ARCHITECTURE PRINCIPIA

ARCHITECTURAL PRINCIPLES OF MATERIAL FORM

GAIL PETER BORDEN

BRIAN DELFORD ANDREWS

Boston   Columbus   Indianapolis   New York   San Francisco   Upper Saddle River
Amsterdam   Cape Town   Dubai   London   Madrid   Milan   Munich   Paris   Montreal   Toronto
Delhi   Mexico City   São Paulo   Sydney   Hong Kong   Seoul   Singapore   Taipei   Tokyo

Editorial Director: Vernon R. Anthony
Editor, Digital Projects: Nichole Caldwell
Editorial Assistant: Nancy Kesterson
Director of Marketing: David Gesell
Senior Marketing Manager: Harper Coles
Senior Marketing Coordinator: Alicia Wozniak
Marketing Assistant: Les Roberts
Senior Managing Editor: JoEllen Gohr
Associate Managing Editor: Alexandrina Benedicto Wolf
Production Project Manager: Maren L. Miller
Senior Operations Supervisor: Pat Tonneman
Operations Specialist: Deidra Skahill
Senior Art Director: Diane Y. Ernsberger
Cover Designer: Candace Crowley
Image Permission Coordinator: Mike Lackey
Cover Art: Gail Peter Borden
Lead Media Project Manager: Karen Bretz
Full-Service Project Management: Penny Walker, Aptara®, Inc.
Composition: Aptara®, Inc.
Printer/Binder: Edwards Brothers Malloy
Cover Printer: Lehigh-Phoenix Color/Hagerstown
Text Font: 10/12 Arial MT

Credits and acknowledgments borrowed from other sources and reproduced, with permission, in this textbook appear on the appropriate page within text.

Copyright © 2014 by Pearson Education, Inc. All rights reserved. Manufactured in the United States of America. This publication is protected by Copyright, and permission should be obtained from the publisher prior to any prohibited reproduction, storage in a retrieval system, or transmission in any form or by any means, electronic, mechanical, photocopying, recording, or likewise. To obtain permission(s) to use material from this work, please submit a written request to Pearson Education, Inc., Permissions Department, One Lake Street, Upper Saddle River, New Jersey 07458, or you may fax your request to 201-236-3290.

Many of the designations by manufacturers and sellers to distinguish their products are claimed as trademarks. Where those designations appear in this book, and the publisher was aware of a trademark claim, the designations have been printed in initial caps or all caps.

Library of Congress Cataloging-in-Publication Data

Borden, Gail Peter.
 Architecture Principia : architectural principles of material form /
Gail Peter Borden, RA AIA NCARB, University of Southern California,
Brian Delford Andrews, RA, University of Arizona.—1 [edition].
     pages cm
  ISBN 978-0-13-157965-1
1. Architecture.   I. Andrews, Brian Delford.   II. Title.
   III. Title:
Architectural principles of material form.
 NA2510.B67 2013
 721—dc23
                                    2012048223

10 9 8 7 6 5 4 3 2 1

PEARSON

ISBN 10:   0-13-157965-7
ISBN 13: 978-0-13-157965-1

About the Authors

Gail Peter Borden, RA, AIA, NCARB
Architecture Discipline Head
Director of Graduate Architecture
Associate Professor of Architecture
University of Southern California

Principal
Borden Partnership llp
www.bordenpartnership.com

Gail Peter Borden attended Rice University, simultaneously receiving bachelor of arts degrees (all cum laude) in fine arts, art history, and architecture. Upon graduation, he won the prestigious William Ward Watkins Traveling Fellowship, the AIA Certificate for Excellence, the Chillman Prize, and the John Swift Medal in Fine Arts. After receiving a Texas Architectural Foundation Scholarship, Professor Borden returned to Rice for his BARCH, also cum laude. He went on to Harvard University's Graduate School of Design to complete a master's of architecture with distinction.

In addition to holding a tenured position as associate professor at the University of Southern California, Borden is the architecture discipline head and director of the graduate architecture program. As principal of Borden Partnership since 2002, his design work has won numerous recognitions including: the Architectural League Prize; the AIA Young Architect Award; *Building Design and Construction* magazine's "40 Under 40" award; and numerous AIA, ACSA, and RADA awards. Borden received artist-in-residence awards from the Chinati Foundation in Marfa, Texas; the Atlantic Center for the Arts; the Borchard Fellowship; and the MacDowell Colony. His teaching has been recognized with an ACSA New Faculty Teaching Award as one of the top emerging architecture faculty. He published his first book, *Material Precedent: The Typology of Modern Tectonics*, in 2010 (Wiley Press) and his second, *Matter: Material Processes in Architectural Production*, in 2011 (Routledge).

As an artist, theoretician, and practitioner, Professor Borden's research and practice focus on the role of materiality and architecture in contemporary culture.

Brian Delford Andrews, RA
Visiting Professor
University of Arizona

Principal
Atelier Andrews
www.atelierandrews.com

Brian Delford Andrews attended Tulane University and the Architectural Association in London. Upon graduation from Tulane, Professor Andrews was awarded the AIA Gold Medal and the Best Thesis Award. He received his master's degree from Princeton University, where he was also awarded the Skidmore, Owings & Merrill Prize and Travel Fellowship.

Andrews is currently a visiting professor at the University of Arizona. He has taught at various institutions including the University of Virginia, Syracuse University, and the University of Nevada, Las Vegas. He was the Robert Mills Distinguished Professor at Clemson University and also the Hyde Chair of Excellence at the University of Nebraska. In addition, Andrews has taught in California and in the Middle East. The University of Southern California has recognized his teaching with numerous awards including an "Outstanding Teaching and Mentoring Award." As a principal in Andrews/Leblanc and Atelier Andrews, his work has been recognized both nationally and internationally. He has won a Progressive Architecture Award and numerous ACSA, Boston Society of Architects, and RADA awards. His work has been exhibited and published widely. Andrews is currently working on a publication about Giuseppe Terragni's *Asilo*.

Professor Andrews' research and practice focuses on drawing and the concept of "Spatial Detritus" in existing society.

01 - Organization Systems — 2
- Centralized — 6
- Linear — 18
- Grid — 30
- Free Plan — 42
- Dispersed Fields / Pods — 46
- Raumplan — 54
- Hybrid — 58

02 - Precedent — 64
- Lineages — 69
- Assemblages — 92
- Comparatives — 106

03 - Typology — 120
- Geometric Typology / Form Typology — 124
- Programmatic / Functional Typology — 168

04 - Form — 230
- Platonic Formalism — 234
- Functional Formalism — 236
- Contextual Formalism — 238
- Performative Formalism — 240
- Organizational Formalism — 242
- Geometric Formalism — 244
- Material Formalism — 246
- Experiential Formalism — 248
- Sequential Formalism — 250

05 - Figure / Ground — 252
- Positive / Negative — 256
- Nolli Map of Rome — 258
- Plan Poche — 260
- Sectional Poche — 262
- Elevational Poche — 264
- Urban Poche — 266
- Military / Defensive Poche — 268
- Monumental Poche — 270
- Structural Poche — 272
- Indeterminate Poche — 274
- Hierarchical Poche — 276
- Material Process Poche — 278
- Programmatic Poche — 280

06 - Context — 282
- Natural Context — 286
- Urban Context — 288
- Historical Context — 290
- Material Context — 292
- Cultural Context — 294

07 - Geometry / Proportion — 296
- Point / Line / Plane — 302
- Three-Dimensional Volume / Mass — 304
- Two-Dimensional Module in Plan — 306
- 2D and 3D Module Form — 324
- Distorted Geometries and Complexities — 326

08 - Symmetry — 328
- Bilateral Symmetry — 332
- Asymmetry — 338
- Local Symmetry — 340
- Material Symmetry — 342
- Programmatic Symmetry — 344

09 - Hierarchy — 346
- Formal / Geometric Hierarchy — 350
- Axial Hierarchy — 354
- Visual / Perceptual Hierarchy — 358
- Hierarchy of Scale — 360
- Hierarchy of Monument — 364
- Hierarchy of Visual Control — 366
- Programmatic / Functional Hierarchy — 368
- Color Hierarchy — 370
- Material Hierarchy — 372

10 - Material — 374
- Material Form — 378
- Material Structure — 390
- Material Detail — 398

11 - Ornament — 410
- Material Construction—Ornament — 414
- Applied Ornament—Religious — 418
- Applied Ornament—Performative—Structural — 422
- Applied Ornament—Performative—Mechanical — 426
- Applied Ornament—Referential—Organicism — 430
- Applied Ornament—Referential—Structural — 434
- Applied Ornament—Historical — 438

12 - Pattern	**442**
Shape	446
Material	452
Color	458
Referential (Historical)	464
Repetition	470
Rhythm	494
13 - Perception	**512**
Light	516
Color	518
Perspectives	520
Material	524
Sound	526
Memory / History	528
Environment	530
14 - Sequence	**532**
Horizontal Sequence—Axial	536
Ceremonial Sequence	538
Experiential Sequence	540
Programmatic Sequence	542
Vertical Sequence	544
15 - Meaning	**548**
Classicism	554
Romanesque	556
Gothic	558
Renaissance	560
Mannerism	562
Baroque	564
Neo-Gothic	566
Neoclassicism	568
Art Nouveau	570
Arts and Crafts	572
Industrialism	574
Modernism	576
Rationalism	578
Brutalism	580
Hi-Tec	582
Postmodernism	584
Structuralism	586
Post-Structuralism	586
Deconstructivism	590
Regional Modernism	592
Globalism	594
Glossary	**596**
index	**600**

Preface

The history of architecture is defined by trends and principles that repeat and reoccur. The evolutionary nature of function and form, and the interrelationship of the way in which building types are conceived, define the meaning of our largest cultural objects. The built world as a mediator of daily life and a snapshot of cultural technology and humanity embodies a moment of time and an associated way of thinking. Every piece of architecture emerges from a context that is specific and rational. This book attempts to unpack the lineage of architecture into primary veins of thinking and re-occurring issues, which remain despite all of our cultural, intellectual, and social advancements. These theories are attempts to resolve known problems that are fundamental to how we think, understand, and make architecture. These principles establish a genealogy of thought manifested through form.

The book is entitled *Principia* as a nod to Sir Isaac Newton's seminal three-volume work *Philosophiae Naturalis Principia Mathematica*, containing explanations of his laws of motion and his law of universal gravitation. Similarly this book focuses its lens of inquiry on the fundamental principles underpinning the history and practice of architecture. Written by designers, for designers, the text is intended to serve as an analytical handbook of the concepts behind these diverse, formal principles as viewed through the history of architecture. It is simultaneously fundamental and advanced.

This book deploys a case study methodology. Using precedents to illustrate and depict the primary aspects of architectural formalism, the analytical diagram and the associated textual description attempt to reveal strands of thought and references embedded within architecture. Through the highlighting of types (functional, formal, geometric, and material, for instance), the catalog of ways of thinking presents, both geographically and chronologically, a broad and comprehensive vision of the evolution of contemporary material form.

Organization
To tackle the breadth of thinking, the book is broken down into 15 chapters. Each takes a specific and critical theme fundamental to architecture and explicates the different modalities and sub-themes within each one. Describing the contextual, formal, and experiential significance of each theme, the book uses both textual and visual illustrations of the project's application.

The book is thematically organized as follows:

01 - Organization Systems
Beginning with the fundamental organization systems of the plan, the relationship of the practicalities of position, placement, and function interrelate to define over-arching principles of order and parti (the organizing diagram of an idea). This realm, governed mostly by geometry and the relative relationships of hierarchies in space, deals with the arrangements and the related social, functional, spatial, and conceptual intentions of a design agenda. Centralized systems (square, circle, Greek Cross, Latin Cross, and radial) deal with self-resolving geometries. Linear systems (single-loaded, double-loaded, and point-to-point) allow for the extrusion of a form. Grid systems (position, form, structure, and module) develop a multivalent system. Dispersed fields (including organized and disorganized systems) establish the interrelationship of discrete objects within a fabric. The free plan of modernity, through technological, material, and structural advancements, allows the elimination of the load-bearing wall, which enables the subdivision of space and fluidity of form to be dependent upon other considerations. Raumplan evolves the organization of spatial zones through the deployment of varied sectional relationships. Hybrid systems combine, splice, mix, juxtapose, and bastardize aspects of all of these organizational approaches.

02 - Precedent
From the focus on organization principles, the book moves towards an overt examination of precedent. Precedent involves the role of history and formal iteration based on referential tactics. Precedent analysis is examined through the lens of vernacular, cultural, and intellectual evolutions and their affect on the iterative production of form. Using a "lineage" methodology, the analysis picks strains of façades, plans, spatial types, and ultimately hybridized assemblies that cross-pollinate multiple precedents to produce a new precedent. The morphological examination allows for the cataloging of referential layers and the rationale for their adoption.

03 - Typology
Directly associated with the use of precedent is the development of typology. Typology deals with the classification of trends, based upon either form or function. It engages geometric typologies that include square, circle, oval, triangle, polygon, spiral, and star. In addition, functional or programmatic typologies include the temple / church, palace, house, museum, library, school, prison, theater, office / high rise, parking garage, and campus. The textual and visual trending of these strains allows for the iterative evolution to be documented and presented.

04 - Form
Form, as the most visually dominant figural shape of a building, is classified based on the primary driver of its derivation. These classifications are broad and varied. Platonic formalism relies upon the purity of geometric relationships. Functional formalism allows for the pragmatics of need and purpose of architecture to derive form. Contextual formalism relies upon surrounding physical and cultural factors to determine form. Typological formalism relies upon referential imagery of function, which is based on a response to historical practices. Performative, or technological, formalism depends upon systems and building construction knowledge to produce formal, environmental, and effectual response. Organizational formalism is dominated by the organization's system (discussed in Chapter 1) as derivational ideals that have universal compositional implications. Geometric formalism uses geometric rationale (traditional and descriptive as well as digitally advanced techniques) to predicate form. A related classification, symmetrical formalism, uses axial geometries to establish reflection lines to establish order. Hierarchical formalism deploys form to produce contrasting gradients that allow for relative readings within a system. Material formalism uses the material and tectonic to define form. Experiential formalism relies upon the perceptive nature of human interaction (that of the viewer or user) to define form. Sequential formalism, a subset of experiential formalism, deploys chronology of movement and perception to define form. Axial formalism, linked to sequential formalism, defines the hierarchy and perception through the privilege of line.

05 - Figure / Ground
Associated with the legibility of form is the dialectical balance of figure and field engaging the graphic and physical ramifications of the interrelationship of positive and negative. This chapter begins with the graphic reading of Nolli's map of Rome, which deploys the idea of poche as a thickened mass. The theme of poche is further examined through position (plan, section, and urban) and function (defense, monument, structure, service, and material).

06 - Context
Context is the relationship of the object to its surroundings. Its classifications include the natural condition; the local, cultural, and broader urban condition; historical context as a condition of cultural, time, and technology; material context as a physical and technological relationship with matter; and cultural context relating to social norms, interactions, and traditions.

07 - Geometry and Proportion
Geometry and proportion refers to the mathematical rules of idealized form. The interrelationship of these principles to architectural form establishes the proportional systems, which include both the two- and three-dimensional modules in elevation and plan, and their implications on space, construction, and perception. Included in the study of these systems are the advanced techniques of distorted geometries and computational complexities that allow for more dynamic spaces with more intricate ornamentation of a structure's skin.

08 - Symmetry
Symmetry focuses on the relationship of form to axis through axial, bi-axial, radial, asymmetrric (denial of the axis), local, material, and programmatic methods.

09 - Hierarchy
Hierarchy focuses on the ordered valuation of objects within a composition through formal, axial, visual / perceptual, scalar, monument, control, geometric, program / function, color, and material methods.

10 - Material
The theme of materiality examines the physicality of the matter with which we build through form (geometry / system / pattern / ornament), the relationship of surface, the engagement of material with tectonic, the perceptual nature of materials, the role of material relative to physics and structure, the engagement of the detail as a conceptual and material factor, the ecology of materials (origin and production), the function of material, and ultimately the meaning and association of materiality.

11 - Ornament
Ornament is the articulation and embellishment of material intended to lend grace and beauty to a building or structure. It has been defined as an element not belonging to the essential harmony or melody. It has been said that ornament is, in fact, all that separates architecture from building. This chapter divides ornament into a number of different types that include material (the physical way in which we make things and the ordering and composition of segmented construction); religious (the iconographic cultural reference of form); performative (the responsive nature of articulated materials toward a specific agenda); structural (the physics of forces and their engagement with functional verse ornamental); referential (creating a relationship to another premise or topic); organic (that tries to evoke specifically the natural condition); and historical (that directly relates to architecture or another identifiable form).

12 - Pattern
Pattern refers to the field effect of aggregated units accomplished through the method of repetition and rhythm, employing material elements such as shape, color, and composition. These variables may occur on varying scales of piece/unit, portion/panel, bay/module, or chunk.

13 - Sequence
The sequential nature of chronology, narrative, and experience of movement through space is engaged through the focused analysis of both horizontal and vertical movement, which include axial, ceremonial, experiential, organizational, and programmatic circulation.

14 - Perception
Perception is the most significant factor in the understanding of architecture and yet one of the most elusive and difficult to govern. The following factors are the primary categories relative to perception: light, color, visual focal point, perspectives (both real and false), material, sound, memory, and environment. Designers employ these factors and their impact on perception of composition to affect the user's experience and ultimately satisfy the programmatic requirements of the building or space. Examining the perceptual qualities of the body and the palette of manipulable factors that influence our five senses, the thematic investigation includes an examination of light, color, focal point, vantage or perspective (Constructed, False, and Multiple), material, sound, memory (history) and environment, and their impact on perception and composition.

15 - Meaning
This chapter focuses on the intellectual content and underpinning theoretical agenda of architectural movements, relative to broader themes and precedents, and their implications on form. An understanding of these movements is critical in comprehending how this book works. The movements discussed follow the primary architectural beliefs chronologically by era, covering classicism, Romanesque, Gothic, Renaissance, baroque, mannerism, neo-Gothic, industrialism, art nouveau, neo-classicism, arts and crafts, modernism, rationalism, brutalism, postmodernism, structuralism, post-structuralism, deconstructivism, hi-tech, regional modernism, pluralism, and globalism.

Layout and Key Features
The specific layout was carefully considered and graphically standardized to produce a consistent analytical lens that engages the diverse themes. In a spread, there is a series of descriptive devices that aid the reader in understanding the theme relative to a vein of thinking and a specific moment within architecture.

Each of the aforementioned chapters has an introductory essay that outlines the primary principles, conceptual evolution, and associated subthemes. Each two-page spread then pursues the illustration and unpacking of these subthemes through a comparative sequence of analytical case studies. Each spread is divided into four columns. The first is dedicated to a textual description followed by three chronologically organized case studies. Each case study is identified by a caption, the first line of which identifies the project name and architect. The second line contains the geographic location and a date of completion. For some urban projects, a larger time span was adopted so as not to minimize the scale and duration of the project. The third caption line then positions the project within the larger conceptual movement in which it was designed and built, followed by a condensed and primary description of the project relative to the theme, and finally the primary materials, illustrating both the level of technology and the associative physical-to-conceptual limitations. The brief description then tries to textually unpack the diagram and reference the significance of the case study relative to the broader principle. Certain case studies occur in multiple chapters of the book and have different visual and textual descriptions as the complexity of their thinking makes them applicable to diverse issues. An effort to use such case studies was intentional to illustrate the sophistication and multi-principled nature of architecture, which is rarely about a single system or reading.

Atop each page is a running band (see example above) that establishes a classification system, allowing the book to be used as a reference tool. This serves as a way-finding system that marks specifically where you are within the text and provides the contextual issues surrounding the case studies being presented. They are an indicator flag of where you are within the larger thinking. In addition to the page number, there are five categories to aid in this classification system: principle, organization, geometry, reading, and scale. The *principle* is the chapter title and the diagrammatic lens through which that section of examination is being viewed. This occurs on both the left and right pages of the spread as it is the fundamental underpinning of the analysis. The *organization* describes the subtheme (typically based in methodology) that establishes the specific thinking within the larger principle. *Geometry* describes the sub-subtheme and typically references the formal methodology of the use. *Reading* refers to the specific drawing type used in the illustration, linking the representational model and vantage to the specific method (for instance, plan, section, elevation, or axonometric). *Scale* refers to the level of operation at which the principle is deployed; for example, the scale of the city, the scale of the building, or the scale of the detail. Occasionally, one of these categorical cells is left intentionally blank. In this situation, the intermediate tier was not needed and thus skipped over. The uniformity of the thematic position is, however, consistently maintained throughout the book.

The unique approach and design of the book includes:

- a comprehensive look at the fundamental themes of architecture;
- the use of comparative precedents;
- case studies from across the history of architecture;
- consistent and clear graphic language;
- a parallel visual and textual unpacking of each principle.

Supplements
A complete image bank is available in PowerPoint format by going to the Instructor Resource Center.

Download Instructor Resources from the Instructor Resource Center
To access supplementary materials online, instructors need to request an instructor access code. Go to **www.pearsonhighered.com/irc** to register for an instructor access code. Within 48 hours of registering, you will receive a confirming e-mail including an instructor access code. Once you have received your code, locate your text in the online catalog and click on the Instructor Resources button on the left side of the catalog product page. Select a supplement, and a login page will appear. Once you have logged in, you can access instructor material for all Prentice Hall textbooks. If you have any difficulties accessing the site or downloading a supplement, please contact Customer Service at **http://247pearsoned.custhelp.com/**.

Principia
The collection of these themes results in a methodical unpacking of the lineages of architectural reference of material form. The simultaneous presentation of the graphic documentation and the analytical highlighting, coupled with the textual description and historical position, establishes a catalog of contemporary principles in relationship to form. This handbook will serve the designer, historian, and layman alike in understanding the complexity of the social, cultural, functional, material, and formal lineage of architecture.

Acknowledgments
Gail Peter Borden: I would like thank my father for his example as an author; my mother for her enduring confidence; and my family, Brooke, Frieda Dorothy, and Gail Calvin, for their continuous support and love.

Brian Delford Andrews: I would like to thank my former professors and students for their friendship and support, my parents for everything, and mostly my daughter Constance for her love, inspiration, and encouragement.

We are indebted to Wili-Mirel Luca for his tireless reliability, passion, and love of architecture, which were all applied to the research and drawing of this book.

Thank you to the following reviewers for your thoughtful comments on this book: Joseph Bilello, Ball State University; Jeremy Ficca, Carnegie Mellon University; Craig S. Griffen, Philadelphia University; Dane Archer Johnson, Ferris State University; Patricia Kucker, University of Cincinnati; Benjamin Perry, JCCC; and Rise Talbot, El Centro College.

Gail Peter Borden
Brian Delford Andrews

2	ORGANIZATION SYSTEMS			
	PRINCIPLE			

01

Organization Systems

01 – Organization Systems: Parti and Order

Primary to the creation and comprehension of architecture is the development or adoption of an ordering or organizational system. Organizational systems are the geometric principles utilized to create the form and layout of an architectural composition. The combination of shape with an associated plan typology generates a building's identity and form. Through time, there are distinct and discrete classification systems that have arisen. The following pages dissect this history through the primary geometric ordering system, focusing both on form and interrelationships of pieces, or parti.

A parti refers to an overall idea or concept that diagrammatically informs a design. It originated as a term used at the Beaux Arts, denoting the essential diagram for a solution to an architectural problem. An architectural parti can take numerous forms and be applied either separately to different aspects of a project, (such as the plan or elevation) or inclusively, impacting many or all aspects of a design. A parti is a way of distilling a diagram down to one word or phrase, thereby allowing easier access to the designer's intent.

The formal organizational systems, or partis, include the following types: centralized, linear, **grid**, dispersed field, free plan, Raumplan, and hybrid. The definitions of these principles and the resulting subclassification of their genealogies establish a handbook of formal types. These types account for a classification methodology that encompasses every building through time. The purity of the parti's deployment, clarity, and visual legibility varies with architect, project, material technology, and era, though the formal organizational premise remains constant.

Organizations systems in architecture are concerned primarily with methods of composition and design. These systems form the basis of most architecture and allow the analysis of buildings and urban plans. Architecture is often defined by an organization system. It connects any building with those of similar systems and aids in their overall understanding. Every building contains and is categorized by some form of organizational system. Either straightforward and immediately recognizable, or hybridized and grafted from a combination of systems, the basic physical organizational systems are finite, and thus the reading of these underpinning principles is discernable and significant. The various systems, their possibilities or characteristics, and the history and precedents behind these systems are essential. These systems derive a catalog of spatial types and functional organizations.

Centralized Historically, centralized systems have been the most common organization system. A centralized system is one that focuses on a central space or object in the plan. Centralized systems come in the form of squares, circles, ovals, triangles, and stars. Often the space can be recognized as a singular, self-resolving, and formally complete entity, such as a church interior or a courtyard. In other examples, it might take the form of a solid structure or object, either as an architectural (building) or urban (city) element.

Linear A linear system is one that organizes elements along a line or axis. A linear system can be single-loaded or double-loaded, or have a point-to-point arrangement. These schemes can be either architectural in scale (such as a simple hallway) or urban in scale (such as a boulevard). A single-loaded system implies that one side is given priority and weighted with ancillary spaces; a double-loaded system uses both sides, whereas a point-to-point system is concerned with the elements being connected at either end.

Grid Grids are very common and often are recognized in both urban plans and structural systems. Grids are ways of deploying **multivalent**, repetitive organizational fields. The grid works equally in all directions through the establishment and repetition of a standard base unit. Present as both positive and negative figures, these systems can be recognized as a grid of **bays** or **modules** within a project. Their intensity of repetition and infinite expandability at either the architectural or urban scale is often adapted to respond to specific and local needs. As a way of either negotiating topography or attempting to emphasize a portion of a design project, deformations, interruptions, and disassociations to the grid establish relative relationships of object to field.

Dispersed Field A dispersed **field** first refers to buildings that separate the components into discrete objects and then deals with the interrelationships of the piece-to-piece, piece-to-field, and the field as a whole. The geometric association of one component to another can be overtly organized into legible patterns, surreptitiously organized through more concealed ordering, or fully disorganized to intentionally disregard a collective order. This system can be employed at both the architectural and urban scale. The typical Roman encampment is a prime example of an organized dispersed field system. It arrays disparately dimensioned objects within a regimented field. A disorganized and dispersed field is typified by the ubiquitous sprawl of suburban development, where independent local decisions outweigh the collective vision of the total composition, resulting in juxtapositions and anomalies.

Free Plan Free plan is a twentieth-century invention that allows a new type of space founded in continuity. In architecture, it allows the walls and spaces to be free of the structural grid. As the champion of this planning, Le Corbusier designed several prominent residential projects using this strategy. On an urban level, the various buildings are free to be placed in situations where they are related, though not by the typical urban qualities of axis and symmetry. The campus of IIT, by Mies van der Rohe, exemplifies this approach.

Raumplan Raumplan (developed and primarily used by Adolf Loos) deploys sectional differentiation between rooms as a way of distinguishing them in addition to their organization or position in plan. The complexity of this model results in limited applications, but allows for visual and spatial hierarchies between adjacent spaces. These hierarchies create intricate dialectics of dominant and subordinate spaces and positions. The Villa Moller and Villa Müller, in Vienna and Prague respectively, are excellent examples of this unusual system.

Hybrid The hybrid system combines two or more of the aforementioned systems into one building plan. Examples include the crossbreeding of both regular and irregular spaces (Guggenheim Bilbao, Frank Gehry), or the synthesis of a traditional, centralized courtyard figure with an internalized free plan (Bo Bardi House, Lina Bo Bardi). There is a long history of these organizational systems; architects throughout time have analyzed the hybrid and attempted to modify, adapt, and synthesize it for contemporary needs and applications. Hybridizations of centralized, linear, grid, and dispersed field of almost every configuration have been in use since architecture began. The Renaissance made large contributions to the refinement and formal usage of all of these methods. Modernism ushered in new spatial and organizational types such as free plan and Raumplan, and accelerated the use of the hybrid systems. The resulting forms and their driving rationale have evolved through associated conceptual intentions and technological capabilities.

It is critical that architects and designers understand these systems not as a series of formulas, but as a classification of principles. All projects are comparable to their historical antecedents, thereby making this knowledge, lineage, and associative sensibility essential.

Centralized

Centrally organized buildings rely upon a **radial geometry** to emphasize a central point. Defined geometrically through the square, circle, Greek Cross, Latin Cross, or other radial geometries, the hierarchy of their geometric system dominates the entire composition. Secondary formal systems are subordinated by the power and dominance of the primary geometry. Easily identifiable and rationally derived, the geometric purity of form provides clarity of understanding to the occupant, allowing construction of a mental map that defines the entire composition. This understanding of the whole allows for a continuous positioning of the self relative to the hierarchy and form of the centralized system. The result is an awareness of one's own location, purpose, and hierarchy relative to the whole.

SQUARE

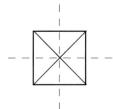

DIAMOND

RADIAL

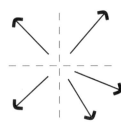

CENTRIFUGAL

ORGANIZATION SYSTEMS
PRINCIPLE 7

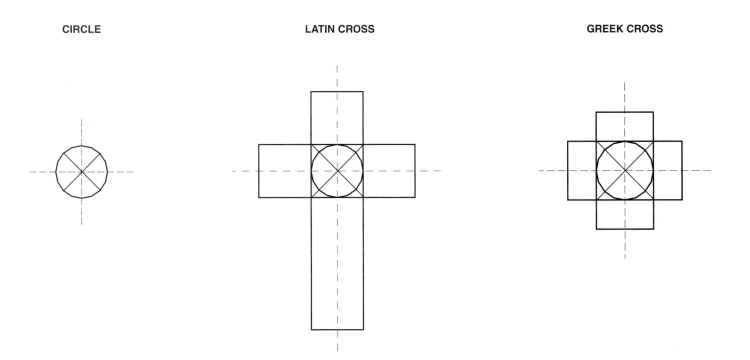

CIRCLE — LATIN CROSS — GREEK CROSS

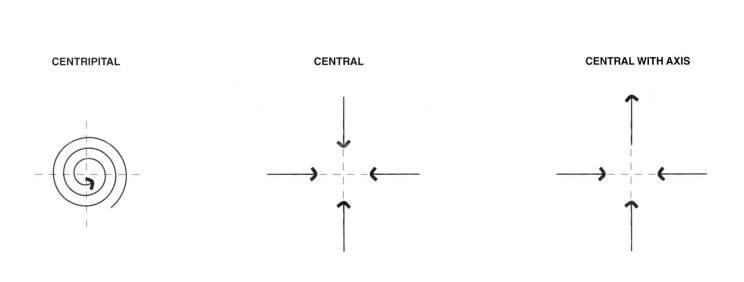

CENTRIPITAL — CENTRAL — CENTRAL WITH AXIS

8	ORGANIZATION SYSTEMS	CENTRALIZED	SQUARE	PLAN
	PRINCIPLE	ORGANIZATION	GEOMETRY	READING

Square

The geometry of the square is defined both by perpendicular and diagonal (forty-five degree) cross axes. Emphasizing axes, center, and faces, the square maintains its geometry through the mapping of the four cardinal directions through the repetitive faces. The equality of the four faces gives a multidirectional, non-hierarchical **façade** that is uniform and repeated, such as is found in the Villa Rotunda and Exeter Library. The geometry of the square is deployed for its simplicity, clarity and equality.

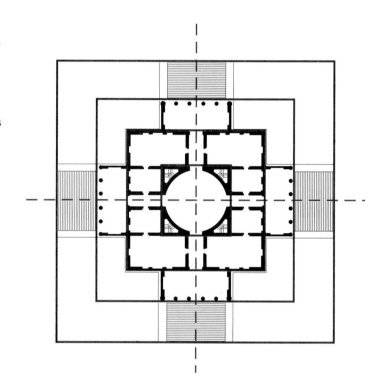

Villa Rotunda, Andrea Palladio
Vicenza, Italy 1571
[High Renaissance, cross axis square with dome, masonry]

The square of Villa Rotunda is concentrically doubled. Established first through the body of the house itself, then amplified through the four-sided repetition of the temple façade. The cross axis of the square's midpoints becomes the axial lines of reflection for the symmetrical façades and the equally delivered central circulation lines of approach and entry for each of the four faces. These distinct faces and axes are oriented extensions of the four cardinal directions: north, south, east, and west.

| ORGANIZATION SYSTEMS |
| PRINCIPLE | **9** |

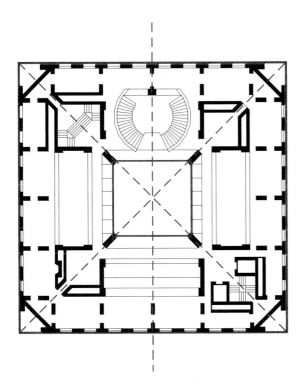 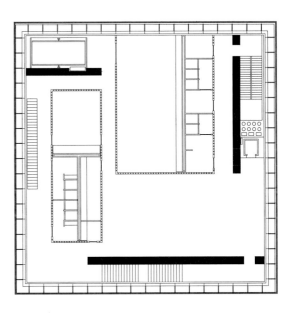

Exeter Library, Louis Kahn
Exeter, New Hampshire 1972
[Late Modern, centralized square, masonry and concrete]

In the Exeter Library, Louis Kahn uses the cube, a Platonic solid, to produce an isolated and enigmatic figure defined by two nested cubes, set one inside the other. The building uses the square to define both the plan and the elevation (with the lower subterranean level defining the bottom of the square). The inner concrete square figure defines the central atrium. Perforated with large circles radiating from the center point of each elevation, the planar enclosure becomes frame-like. The top of the inner concrete frame is defined by massive crossing beams connecting the diagonal corners. The emphasis of center through central and diagonal axes produces a centralized hierarchy on the void of the atrium that re-emphasizes the square geometry.

Kunsthaus Bregenz, Peter Zumthor
Bregenz, Vorarlberg, Austria 1997
[Modern materialism, stacked square, glass and concrete]

The Kunsthaus Bregenz has an exterior size and form similar to the Exeter Library, but uses a very different spatial and material intention. Square in plan and elevation, the figure is repetitively patterned with a blank material-based module. The inner plan maintains the square figure of the perimeter, subdivided by three carefully positioned concrete walls that introduce a dynamism of directionality and break the uniformity of the floor plate.

| 10 | ORGANIZATION SYSTEMS / PRINCIPLE | CENTRALIZED / ORGANIZATION | CIRCLE / GEOMETRY | PLAN / READING |

Circle

The circle as a centralized organizational system allows for the perfection of the geometry of the radial form to produce uniformity. The figure innately creates deference to the hierarchy of the center point. The form defines simultaneously the figural resolution and the interior space. The geometric, self-resolving form provides a space without hierarchy beyond the center point. This allows for radiating multidirectional equality. The nature of the circle's geometry intrinsically prioritizes the center point as a place and position of dominance. When occupied by the viewer, this central position establishes a point of control. When occupied as a focal point by an object or user, the circular geometry establishes a perimeter and a rippling of spatial layers that ring and reinforce the hierarchy. This spatial result is intrinsic to the circular geometric type.

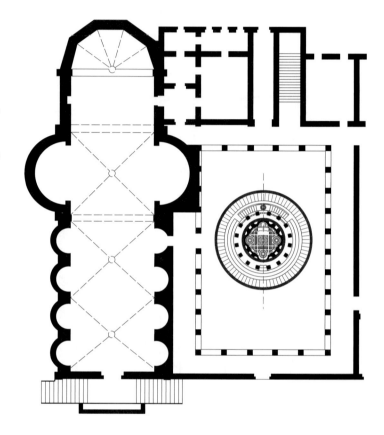

The Tempietto, Donato Bramante
Rome, Italy 1502
[High Renaissance, freestanding circular **martyrium**, masonry]

*In the Tempietto, Bramante uses the small scale of a martyrium to execute a simple, centralized, circular plan. A collaborative, **proportional** and detailed system reinforces the concentric organization. From the overall figure, through the ringing steps, perimeter columns, and hemispherical dome, the collective composition reinforces the clarity, elegance, and simplicity of the fundamental circular geometry.*

	ORGANIZATION SYSTEMS	
	PRINCIPLE	11

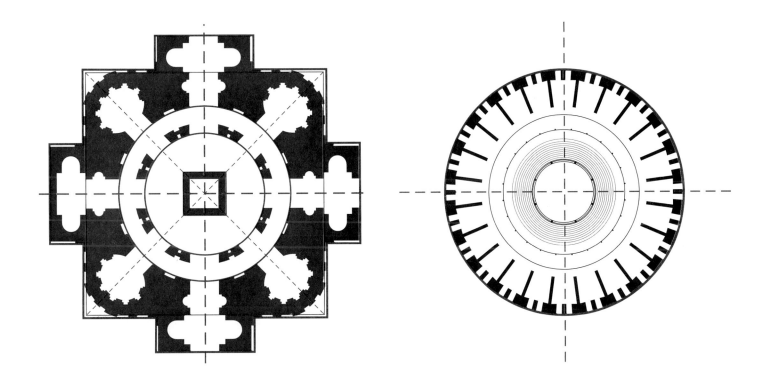

San Giovanni de Fiorentini, Michelangelo Buonarroti
Unbuilt 1560
[Mannerist, centralized circle-square-Greek Cross, masonry]

*In the unrealized design of San Giovanni de Fiorentini, Michelangelo simultaneously deployed the superimposed figures of circle (to define the primary inner **drum** space), square (to resolve the corner condition), and Greek Cross. The latter was used to create the cross axis necessary to define entry, reiterate the iconic, representational cruciform figure, and reinforce the relation of nave, transept and altar. Each figure, laid on top of one another, restates and reframes the one below. The use of multiple geometries allows for each figure to collaborate, while providing the local formal and functional strengths of their individual qualities.*

Panopticon, Jeremy Bentham
Unbuilt 1785
[Neoclassical, radial plan for visual surveillance, masonry]

Based upon a fundamental desire to establish visual control, the organization is defined by a radial configuration. As a prison design, the scheme locates the guard at the heart of the composition, ringing the perimeter with the individual cells. The visual domination established through the configuration allows the central position to see all of the perimeter positions. This puts a single control point in a dominant position leaving the prisoners in a constant state of surveillance, requiring constant good behavior as one never knows when one is being scrutinized. The circular configuration, through its geometric construction of center point, radius, and perimeter circle, works to establish the viewer, sight line, and the viewed.

Latin Cross

The Latin Cross is defined by the extension of the primary axis (typically the east-to-west orientation). This simple hierarchy establishes a discrete nature to each of the four arms. The western arm extends in length, differentiating itself from the other three and establishing the entry. The hierarchy of the elongated nave creates a gathering space for the audience, allowing the eastern tip of the same axis to establish itself as a backdrop to the transept and a secondary place of hierarchy as a position for the choir. The transept establishes two lateral arms at the crossing axis, allowing for a secondary figure that is minor in relation to the nave, but essential in enforcing the significance of the central position of the altar. The crossing is further given hierarchy through the position of the dome and its diverse forms, defined by structural and spatial considerations. The overall combination of the figure replicates the primal Christian figure derived from the crucifixion.

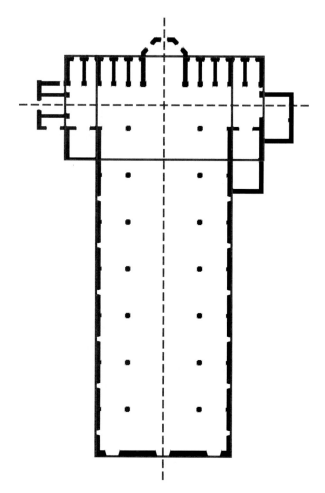

Santa Croce, Arnolfo di Cambio
Florence, Italy 1442
[Late Medieval, Latin Cross church, masonry]

Santa Croce in Florence is the largest Franciscan Church in the world. It was originally designed by Arnolfo di Cambio and used an Egyptian Tau Cross Plan (the symbol of St. Francis). This type of plan formed the beginning of the transept, which then resulted in the Latin Cross. The plan recognizes the importance of the position of the altar. The original ten chapels, situated on either side of the altar, add to the importance of this second axis. The proportional relationships that are so critical in later church plans are not as dominantly evident In Santa Croce.

	ORGANIZATION SYSTEMS	
	PRINCIPLE	**13**

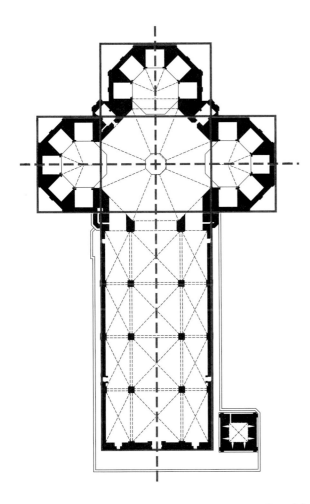

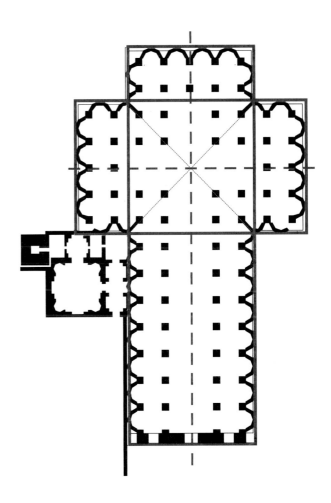

Santa Maris del Fiore, diCambio and Brunnelleschi
Florence, Italy 1462
[Gothic, Latin Cross church, masonry]

San Spirito, Filippo Brunelleschi
Florence, Italy 1482
[Renaissance, centralized Latin Cross church, masonry]

This Gothic Cathedral of Florence is based around the large octagon at the crossing of the nave and transept. This octagon was built to support the immense dome that was later engineered by Brunelleschi. The three apses that are connected to the high altar are identical; the nave makes up the fourth side and extends to the west façade. The plan is a modified Latin Cross, where the transept and choir space are replaced by large apses. Though the proportions that later become so critical are not yet obvious, the underpinning organizational figure is distinctly defined by the Latin Cross. The notoriety of Brunelleschi's dome often overshadows the clarity of this early example of the Latin Cross.

The church of San Spirito is an excellent example of a Latin Cross church. The crossing of the nave and the transept is carried out in a complete and highly ordered system to establish the position and hierarchy of the high altar. As a system based on the square and its subsequent multiples, the left and right naves are constructed of squares, each of which relates to a chapel. The square is then doubled and quadrupled to make a larger series of squares that comprise the middle nave and the transept. One of the end results of this most ingenious scheme is the interior portico that runs around the entire interior of the church. Here Brunelleschi adeptly introduced the Renaissance system of proportioning into church planning.

Greek Cross

The Greek Cross is a close evolutionary iteration of the centralized circle. Defined by equal cross-axial transepts, the perpendicular, intersecting geometry establishes an equalized central space determined and prioritized by the positional dominance. The multidirectional radiation from the center typically corresponds to the four cardinal directions. These positions establish a calibrated geographical context while simultaneously producing a dominance of the overlapping center. The establishment of the hierarchy of the form and its centralized position resonates with the religious symbolism and hierarchy of both ceremony and experience. The centralized geometry of the internal organization makes it difficult to prioritize one axis and associated façade over another, but internally allows for an efficient and singular focus on a highly privileged point.

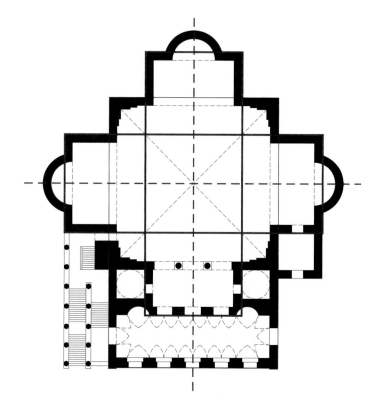

San Sabastiano, Leon Battista Alberti
Mantua, Italy 1475
[Renaissance, Greek Cross votive church plan, masonry]

Many of the churches designed during the Renaissance originally intended to deploy the organizational geometry of the Greek Cross in plan. The centralized Greek Cross, with all four arms being equal, related more to the pure geometric ideals of the Renaissance. Typically this desire was not carried out due to the functional concerns of housing the congregation. San Sabastiano was one of the first planned Greek Cross churches. It has three equal arms, and a fourth that connects to a narthex that stretches across the façade. The fact that San Sabastiano was a Votive Church, which deals primarily with offerings and is not used as a typical church, possibly explains why Alberti was able to carry out the Greek Cross plan.

	ORGANIZATION SYSTEMS	15
	PRINCIPLE	

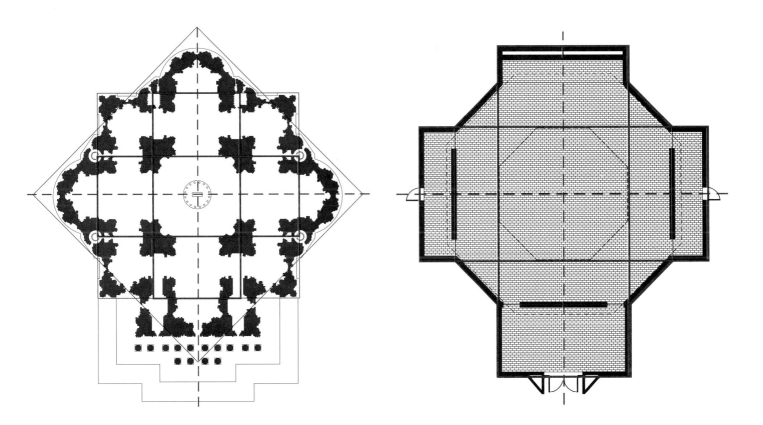

St. Peter's Basilica, Michelangelo Buonarroti
Rome, Italy 1547
[Renaissance, dome and Greek Cross plan, masonry]

When Michelangelo took over the planning of St. Peter's Basilica, he was confronted with the remains of numerous architects. Michelangelo, as most architects of the late Renaissance, desired a Greek Cross plan. The reasons for this are twofold: first, the Greek Cross was more in keeping with the pure geometrical ideals of the Renaissance, and second, the Greek Cross plan allowed the dome to be seen better from the piazza in front of the basilica. Bramante had originally designed St. Peter's Basilica as a Greek Cross, though the piers were not structurally up to the task of supporting the resultant dome. Michelangelo reconfigured the plan to provide adequately sized piers and streamline the overall building plan within a square. The resulting composition was more open, allowing a spatial fluidity that liberated the stagnation of Bramante's compartmentalization. Maderna later added the extended nave, which disorients the centralized nature of the organization and blocks the view of the dome from the piazza.

Rothko Chapel, Philip Johnson
Houston, Texas 1971
[Late Modern, Greek Cross, masonry]

The Rothko Chapel, as a Modernist interpretation of the traditional Greek Cross, transposes the form to a non-denominational space designed to house a series of abstract paintings by Mark Rothko. As a simultaneous mediating extension of the de Menil collection and the University of St. Thomas, architect Philip Johnson continued the masonry and steel palette of the campus to a free-standing centralized form. Emerging from the specially designed, experiential configuration of Rothko's somber paintings, the central hall and its form establish and engage the purity and multi-directionality of the inner-faceted perimeter produced by the Greek Cross. The resulting composition creates an architecture that has an external formal presence as an independent figure; internally, the composition establishes a spatial configuration that directly reinforces the staging of the artwork.

16	ORGANIZATION SYSTEMS	CENTRALIZED	RADIAL	PLAN
	PRINCIPLE	ORGANIZATION	GEOMETRY	READING

Radial

Radial organization hybridizes a combination of linear organizations with the principles of the crossing nodal condition inherent in the Latin or Greek Cross. Established by the hierarchy of a discrete point, the radial (inward converging or outward expanding) extensions define a relationship between point, line (as defining either the edge or an axis), and field. The central point of crossing can be viewed as a point of convergence with centripetal forces directing inward, or a point of radiation becoming a centrifugal dispersal projecting outward from the point of origin. Each of these formal readings can be applied based upon the programmatic and experiential necessities of the building typology. Its geometry has a clearly established hierarchy, but innately sets a conflict between center, edge, and the hierarchy of entry.

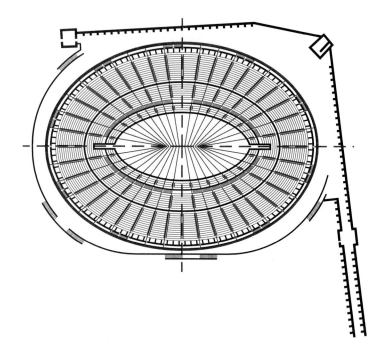

Pompeii Amphitheater
Pompeii, Italy 70 BCE
[Classical, oval shaped outdoor theater, masonry]

The radial configuration deployed here is one typical of performance spaces. Defined by a central stage, the seating radiates outward, providing both visual and auditory equality to the audience. Defined by practical principles that form the geometric plan and the terraced section, the resulting figure optimizes the individual experience through the collective form.

	ORGANIZATION SYSTEMS
	PRINCIPLE **17**

 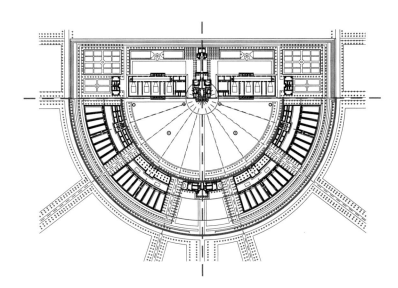

Palmonova, Vincenzo Scamozzi
Palmonova, Italy 1593
[Renaissance, star shaped fortress town, masonry]

Palmonova developed with a dominant radial geometry. Emanating from the sight lines of the ramparts, the repetitive streets are both radially and concentrically layered. The entire town is defined by the planametric decision of this organizing geometry. Originating from a hexagonal town square, the midpoints of each face determine the primary radial street layout. The organization and governance of the urban planning makes the town a unique, singular figure despite the localized variations in its architecture.

Royal Saltworks, Claude Nicolas Ledoux
Arc-et-Senans, France 1779
[Neoclassical, semi-circular industrial compound, masonry]

The built portion of the Royal Saltworks is a semicircular, utopian development. Containing factory facilities for the processing of salt, dormitories, military outposts, gardens, and administration buildings, the compound represents an ideal self-sustained community. Designed as a complex governed by its geometry and controlled through the axes of surveillance, the form establishes the hierarchy of its configuration. Buildings on the circumference are subordinate to and controlled by the central power of the Director's House. The visual domination is facilitated by the radial configuration, approached on the mid-point, establishing an axis of symmetry.

Linear

Linear organizations develop a narrow cross section that is extruded along an elongated axis. Linear types are defined through their planametric configuration. They are established through the relative positional relationship of the programmatic destination spaces in relation to the circulation. These types include single-loaded corridor, double-loaded corridor, and point-to-point configurations. Each configuration establishes differing local and overarching parameters that result in dramatically varied formal results. These forms, often categorized as "bar" buildings, produce a fundamental and ubiquitous building type.

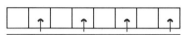

STRAIGHT

SINGLE-LOADED

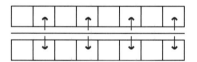

STRAIGHT

DOUBLE-LOADED

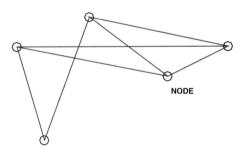

NODE

POINT-TO-POINT

| | ORGANIZATION SYSTEMS PRINCIPLE | 19 |

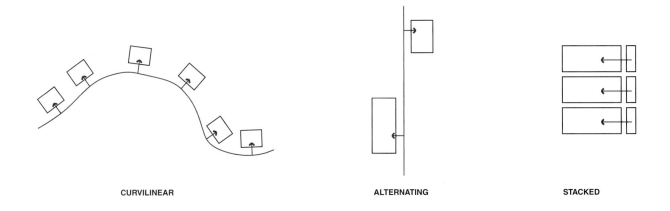

| CURVILINEAR | ALTERNATING | STACKED |

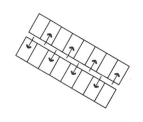 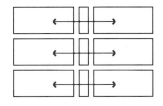 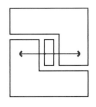

| ALTERNATING | STACKED | VARIABLE STACKED |

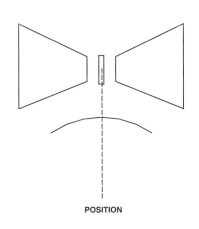 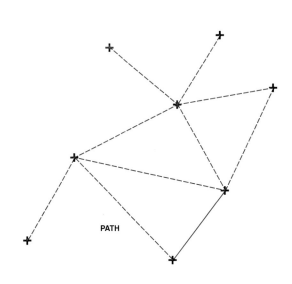

| POSITION | PATH |

| 20 | ORGANIZATION SYSTEMS
PRINCIPLE | LINEAR
ORGANIZATION | SINGLE-LOADED
GEOMETRY | PLAN
READING | ARCHITECTURE
SCALE |

Single-Loaded—Architecture

The architectural single-loaded corridor establishes a parallel linear circulation pathway with associated side spaces. The configuration clearly individuates the area of movement from the space of destination. The linear nature of the plan is efficient for repetitive typologies such as housing, and allows for a narrow cross section and thinness of building to allow light and air to laterally move through the spaces.

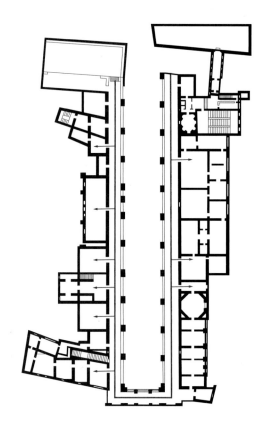

The Uffizi, Georgio Vasari
Florence, Italy 1560
[Renaissance, urban wrapper defining linear courtyard, masonry]

The Uffizi is a single-loaded corridor that in form and scale establishes a perimeter to an urban space. This void space is carved out of the dense urban fabric of the city and is uniformly articulated through the repetitive, stacked lining pathways of circulation. The inner ring of discrete chambers that house the museum collection is accessed through this edging corridor. The result is a constant urban edge that establishes an intense and unifying regularity to a lining façade that homogeneously cloaks a collection of discretely scaled gallery rooms behind. The linear circulation creates an edge buffering between the void of the city and the interior rooms of the gallery.

ORGANIZATION SYSTEMS
PRINCIPLE 21

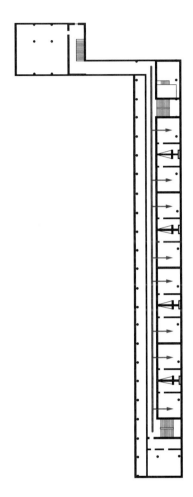

Narkomfin, Moisei Ginzburg
Moscow, Russia 1932
[Modern, linear plan with interlocking section, concrete]

As an early Modernist housing project, the Narkomfin is a quintessential single-loaded corridor. Defining one edge of the composition, the linear circulation pathway parallels the stacked, repetitive units of the residential units. The corridor is linked at the ends by stairways of vertical circulation, allowing a repetitive, cellular plan configuration to be bracketed by circulation. The strict, repetitive formalism establishes the importance of functionalism in the Modernist composition.

Gallaratese Housing, Aldo Rossi
Milan, Italy 1974
[Postmodern, linear repetitive housing, concrete]

The Gallaratese Housing Block is a 200-meter-long building that is accessed through a single-loaded, exterior circulation gallery. The language of Rossi's building borrows from the Italian Rationalist movement, adopting its streamlined formalism rooted in function. It is a high point of the single-loaded corridor scheme and makes no apologies for its simplicity, clarity, and cellular repetition.

Single-Loaded—Urban

The urban single-loaded organization uses the same principles as the architectural single-loaded corridor but scales them to the **grain** of the city. Still relying on the parallel but discrete zones of circulation and zones of rest, the increase in scale allows for its pragmatic expansion. The scope of room increases to the size of building while hall inflates to the size of street. The configuration is often dependent on a large open space defined either by an orchestrated void in the urban fabric (such as a square, park, or piazza) or by the adjacency to a natural feature (such as a boardwalk, beachfront, or river).

Piazza San Marco
Venice, Italy 1400s
[Italian Renaissance, perimeter uniform colonnade, masonry]

The perimeter circulation of the colonnade at the Piazza San Marco is an urban single-loaded plan. It deploys a repetitive continuity to establish uniformity to the perimeter façades of the Piazza. The highly ordered elevations bridge the massive scale of the space to create a singular urban room. Additionally, they establish continuity of the disparate spaces and functions embedded within the mass of the cloaked urban fabric. The corridor is single-loaded as the built fabric lines one edge; the open field of the piazza defines center. The uniformity possible through the consistency of circulation allows the façades, and their repetition, to address a larger agenda of monumentality and continuity at the urban scale.

	ORGANIZATION SYSTEMS	23
	PRINCIPLE	

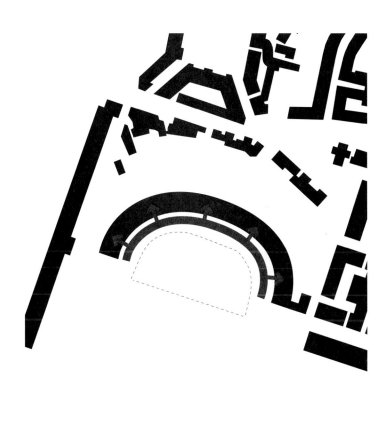 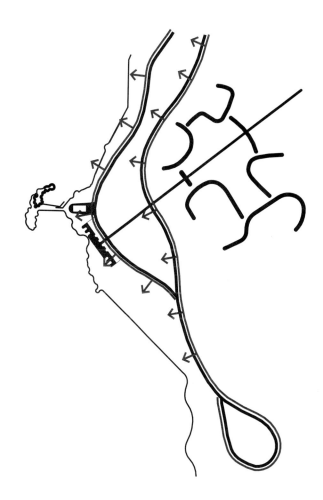

The Royal Crescent, John Wood the Younger
Bath, England 1774
[Georgian, crescent shaped repetitive housing units, masonry]

Much like the colonnade at Piazza San Marco, the Royal Crescent by John Wood the Younger deploys a single-loaded repetitive system. Its unitized repetition establishes continuity and identity on the urban scale. The simple "crescent" gesture is established through the concise system of the bayed façade and adjacent pedestrian circulation. Gathering the row house bays of the individual residential units, the single-loaded corridor serves as a necklace tethering the individually owned and occupied units into a collective whole. The resulting figure creates a solid figure of assembled pieces.

Plan for Algeria, Le Corbusier
Algeria, Africa 1933
[Modern, hyper linear plan, concrete]

Le Corbusier's unbuilt plan for Algeria deploys the single-loaded corridor as an urban parti. Extruding the functions of the city into a continuous bar, the automobile establishes the uppermost level as the "corridor" of circulation. The city then builds downward beneath the roadway to create an extruded cross section that draws a constant profile along the buildings' length. The resulting curvilinear bar form hugs the topography of the natural context, creating a worm-like building and city founded in the organization of the single-loaded corridor.

24	ORGANIZATION SYSTEMS	LINEAR	DOUBLE-LOADED	PLAN	ARCHITECTURE
	PRINCIPLE	ORGANIZATION	GEOMETRY	READING	SCALE

Double-Loaded—Architecture

The double-loaded corridor is defined by a central linear circulation with access points on both sides. As one of the most efficient organizational methods emerging from contemporary building codes, it inherently provides for discrete paths of access on both ends. The double-loaded nature allows for an optimization of space which reduces the amount of circulation in the ratio of path to place. The efficiency of the configuration leads to naturally associated linear forms.

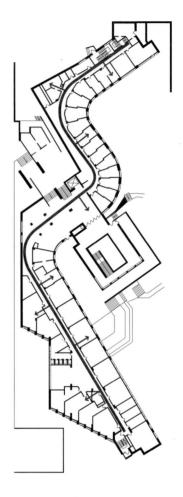

Baker House, Alvar Aalto
Cambridge, Massachusetts 1948
[Modern, double-loaded curved corridor, masonry]

The Baker House dormitory, by Alvar Aalto, extends the efficiency of the single-loaded corridor by mirroring its configuration. Defined by a central corridor that feeds spaces on both sides, the corridor is centrally located with equal residential spaces on either side. The organization optimizes the spatial configuration with vertical circulation tying into either end of the center corridor. The plan figure curves and bends to produce a geometry that responds to and references the adjacent river's organic form, optimizing each room's view and exposure.

ORGANIZATION SYSTEMS PRINCIPLE **25**

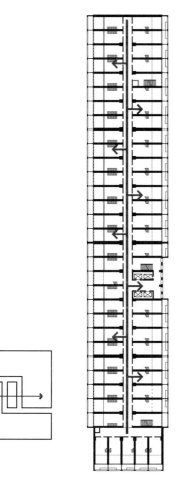

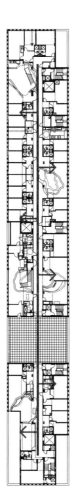

Unite d'Habitation, Le Corbusier
Marseille, France 1952
[Modern, double-loaded corridor, concrete]

Simmons Hall, Steven Holl
Cambridge, Massachusetts 2004
[Postmodern, double-loaded corridor, glass and metal]

The Unite d'Habitation deploys a double-loaded corridor in collaboration with an interlocking "L" section. Defined by two sectionally combined figures, the central corridor occurs on every third floor. Depending on the unit, the organization provides alternating access to either the upper stem or the lower stem of the two-story unit. The result is a minimization of circulation space and a maximization of both quantity and quality of the residential units, which provide two stories of usable space with double-height spaces.

Simmons Hall, located adjacent to Baker House on the campus of the Massachusetts Institute of Technology, uses a double-loaded corridor configuration as its horizontal organization, with residential units on either side of the central corridor. The uniqueness of the configuration comes through the vertical erosions of what Steven Holl terms "mixing chambers" (illustrated by the organically formed walls in plan), which provide atrium-like spaces as formal vertical circulation that aggregate and condense the horizontal movement. The moments of densification and overlap become opportunities for social interaction and communality.

Double-Loaded—Urban

The urban double-loaded corridor offers the efficiency of a central circulation trajectory flanked by layered spaces of commerce, office, and residential. Some of the most potent urban conditions are defined by concentrated shopping and organized aggregation. The boulevard, shopping street, public mall, or promenade typify and activate an intensity of social density, urban activity, and mutually supported life.

The Avenue de Champs-Élysées
Paris, France 1724
[Neoclassical, boulevard]

The urban proposition of the double-loaded corridor applies to nearly any street, but merits particular distinction when considered as an urban agenda. The boulevard as a spatial type demands a cohesive sense of scale, plantings, and circulation zones to breed consistency despite the variety of functions and storefronts. The Champs-Élysées is an example of a commercial strip defined by a central multi-lane car thoroughfare and large sidewalks that create a massive pedestrian zone lining the storefronts. As an aggregation of the world's finest retailers, the entire boulevard has become a shopping zone connected by the infrastructure of the street and the self-similarity and physical adjacency of **programs** *(shops). The double-loaded nature works on a larger scale as the boulevard connects the Arc de Triumph to the Grand Arch in La Defense, which establishes its presence as equal to the monuments that it connects.*

	ORGANIZATION SYSTEMS	27
	PRINCIPLE	

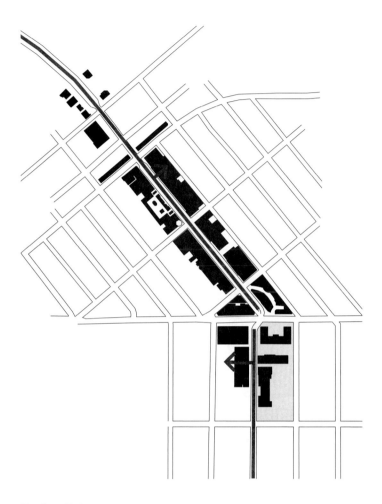 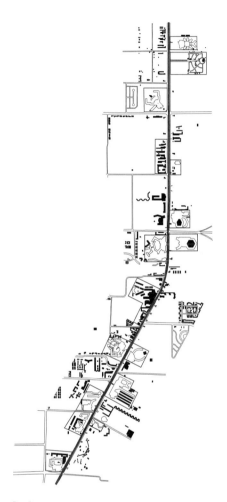

Rodeo Drive
Beverly Hills, California 1970s
[Postmodern, eclectic styles]

Rodeo Drive is a world-renowned and iconic hub of commercial retail. The form is a typical street configuration, simply lined with highly iconic, over-the-top luxury brands and their associated spaces. The double-loaded circulation is largely pedestrian with vehicular traffic used for touristic "cruising." The urban fabric, defined by elaborate storefronts, establishes the edges of the double-loaded corridor. Each storefront establishes its relationship to the street by developing visual appeal through identify and enticement, generating both image and sales.

Las Vegas Strip
Las Vegas, Nevada 1970s
[Postmodern, linear strip eclectic styles]

The Las Vegas Strip developed as a series of nodal hotels and casinos arranged along a primary automobile artery. Though each originated as an independent, self-contained world, they are connected by their shared street and the traffic it bears. The expanding complexes engage the pedestrian and vehicle into their spectacle. Their power comes not just from the uniqueness of their offerings, but from their relative adjacency, density, and collection. The double-loaded "strip" becomes the urban collector, stringing the casinos together and creating a continuous urban experience.

| 28 | ORGANIZATION SYSTEMS PRINCIPLE | LINEAR ORGANIZATION | POINT-TO-POINT GEOMETRY | PLAN READING | URBAN SCALE |

Point-to-Point—Urban

An urban point-to-point configuration refers to an urban planning methodology where a series of nodal points, defined by monuments, geography, or historical significance are connected by radiating axial boulevards. These connecting lines establish view corridors and collectively generate a web of hierarchical circulation. The resulting urban fabric is one of incremental spatial connectivity that heightens the understanding of one's relative position in space. The civic and monumental nodes are set in heightened contrast to the infill fabric of the subordinate fabric buildings. This juxtaposition provides a collective civic perception of the object (monument) relative to the field (urban fabric). The "point" stands out as a distinct and hierarchical element.

Rome
17th Century Nolli Plan
[Baroque, linear streets connecting piazzas]

Motivated by a desire to connect the seven pilgrimage churches of Rome, Pope Sixtus V (1520–1590) designed a plan for a new organization of the city itself. Implementing the Renaissance ideals of the primary straight street contrasted by the urban fill, he implemented a series of nodes (marking the crossing of the axes with obelisks), which served as cornerstones of new thoroughfares and circulation patterns. The system not only connected the critical points of the Church and city, but also produced a cohesive urban fabric though the carved-out system of circulation, which the city still enjoys today. One of the clearest examples of this urban plan is seen in Piazza del Popolo, where three different axes converge at the obelisk in the center of the Piazza.

| | ORGANIZATION SYSTEMS PRINCIPLE | **29** |

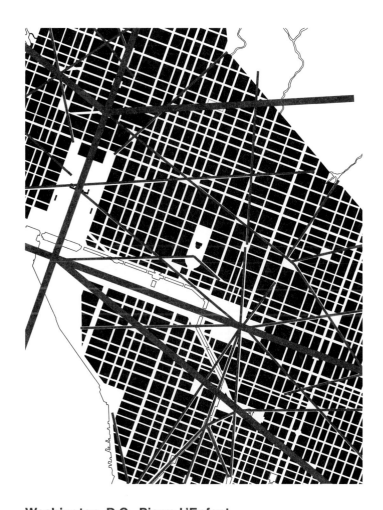 

Washington, D.C., Pierre L'Enfant
1791
[Neo-Baroque, broad avenues radiating from circles and squares]

Paris, Georges-Eugène Haussmann
1870
[Second Empire, connecting boulevards]

As the nation's capital, Washington, D.C. is defined by its collection of monuments, museums, and governmental buildings. The city's urban design was laid out with a Baroque methodology that identifies hierarchical nodes within the city fabric and connects them with grand radial boulevards. The intensity of the point-to-point visual and physical connection of the city's civic infrastructure is juxtaposed against the regular gridded urban fabric. This condition is further emphasized by the standard height of the fabric buildings, capped by legislation in 1899 to give consistency to the collective. As a result, dominance is conveyed on the U.S. Capitol Building and the Washington Monument through their contrasting scale. The consistency of the resulting urban fabric allows for the clear articulation of formal subtractions, or removals, that do not subscribe to the orthogonal geometry of the street grid.

Haussmann's influence and impact on the Paris city fabric is dramatic and omnipresent through the contrasting presence and uniformity of the connective boulevards. Designed as subtractions from the overcrowded Parisian fabric, Haussmann carefully identified a series of trajectories that linked the major monuments and cultural institutions with grand boulevards. He simultaneously threaded these subtractive avenues through the most derelict, crime-ridden, and "lower class" neighborhoods, thereby changing not only the physical but the social fabric through his compositional transformation. By establishing a street width, building height, module of bay, and horizontally layered coursing—as well as roof levels, setbacks, sidewalk configurations, and plantings—the dimension and compositional effect is formal, cohesive, and grand.

Grid

The grid is a fundamental figure produced through the multidirectional field of arrayed rows and columns. The system produces amplified density at the crossing lines, which establishes a patterned array of nodes while simultaneously producing a field of adjacent orthogonal islands of space. Grids are flexible in the dimension of the spacing in both the X and Y direction, allowing for large formal diversity. The variable spacing of the increment allows for variety, which, through equidistant spacing, produces a square grid. A more elongated spacing can produce a rectangular Roman grid. The relative reading of the grid can emerge from the emphasis on point (structure), divergent patterns (variation within the pattern and spacing of the grid), module (repetitive spacing of unit and increment), and the relationship of multiple grids and their relative positions (multiple overlain or shifted grids). All of these operate dynamically at both the urban and the architectural scales. The grid can be used to define building form, structural module, space planning, ornament and pattern, or simply underlying geometric bay systems.

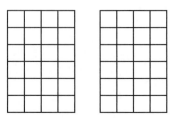

RELATIVE

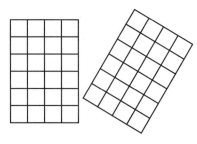

ROTATED

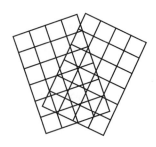

OVERLAID

POSITION

ORGANIZATION SYSTEMS
PRINCIPLE 31

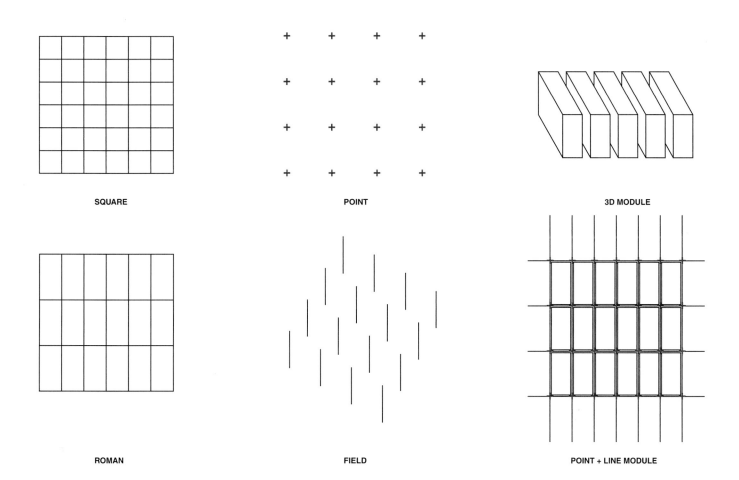

32	ORGANIZATION SYSTEMS	GRID	POSITION	PLAN	ARCHITECTURE
	PRINCIPLE	ORGANIZATION	GEOMETRY	READING	SCALE

Grid Position—Architecture

On the architectural scale, the grid has been heavily deployed as a form generator. The dominance of the field establishes a focal importance to the relative position of an entity within a larger context, which sets up a dialogue based on interrelationships. The authority of the grid establishes the field-to-field relationship as one that privileges relativity through the geometric differentiation of position and orientation relative to the larger field. The rotation can be applied to the field, the frame, and/or the object. The relative relationships established by the regularized increment allow for either the reading of the grid itself or the objects within it.

Berlin Philharmonic Hall, Hans Scharoun
Berlin, Germany 1963
[Eclectic Modern, rotated geometries, metal and concrete]

The Berlin Philharmonic Hall is the premier performance hall for Germany's capital city. The building establishes its significance through its uniqueness of form. Defined through a series of aggregated grid geometries, the relativity of their internal organizations produces a collective formal dialogue. The experiential performance is orchestrated not just by the highly composed local grid forms, but also by the spaces and interactions generated between these grid forms.

| ORGANIZATION SYSTEMS | 33 |
| PRINCIPLE | |

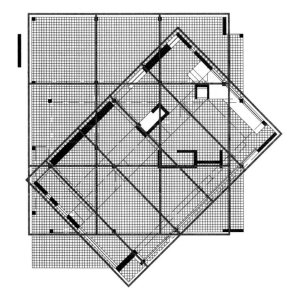

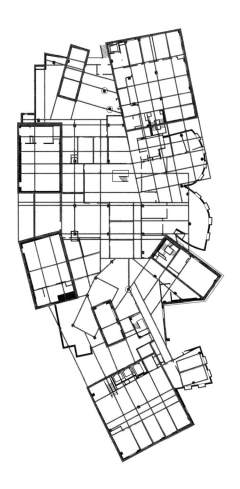

House III, Peter Eisenman
Lakeville, Connecticut 1971
[Deconstructivist, rotated geometries, wood and stucco]

House III operates on pure geometry. Defined by two grids, their shift and overlay create a condition of juxtaposed, superimposed, and **palimpsest** *systems. The interrelationship of the grids allows for the discrete yet dynamic engagement of the two systems to produce a third condition of interface and response. The localized decisions between the hierarchies become the compositional directives closely associated with Eisenman's post-functionalist agenda.*

Stata Center, MIT, Frank Gehry
Cambridge, Massachusetts 2004
[Deconstructivist, rotated geometries, metal and masonry]

The Stata Center extends Gehry's formal principles of the building as an aerosolized "village" of isolated forms, in collaboration with his rotational formalism. Each piece defines a localized geometry; their rotation, deformation, and interrelation establish relativity between the pieces that provides for both a formal and experiential relationship. The individuated geometries of the discrete pieces allow for the massiveness of the building to be broken down in scale while presenting a dynamic visual composition both locally and collectively.

Grid Position—Urban

On the urban scale, the grid establishes both the circulation and the block pattern of the urban fabric. The position of the grid becomes a responsive mechanism relative to contextual conditions or hierarchical intentions. The primary scale of the establishing field creates the urban rhythm, dimension, and infrastructure. The localized responses to the natural geography and topography, varied geometric figures produced by overlaps and disruptions in the grid, along with programmatic nodal conditions, all present opportunities for specificity and neighborhood identity.

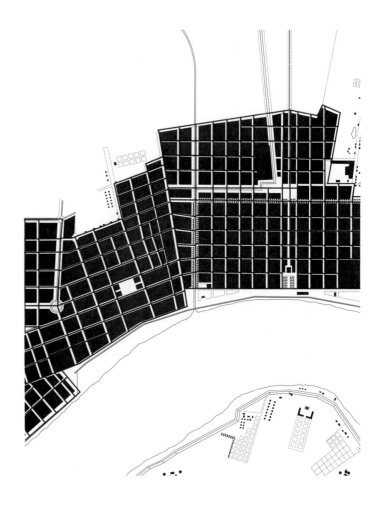

New Orleans, Louisiana
1718
[French Colonial, radial sections adjusting to river]

In New Orleans, the urban fabric is organized through a series of grids. Each grid rotates to address the natural curvature of the Mississippi River, resulting in disjunctive seams. The internal rigidity and dominance of the grid's geometry are confronted with the need of its idealized system to negotiate the massive organicism and curvature of the meandering river. The geometry has to adjust. The grid is broken into wedge-like fields in order to maintain its parallel presence to the riverfront, resulting in a series of seams that resolve the grids and create radial connectors to the river.

	ORGANIZATION SYSTEMS
PRINCIPLE	**35**

Baltimore, Maryland
1869
[Jeffersonian grid, topographic patchwork configuration]

Baltimore deploys a grid to organize a standardized and regimented system of development. The natural topography that transitions from waterfront to rolling hills and valleys results in the emergence of a series of geographically undevelopable zones. These "natural" swatches become disruptions within the urban fabric. As broken voids, the grid assumes a patchwork nature to adjust and respond both functionally and formally. The result is an integration of a superimposed geometric organization on the natural context of local geography and ecosystems.

Athens, Greece
1909
[Neoclassical, radial grids of extension geometry from Parthenon]

As a city founded in the Hellenistic period, Athens has a long evolutionary development. Built at the base of the Acropolis (one of the most significant global, architectural, and cultural monuments), the impact of its presence is significant. As a geographic, topographic, and architecturally dominant figure, the Acropolis establishes a massive and looming presence in the city fabric. The physical and cultural development of the surrounding urban fabric developed as layers of radiating grids. The gridded fabric in relation to the iconic presence of the Acropolis establishes wedges of grid fields that locally relate and respond. The collective city fabric is dependent on the instigation of a singular hierarchical moment.

36	ORGANIZATION SYSTEMS	GRID	FORM	PLAN	URBAN
	PRINCIPLE	ORGANIZATION	GEOMETRY	READING	SCALE

Grid Form—Urban

The relationship of columns and rows within the grid type itself establishes variable formal types. The relative ratio of the grid unit establishes the grain, directionality, and perception. The field can be repetitive or evolutionary (changing across itself). The shape of the grid dramatically impacts the scale of the fabric and the urban feel of a city.

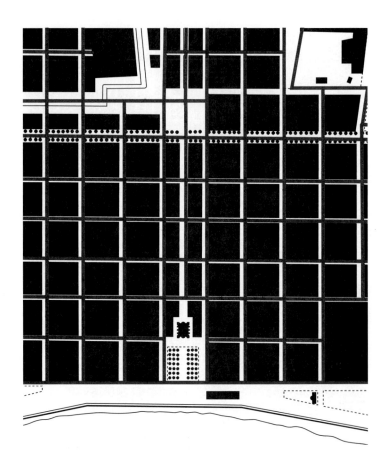

New Orleans, Louisiana
1770
[French Colonial, square grid]

New Orleans' French Quarter deploys a square grid. Its equal-sided nature establishes the base module of a **multivalent field**. *The lack of grain (a hierarchy or differentiation of length to width of the base module) and resulting directionality provided makes for a ubiquitous condition with equality in the field. The result is an evenness to the field that requires hierarchy to be established by other means. In New Orleans, the river, squares, parks, and radial seams between grids establish the hierarchies of the system by creating localities of form, identity, and community.*

	ORGANIZATION SYSTEMS	**37**
	PRINCIPLE	

New York, New York
1811
[Jeffersonian grid, rectangular grid]

The New York City grid is a Jeffersonian grid. Defined by its differential grain (a minor east-to-west width, and an elongated north-to-south length), the resulting form is a directional field (formed by the differing width to length) that allows for an understanding of personal position and movement within a larger field. The grid form also establishes a varied orientation and local condition based on both position within the block and location of block within the urban field. The elongated block negotiates the natural proportions of Manhattan, adjusting the aspect ratio of the urban grid to correspond to the local geography.

Bari, Italy
1893
[Neoclassical, transformational grid]

Bari is a city that evolved its urban form through the chronological development of its planning. It represents urban growth that engages the hybrid systems of diverse eras, scales, speeds, and their associated organizational systems. Beginning with a medieval fabric that meanders and develops with a localized decision-making process, the urban fabric breaks and evolves to a conventional grid. The modern era, governed by the efficient and organized movement of the automobile, required a standardized and rationalized condition of urban development. The contrasting juxtaposition of the informal medieval street with the rational grid within a single city fabric results in an overtly layered condition of chronologies and planning methods.

Grid Structure—Architecture

In nearly any composition, the grid is used as a system that embraces repetition and thus standardization and uniformity. The efficiency of the grid can be extended to include compositional deployments, functional boundaries, framing elements, and planar figures in both plan and section. The grid can be deployed as a controlling field to locate structural positions. When this occurs, the structural grid establishes a meter and organization for the larger composition that the architect must reconcile. Its density and regularity can create the field of a **hypostyle hall** (a repetitive field of tightly positioned columns), or a more loosely placed field (as defined by structurally performative, materially rationalized, and optimized structural frames).

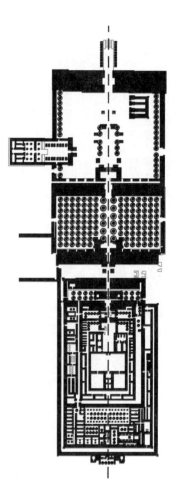

The Great Temple of Amun at Karnak
Luxor, Egypt 1306 BCE
[Egyptian, hypostyle hall, masonry]

The Temple of Amun at Karnak is a quintessential example of a hypostyle hall. Defined by its field of densely located columns, ordered in a field, to establish an abstract forecourt that is grand in scale and multidirectional, the visual power in the layering of the sequence produces a hierarchy and drama to the experience. At Karnak, there is a varied series of hypostyle halls. The interrelations of mass of column, density of grid, and scale of units (relative to the anthropomorphic dimension) produce an otherworldliness of monumental proportions. The heroic qualities impress upon the visitor the power, significance, and dominance of the temple and those being honored.

	ORGANIZATION SYSTEMS
	PRINCIPLE **39**

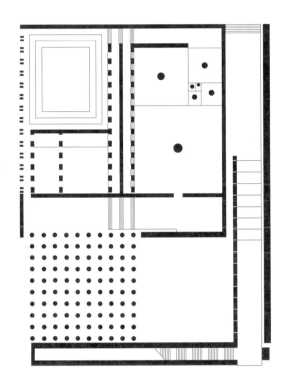

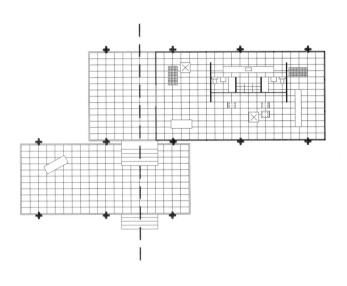

Danteum, Giuseppe Terragni
Rome, Italy 1942
[Italian Rationalist, hypostyle hall, masonry]

In the Danteum, an unbuilt project by Giuseppe Terragni, the hypostyle hall is a grid of masonry columns devoid of ornament and stripped of capital and base. The gridded field establishes a visual density through the archetypal tradition of the hypostyle hall, deployed with Modernist sensibility. The organization of the grid sets up an infinite, expansive pattern that references the woods in "The Divine Comedy." The basic geometry of the grid establishes the order, perspective, repetition, and seriality that define the power of the chamber.

Farnsworth House, Mies van der Rohe
Plano, Illinois 1951
[Modern, free plan – universal space, steel and glass]

*The regularity of the grid is relentlessly deployed by Mies van der Rohe with a reductive power in the Farnsworth House. The dominant governing of the grid, which can expand infinitely in all directions, and the transparent glass boundary create an expansive extension and connection with the surrounding natural landscape. The bold, white-painted structural steel columns create a modularized framework that dimensions and defines the rational "local space" within a broader infinite space. The structural grid, unbounded and extended by the slipping floor and roof plates, is further sub-gridded and scaled through the module of the **travertine** floor. The articulation of the details allows for the built system to project beyond its physical presence through the systemization of material and experience.*

40	ORGANIZATION SYSTEMS	GRID	MODULE	PLAN / ELEVATION	ARCHITECTURE
	PRINCIPLE	ORGANIZATION	GEOMETRY	READING	SCALE

Grid Module—Architecture

The grid as an architectural module uses the repetitive field to establish latent or overt modular patterns within the broader composition of the plan, section, or elevation. As a two-dimensional or three-dimensional methodology, the system can establish the geometric governance of the composition and determine the larger repetitive proportional systems. As a repetitive field, the localized bay or unit can emerge as a module within the repeating sequence that then orders and governs the resulting architectural form and its reading.

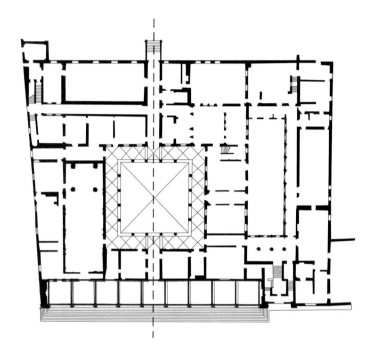

Ospedale degli Innocenti, Filippo Brunelleschi
Florence, Italy 1445
[Italian Renaissance, modular arcade, masonry]

At the Ospedale degli Innocenti, Brunelleschi utilized the square as a method of giving this building a clear and concise organization and ensuring that its relationship to the piazza was unambiguous. The parallel bays of nine squares form the **loggia** *that spreads across the façade of the Ospedale, thereby establishing a transparent relationship with the piazza. The modules are placed exactly perpendicular to the axis of the piazza, which through its extension establishes the purity of form in the piazza. This idea of regular modular proportion ultimately leads to the majority of design principles used during the Renaissance.*

ORGANIZATION SYSTEMS
PRINCIPLE 41

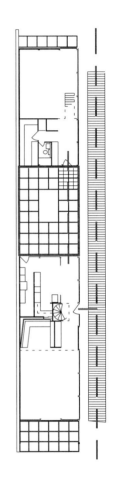

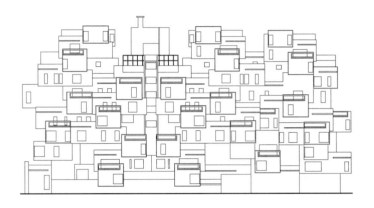

Eames House, Case Study 8, Ray and Charles Eames
Los Angeles, California 1949
[Modern, repetitive module, steel and glass]

In Case Study House 8, Ray and Charles Eames are interested in the design of mass-produced elements. The final built project employs a repetitive grid module made possible through the use of standardized components. Founded in the modularity of steel, delicate trusses and columns establish a repetitive bay system that is equally subdivided with careful standardization. The result is a series of composed and layered grids that derive the module of aggregated pieces. Intrinsic to these standard methods of construction, but artfully and cleverly deployed, the repetitive unit establishes a systematically composed whole.

Habitat 67, Moshe Safdie
Montreal, Canada 1967
[Late Modern, modular unit aggregation, concrete]

In Habitat 67, Moshe Safdie extends the use of the module into a fully three-dimensional capacity. Through spatial, formal, and fabrication-based methodologies, the module dominates the localized piece and collective composition alike. The precast, prefabricated unit, stacked and aggregated to provide interrelationships, becomes a local and collective scale maker. The grid of the module becomes fully three-dimensionally enacted with predictable systemization but localized individuality. In Habitat 67, the module is fully deployed compositionally, **tectonically**, organizationally, and spatially.

Free Plan

The free plan emerged from the structural and material innovation that allowed for the shift from the solidity and dominance of the load-bearing wall to a columnar system. No longer dependent on the positional necessity of span, weight, and gravity, the open composition of the floor plan allows for a fluidity of space, flexible in form and effect. To emphasize spatial continuity, interior subdivisions are minimized, curvilinear forms are used, and programmatic blurring and boundary flexibility provide for a spatial decompartmentalization. In addition, the development of the independent interior core (housing, circulation, and utilities) allows for an often furniture-like formal reading, detached from the ceiling plane and often materially varied in surface. The freedom of the plan is now compositionally governed, fundamentally transforming the spatial and formal capabilities of architecture.

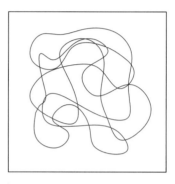

FREE MOVEMENT

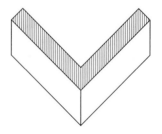

WALL MASS

| | ORGANIZATION SYSTEMS PRINCIPLE | **43** |

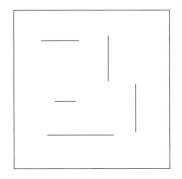

FREE WALL POSITION

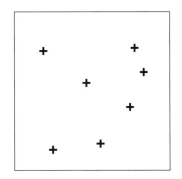

FREE STRUCTURAL POSITION

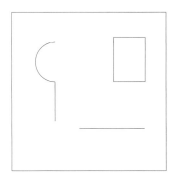

FREE FORM

STRUCTURAL COLUMN

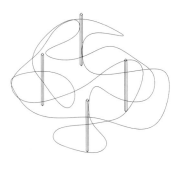

FREE MOVEMENT

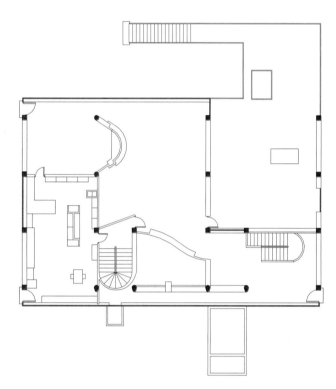
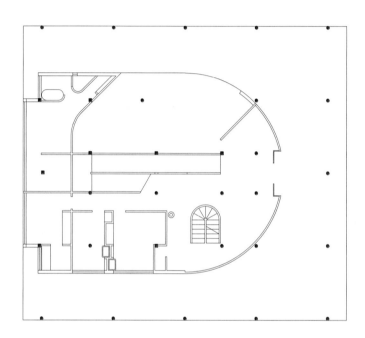

Villa Stein at Garches, Le Corbusier
Garches, France 1927
[Modern, free plan, concrete and masonry]

The Villa Stein at Garches remains one of the seminal examples of Corbusier's free plan. Executed prior to Villa Savoye, it establishes the principle of the free plan, albeit primarily on the **piano nobile**. *Unlike the Villa Savoye, Villa Stein is not removed from the ground plane; however, the concept of the free plan is immediately made obvious. Upon entering you are confronted with four freestanding columns that articulate the separation of structure from surface. On the main public floor there are numerous columns that disengage from the wall, float in the space, and allow the non-load-bearing walls to move around them. As a result, the space is orchestrated and controlled by the partition walls, allowing a complex composition independent of the structural requirements.*

Villa Savoye, Le Corbusier
Poissy, France 1929
[Modern, free plan, concrete and masonry]

Villa Savoye, as perhaps the most iconic Modern house, deploys Le Corbusier's five points: free-form façade, ribbon window, **pilotis**, *roof garden, and the free plan. These five points are interconnected and interrelated. The pilotis are columns that elevate the building, detaching it from the ground plane and allowing the removal of the structural responsibility of the wall. On the exterior, this allows for a free-form façade, independent of weight and force and now independently composable. To emphasize this quality, the ribbon window is employed to illustrate the edge-to-edge connectivity. Running across the full façade, there is no continuity of the opaque surface, thus the segmentation and independence of the façade from carrying loads is further expressed. In the interior spaces, a similar dialogue occurs. With the walls being released from the responsibility of weight, their form and position are flexible, allowing for gestural, compositional, or fully open plans. These quintessential characteristics are the hallmarks of the free plan.*

| ORGANIZATION SYSTEMS |
| PRINCIPLE | 45 |

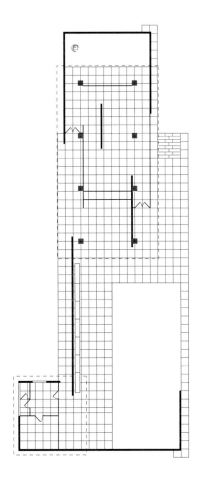
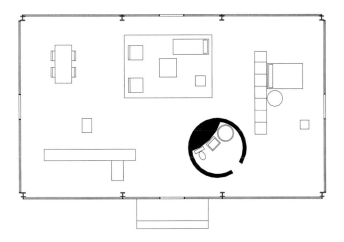

Barcelona Pavilion, Mies van der Rohe
Barcelona, Spain 1929
[Modern, free plan, steel, glass and masonry]

The Barcelona Pavilion uses the free plan to define a specific moment within an infinite space. The dissolve of boundary, through the compositional effect of slipping wall planes, dramatically extends the space blurring any demarcation of interior and exterior. The fluid flow of space from function to function and from inside to out allows the masterful and highly evocative ambiguity of edge. Transparency, coupled with a hyper-material celebration of wall (horizontal book-matched marble) and column (chromed cruciform) objectifies the discrete elements and further removes any reading of them as either traditional or static. The spatial and organizational result is a highly fluid free plan.

Glass House, Philip Johnson
New Canaan, Connecticut 1949
[Modern, free plan, steel and glass]

In The Glass House by Philip Johnson, the free plan is accomplished through the singularity of the space. A large, single room divided only with one "core" containing the bathroom and fireplace, which allows for a continuity and consistency to the space. Functional subdivisions are made through furniture position, orientation relative to the site, the position of the core, and the physical presence of social rituals. The effect of the free plan is further emphasized through the uniformly transparent façade. The glass dissolves the boundary between inside and out, and extends, orients, orchestrates, and composes the adjacent interior zones. The material allows for a merger between house and site, commingling and interlacing their interdependence and experience.

| 46 | ORGANIZATION SYSTEMS — PRINCIPLE | DISPERSED FIELDS / PODS — ORGANIZATION | | |

Dispersed Fields / Pods

Dispersed fields are defined by a series of pods organized within a collective system. This array of individually articulated forms aggregates to define a single composition. Their individual figuration and form maintain an equal importance to the interstitial spaces defined by their relative positions. As repetitive modular units, systematized and regimented (organized fields), or individuated and iconically independent in their formal articulation (disorganized fields), the conceptual spatial configuration is legible as a single collective configuration.

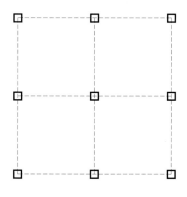

POINT

ORGANIZED

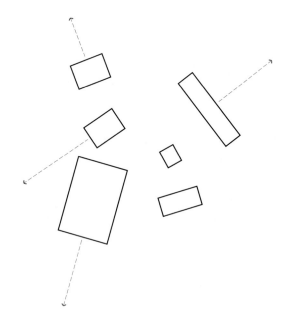

DISPERSED CONTEXTUAL POSITION

DISORGANIZED

| ORGANIZATION SYSTEMS PRINCIPLE | 47 |

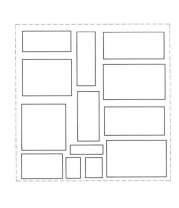
FORMED BOUNDARY

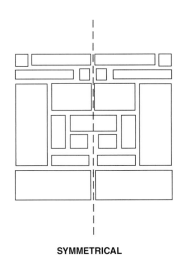
SYMMETRICAL

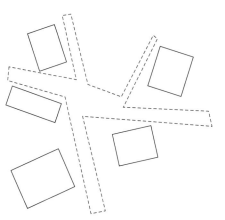
**DISPERSED CONNECTED POSITION
[CIRCUIT + PATH]**

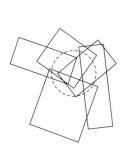
SUPERIMPOSED AND OVERLAP

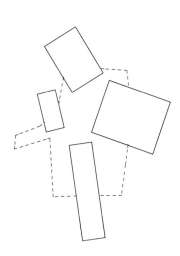
CONNECTIVE TISSUE FORM

| 48 | ORGANIZATION SYSTEMS
PRINCIPLE | DISPERSED FIELDS / PODS
ORGANIZATION | ORGANIZED
GEOMETRY | PLAN
READING | ARCHITECTURE
SCALE |

Organized Fields / Pods–Architecture

The organized configuration of pods comes out of three primary considerations: modularity as determined by the mass-produced serial object; varied familial relationships with proportional associations in either the form or the spaces produced between that collectively create ordered spatial effects and perceptual qualities; and intense geometric relationships that produce ordered connections and inter-formal intensities.

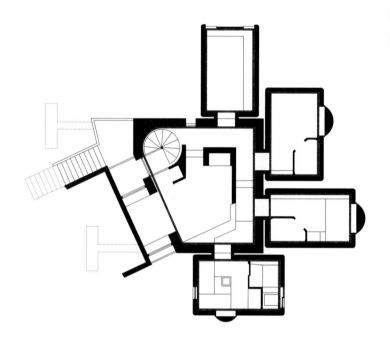

Capsule House K, Kisho Kurokawa
Tokyo, Japan 1972
[Metabolist, individuated units, concrete, metal and wood]

The Capsule House is defined by its repetitive pods. Four identical metal modules cantilever off a central concrete mast; their varied orientation and program provide identity to the repetitive whole. Maintaining an identical shape and dimension, each of the units has a highly modern exterior that emphasizes standardization and dramatizes their "plug-in" nature. The varied position and orientation allow the highly standardized units to identify themselves, finding individuation through their relative position and the functional subdivision.

	ORGANIZATION SYSTEMS
	PRINCIPLE **49**

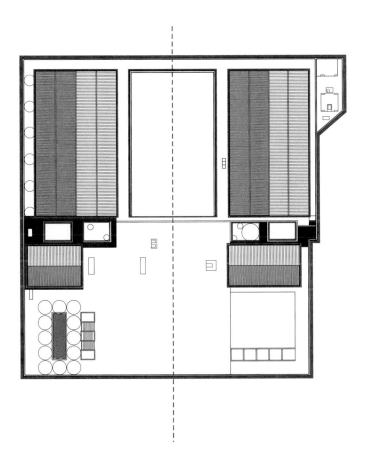
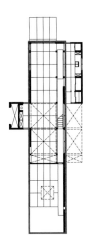

Manzana de Chinati, Donald Judd
Marfa, Texas 1974
[Minimalist, symmetrical array of buildings, masonry and metal]

The Manzana de Chinati, alternately known as "The Block," is an urban-walled compound designed and lived in by the minimalist sculptor Donald Judd. Isolated by a high perimeter wall, the house comprises an entire block in downtown Marfa, Texas. The framed composition creates an inner world that allows for a controlled composition. Dealing with existing buildings, moved buildings, and ground-up construction, Judd compartmentalized the functional components and established a series of discrete objects within a highly organized field. Though symmetrical in building position, localized deviations in function, interior subdivisions, the position of installed artwork, and moments of discreetly composed asymmetry produce moments of discovery and individuation. The house emerges from this relentless celebration of geometry. The result is an objectified, yet highly considered composition that blurs between art and architecture.

House 22, Bryan MacKay-Lyons
Oxner's Head, Nova Scotia 1997
[Postmodern Regionalism, axial house dispersed in line]

In House 22, Bryan MacKay-Lyons develops a scheme where the house engages the physical site and topography of the narrow peninsula as well as the extended views to the water. By developing a tube-like house with large amounts of transparency on the lateral ends, he visually extends the house directionally outward. By then repeating the same form and material, but with a shorter extrusion dramatically positioned as an extension of the house (though clearly devoted to the other waterfront), the house becomes a sequential layering of water view, house, land view, house, water view. With the bold primary move, the small square footage dramatically expands into the landscape, marking itself and the purity of its axially organized diptych pods.

| 50 | ORGANIZATION SYSTEMS
PRINCIPLE | DISPERSED FIELDS / PODS
ORGANIZATION | ORGANIZED
GEOMETRY | PLAN
READING | URBAN
SCALE |

Organized Fields / Pods—Urban

Organized fields and pods on an urban scale refer to segmented compositions that are legible as singular developments. The field of the composition uses either the node of the organizational grid crossing or the infill of the organizational system (typically a grid) as a figure ground. As hierarchical systems, the fragmentation sets up an organized field that establishes either functional gradients or formal **datum**.

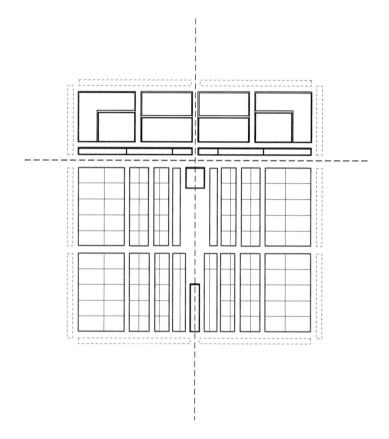

Roman Encampment
Various locations ca. 70
[Roman, organized dispersed field]

The Roman Encampment is the consummate illustration of an organized urban field. As a temporary camp or the foundational establishment of a town, the organizational deployment was incredibly rational and regimented. Encampments were laid out based on the Cardo (north – south) and Decumanus (east–west), orienting the primary cross axis along these cardinal directions. Carefully planning and organizing the disparate functions along positional and hierarchical guidelines, the resulting urban composition was highly repeatable, internally organized, and rationally dominated.

| | ORGANIZATION SYSTEMS PRINCIPLE | 51 |

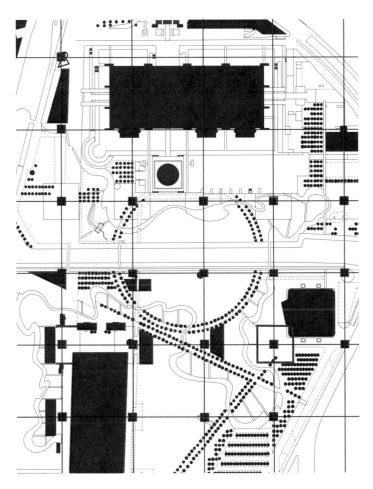

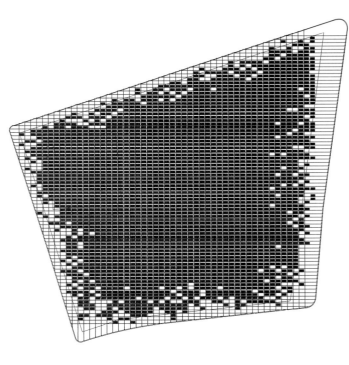

Parc de la Villette, Bernard Tschumi
Paris, France 1982
[Deconstructivist, arrayed gridded field]

Berlin Holocaust Memorial, Peter Eisenman
Berlin, Germany 2005
[Postmodern, repetitive field, concrete]

Parc de la Villette uses an urban-scaled grid to organize itself. A series of deconstructed red cubes is located at the intersecting nodes of the grid. Regimented in their repetitive positions, but individually composed, the diversity of the units is held together through their commonality of position, color, and base cubic geometry. The result is a family of siblings that relate to one another and collectively define an organized field across the diverse landscape of the park. The organizational superstructure allows for the reading of the larger landscape as a connected field with discrete interstitial zones that include and organize diverse functions and existing buildings. The result is a dispersed field that is locally individuated, yet collectively hyper-organized. The dispersed field establishes a formal infrastructure that regiments the landscape and defers to a chronological, cinematic, memory-based perception to accomplish its experience.

At the Berlin Holocaust Memorial, Eisenman uses a repetitive organized field to emphasize and demonstrate the scale and scope of the holocaust. A rectangular monolith is developed and then arrayed across the vast expanse of the site. Varying in height, the relentlessness of the repetitive field establishes a monumental experiential quality that subtly differentiates across the array to create standardization and variation at the same time. The iteration of the monument references the scale, scope, and impact of the events it is commemorating. The collective field is a dominant, repetitively organized scheme that through its scale becomes urban in scope, impact, and experience.

52	ORGANIZATION SYSTEMS	DISPERSED FIELDS / PODS	DISORGANIZED	PLAN	ARCHITECTURE
	PRINCIPLE	ORGANIZATION	GEOMETRY	READING	SCALE

Disorganized Fields / Pods—Architecture

The disorganized composition allows for more experientially based configurations as they respond and allow for localized adjustments. Determined through the conditional engagement with site through context, view, light, wind, topography, etc., the localized response through position and form allows for specificity of the part while negotiating the interrelationships of the collected pieces.

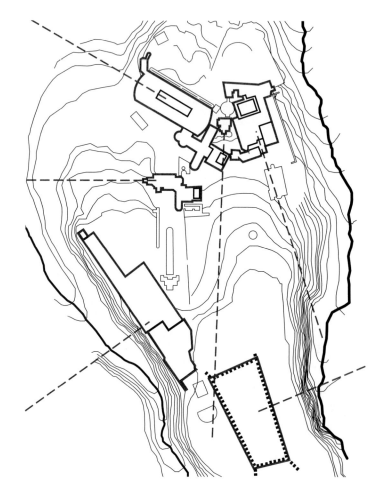

Hadrian's Villa
Tivoli, Italy 120
[Classical, conglomeration of shifted buildings, masonry]

Given its placement in the countryside (not the city), Hadrian's Villa in Tivoli is an example of a disorganized dispersed field built with overlapping and incongruous geometries. Many of the architectural decisions were based on the topography of the site as well as the various axes and spaces and how they were to align. It is considered disorganized in that it was not orthogonal, but instead formed a series of seemingly unrelated angles. The genius of the plan was the ability of these axes and angles to form a comprehensive whole. The villa was extremely influential during the rise of postmodernism in the 1970s and 1980s for its flexible variability of form.

	ORGANIZATION SYSTEMS	
	PRINCIPLE	53

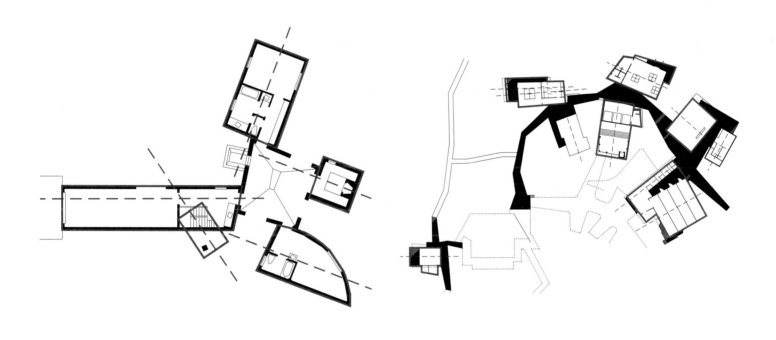

Winton Guest House, Frank Gehry
Wyzata, Minnesota 1987
[Deconstructivist, diverse aggregated forms, diverse materials]

The Winton Guest House uses highly postmodern methodologies to establish the iconography of the individual components through form, function, and material. Each programmatic piece is given a specific formal response and a material enclosure to reinforce that form. The collective assembly is determined through collagist methods that compositionally aggregate and juxtapose the diverse forms. The resulting figure is identifiably both the parts and the whole.

Atlantic Center for the Arts, Thompson Rose
New Smyrna Beach, Florida 1997
[Postmodern, linearly linked varied pavilions, wood, concrete]

The Atlantic Center for the Arts is a collective campus set in the primal landscape of eastern Florida. As a prestigious artist residency program, the center caters to artists working in visual media, dance, performance, music, and written word. The architectural response was to provide discrete pavilions for each of the disciplines. The diverse pavilions take their identity from their relative positions (to both the landscape and one another), and their internal functional requirements. They are connected through a series of locally calibrated walkways that govern circulation while facilitating exterior spatial pockets for pause and interaction.

Raumplan

Raumplan deploys a sectional variation to define programmatic separation. The vertical differentiation allows for a split leveling that provides visual layering and interconnectivity between the spaces. The vertical hierarchy allows for a spatial condition established in the control and dominance of position, function, and place. The unique nature of this organizational system, developed by Adolf Loos, and the functional difficulties inherently associated with the sectional premise result in a limited scope of usage, despite the power of its spatial configuration.

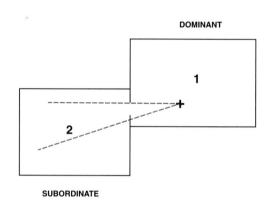

SECTIONAL VISUAL CONTROL

ORGANIZATION SYSTEMS PRINCIPLE	**55**

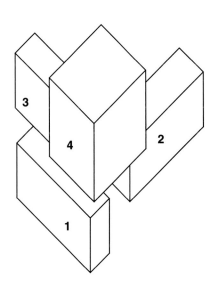

VOLUMETRIC ACENSION

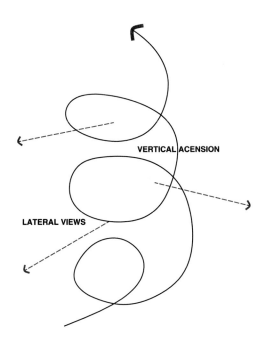

LOOPING EXPERIENCE

56	ORGANIZATION SYSTEMS	RAUMPLAN		SECTION	ARCHITECTURE
	PRINCIPLE	ORGANIZATION		READING	SCALE

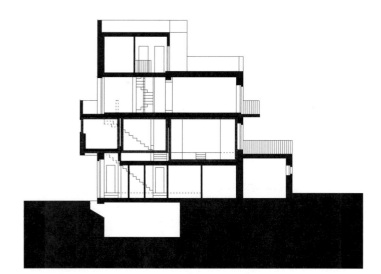 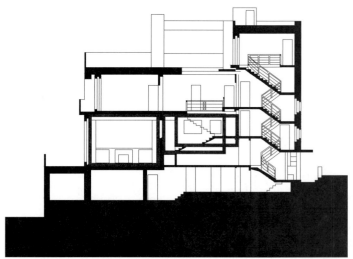

Villa Moller, Adolf Loos
Vienna, Austria 1927
[Modern, Raumplan, sectional arrangement of spaces, masonry]

The Villa Moller, like most of Loos's residential buildings, only uses the Raumplan on the public floors. In this particular Villa, Loos utilized the Raumplan beginning with a central room on the second floor of the house. This room essentially has no function other than to be a starting point for the other rooms that surrounds it. From this foyer, one can ascend to the breakfast nook and library, descend to the music room, or move into the dining room, which is also connected to the music room through a disappearing stair. Loos used the various functions and relationships of these rooms and this foyer as a way of laying out the primary public rooms of the house and establishing visual and physical interface of hierarchical spaces.

Villa Müller, Adolf Loos
Prague, Czech Republic 1930
[Modern, Raumplan, sectional arrangement of spaces, masonry]

The Villa Müller uses the Raumplan in a different way to accomplish the vertical differentiations between the various rooms, employing a series of stairs that spiral through the house. The stairs address each room, beginning at the entry foyer, ascending to the main living room, then the dining room, and finally ending in the ladies room, or dammenzimmer. This final room of the sequence is itself a three-dimensional diagram of the house, similarly redeploying multiple levels within the room.

	ORGANIZATION SYSTEMS	57
	PRINCIPLE	

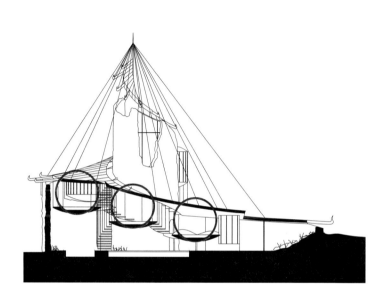

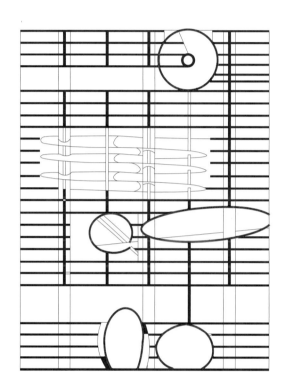

Bavinger House, Bruce Goff
Norman, Oklahoma 1955
[Expressionist Modern, suspended trays, metal and masonry]

The Bavinger House develops as one 96-foot-long spiral with sectionally terraced levels arrayed within. The different functions are both bounded and identified within a single level. The open terrace provides a sectional interrelationship between each of the levels, allowing for downward views, but providing visual privacy as one looks up. The organic form is clearly Wrightian, but the unique spatial configuration is decidedly Loosian.

Très Grande Bibliotheque, Rem Koolhaas, OMA
Paris, France 1989
[Postmodern, cubic book field with subtracted spaces, unbuilt]

Rem Koolhaas and OMA use the Raumplan in this unbuilt, yet still iconic, project through a subtractive spatial method. By first deploying a solid cube, defined as a "full" **poche** *of the book stacks, the primary congregation spaces and circulation pathways are carved out of the mass. The removal of the sectionally staggered and individually figured voids produces a terraced, spatial Raumplan. Identity of figure comes from the purity of individual form. Connected by circulation systems (a vertical field of elevator towers and divergent, sequential promenades of stairs and escalators), the disparate figures are separated with varied margins through the thick poche of the book stacks, limiting the spatial overlap and visual interpenetration typical of Raumplan.*

Hybrid

Perhaps the most common condition is not the employment of any one organizational system taken through a project, but the combination of multiple organizational systems, deployed for their localized efficacy. Synthesizing aspects of multiple systems or splicing varied systems into one another, the hybrid is perhaps the most common method of the pluralist, postmodernist style that emphasizes form over the clarity of plan. The integration of varied systems allows for internal contrast of spatial types, collagist formal composition methodologies, reinterpretation of traditional types through contemporary cultural conditions, varied material and technological capabilities, and diverse spatial agendas. The resulting hybridizations allow for spatial complexities, nested compositions, and diverse (yet juxtaposed) decedents of the primal organizational methods.

| ORGANIZATION SYSTEMS |
| PRINCIPLE | **59** |

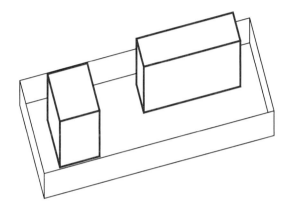

BOX IN BOX

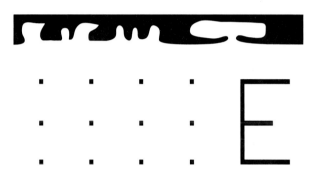

POSITIVE NEGATIVE

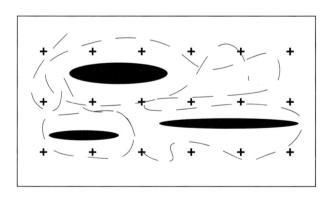

FREE PLAN WITH VOIDS

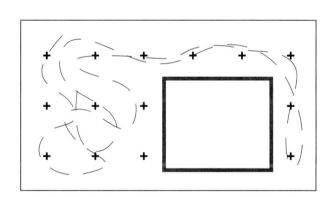

FREE PLAN AND COURTYARD

60	ORGANIZATION SYSTEMS	HYBRID		PLAN	ARCHITECTURE
	PRINCIPLE	ORGANIZATION		READING	SCALE

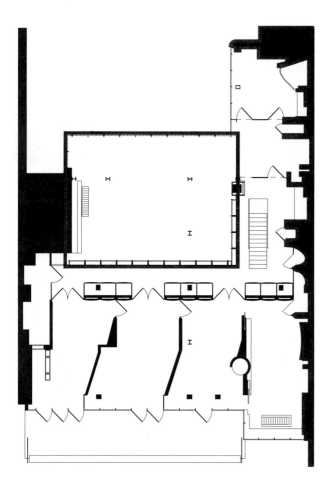

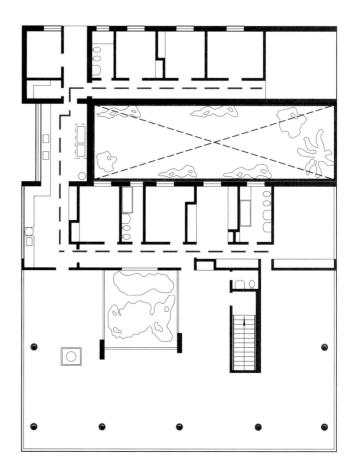

Maison de Verre, Pierre Charreau
Paris, France 1931
[Modern, additive and subtractive spaces, glass and metal]

Maison de Verre uses both an additive and subtractive planning methodology. The side party wall is thickened with carved removals that become service-based support spaces, which are defined by a curvilinear organization. The main double-height volume is defined by a volumetrically additive and articulated riveted-metal frame that shows both its segmentation and componential construction. The two methods represent entirely contrasting spatial methodologies and formal resultants. Their hybridization and juxtaposition heighten their contrast and establish an increased power to each of their disparate methodologies.

Bo Bardi House, Lina Bo Bardi
Sao Paulo, Brazil 1951
[Modern, free plan and courtyard, concrete and glass]

The Bo Bardi House fuses the idea and space of the free plan with the more traditional aspects of the courtyard house. Bo Bardi employs the free plan in the large, public, glass rooms of the house and then contrasts these with the cellular private and service rooms, which both look onto a courtyard. This juxtaposition is emphasized by the fact that the free plan portion of the house is raised on pilotis and the courtyard side is grounded into the earth. This allows for an extreme and captivating reading of the house.

| | ORGANIZATION SYSTEMS PRINCIPLE | **61** |

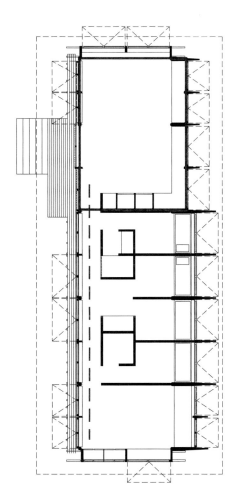
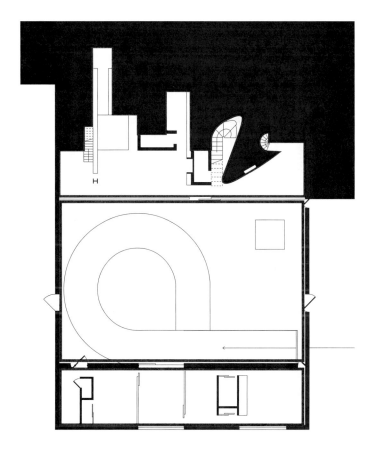

Marika-Alderton House, Glenn Murcutt
Northern Territory, Australia 1994
[Postmodern, free plan with corridor, metal and wood]

The Marika-Alderton House, even in its very small footprint, uses an evolutionary spatial planning. The public spaces are organized in a single free plan room that allows for diverse configurations and interconnections of its disparate programs. In opposition, the private zone is highly regimented, deploying a single-loaded corridor to connect to and access the clearly bounded and specifically segmented bedrooms and bathroom. The combination of the two allows for their juxtapositions: public vs. private, day vs. night, fluid vs. fixed, and open vs. closed.

Bordeaux House, Rem Koolhaas
Bordeaux, France 1998
[Postmodern, additive and subtractive spaces, diverse materials]

In its organizational structure, the Bordeaux House subscribes to the same principles as the Maison de Verre. Koolhaas juxtaposes the heroic scale of the large open space of primary living space with the subtractive (carved out and removed) and independently formed service spaces. The relationship of the free plan to the carved organic form of the poche establishes contrasting spatial scales and formal environments.

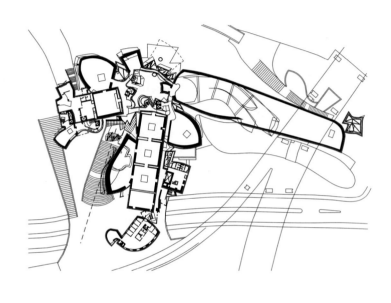

German Architecture Museum, O.M. Ungers
Frankfurt, Germany 1984
[Postmodern, building within a building, diverse materials]

This project is a renovation of an older, existing building. The exterior of the original building is essentially left untouched, whereas into the interior Ungers effectively inserts a new building. The original classical building is primarily constructed of masonry while the new postmodern building is completely white and without material character. This difference in material allows for the two to contrast, whereas the similarity of design principles ensures that the two buildings work in unison.

Guggenheim Bilbao, Frank Gehry
Bilbao, Spain 1997
[Expressionist Modern, regular and figured spaces, metal]

The Guggenheim Bilbao uses diverse organizational and spatial types. Like many of Gehry's schemes, the program is subdivided into discrete elements based on function and gallery type, and then defines the form of the individual components from a localized functionality and experiential effect. The diversity of the forms, rotated and positioned, then draped with an encompassing **cladding**, *creates an even further divergent spatial type. From the orthogonal to the highly curvilinear, the forms and functions deploy varied organizational and hierarchical tactics, resulting in collagist forms and experiences governed by visually expressionist methods.*

ORGANIZATION SYSTEMS
PRINCIPLE
63

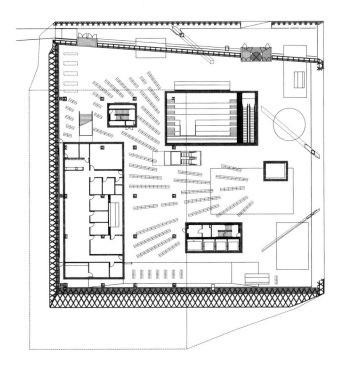

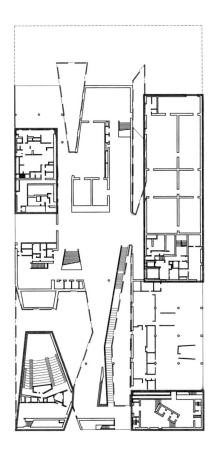

Seattle Public Library, Rem Koolhaas
Seattle, Washington 2004
[Postmodern, continuous free plan, concrete and glass]

The Seattle Public Library is innovative in both form and program, from the reframing of a series of traditional programs to the addition of entirely unique programs to the library, such as the transformation of periodicals into a "living room." These innovative functional ideas manifest in unique environments that collect in a functionally driven, aggregated form. The sectional arrangement is defined by a linear organization along a distinct experiential promenade. Layering the sequence to ease and orchestrate an interpenetration of spaces, the innovation of this organizational system extends into the subtlety of each piece. For example, the book stacks slowly ramp and fold upward along a continuous floor plate, allowing for a singularity of the space despite its sectional layering. The collective and individual organizational methodologies collaborate in their diversity to produce a form that is driven by the diagram of the program.

De Young Museum, Herzog and de Meuron
San Francisco, California 2005
[Postmodern, free plan with courtyards, metal and glass]

The de Young Museum uses a simple **mat building** design governed by a regular structural grid and relatively conventional orthogonal galleries. Its innovation lies in the carefully articulated development of surface through the variably perforated and dimpled skin. The three-dimensional dynamism emerges through a method of introducing diversity and locality into the massive horizontality of the plan. By pulling diverse strands apart to create internal courtyards, the bending forms create divergent geometries and open the mass of the mat to outdoor spaces. The resulting courtyards serve as interruptions in the field. As locally sculpted conditions, they introduce a second hybridized organizational system into an otherwise conventional plan.

64 PRECEDENT
PRINCIPLE

Precedent

02 – Precedent

Precedent is one of the most important and widely used design methodologies. It centers on the concept of utilizing ideas or forms that are found either in other buildings or in other fields. It references the idea of what has come before and, if properly deployed, takes the original idea in an entirely new direction. The significance of precedent is that it is always with the design in terms of experience and memory, and also with the viewer as a framework for evaluation. Thus the role of precedent is intertwined with the role of history.

Founded in the interrelationship of history and design, precedent relies on our ability to look at built structures of a past time and unpack the underpinning principles that dictated their form and purpose. We then translate these principles into active design tools and use them to produce a unique composition. The result may be directly dependent upon the historical reference or only obliquely attached.

The following list classifies the broad strokes, the diversity of precedents, and their references. Bracketing them into categories that look at: vernacular construction relative to natural environmental forces; cultural vocabulary associated with place and tradition; intellectual, theoretical, or philosophical trajectories that relate to methodological and abstract formal agendas; material and construction techniques that emerge from craft, tectonics, and making; and historical precedent that typically emerges out of typology and program.

Vernacular Precedent Vernacular precedents are typically concerned with historical buildings that emerge out of a specific time and place and their influence on other buildings that are executed later. Unlike typology, which is primarily concerned with programmatic or functional types, vernacular precedents can and often do cross over to include numerous types, focusing more on technique and response to place and environment. A vernacular precedent usually focuses on the language of a project and its rationale in addition to its direct form or space. This establishes an importance of formal language as a critical methodology to the understanding of precedent.

Cultural Precedent Cultural precedent references the use and cultural customs of a building and form. Buildings throughout history have always been connected to rituals and events specific to the communities that built them and the constituents they served. The associated forms are often critical participants in the formation of architecture. Architects can utilize rituals as a way to engage designs that are related to these traditions (either directly or as evolutionary iterations). Cultural ideas of religion, education, defense, literature, and film represent a portion of the possibilities evident in this extremely rich vein of design possibilities.

Intellectual Precedent Intellectual precedent, while not intrinsically visual, remains one of the most powerful and useful tools available to a designer. Unlike other precedents that are concerned with concrete conditions, intellectual precedents are typically related to the underpinning ideal of a precedent. This movement away from form and toward idea is no less powerful, but often has more flexible formal interpretation and application. Often, architectural movements (discussed specifically and chronologically in – Chapter 15, "Meaning") are a direct way of using an intellectual precedent. For example, an architect's decision to create a minimalist design for a project is based on an intellectual rather than a form-based argument.

Material Precedent As the name implies, material precedent deals with the concept of materiality and its historical application. Materials are fundamental to the making of architecture. They provide a language of how things are made and also affect the composition and development of form. The root of architecture is in the technology of how we build. The relationship of material precedent to craft (that relates to cultural precedent), or performative characteristics such as environment (that determine vernacular precedent), illustrates its associative and interlaced nature. Materials have always carried with them certain meanings that are employed through associative techniques of tooling, methods of construction and assembly, and the resulting forms. For example, marble carries with it different meanings, crafts, and forms than wood or plastic. Building on these traditional associations, material precedent is also engaged in the continual evolution of new materials and new fabrication techniques. As evolutionary assembly processes emerge and established materials are challenged, these material associations with the traditions of manufacturing, fabrication, and construction are challenged, defining new boundaries of material precedent.

Historical Precedent Historical precedent is the most overt and common of the precedent types. It encompasses a broad territory including all of the aforementioned precedent types. Historical precedent specifically refers to using history as a design source. It is rooted in functional typology (which is based in how a building is used), but branches out to all architectural considerations, such as natural systems, formal traditions, and contextual methodologies. Historical precedent looks at everything that has come before, building and evolving from its understanding and continuing the advancement of architecture.

There are of course bridges that hybridize and synthesize these categories. For instance, a southern American farmhouse has a response to climate (vernacular precedent), emerges out of local cultural norms (cultural precedent), is derivative of the inhabitants' nationality and traditions (intellectual precedent), uses indigenous craft and materiality (material precedent), and has direct historical associations with the house and farm type (historical precedent). Described in more detail, each of these categories represent the diverse aspects of history that can serve as referential models when associated with their lineage and stand on the shoulders of precedent.

Using precedent as a design tool requires extensive knowledge of the history of buildings, not only from a functional standpoint, but also from a cultural one. It is this knowledge that allows architects to survey and embrace appropriate models for their designs. Each precedent carries with it distinct political and cultural factors, and the architect must always remain mindful of these contextual influences. Precedent remains one of the most useful and powerful methodologies as it offers a wealth of information that provides opportunity either overt or subtle. The potential combinations and hybridizations of precedent allow for a multiplicity of levels within which a designer can create architecture. Precedent has long been a fundamental tool in education, design theory (from Vitruvius to Alberti to Rowe), and practice. Even as architects continue to look for new ideas and tactics in their exploration of form and meaning, history and the reference of that which has come before will be essential.

Lineages

Throughout history, architects have studied the evolution of form. Engaging the legacy of history through architectural types and projects that have addressed similar design agendas is fundamental to the design process and defines distinct lineages. The most direct application of precedent is working within a particular functional typology. These typological classifications create lineages that run as veins throughout the history of architecture. The specific iterations and evolutions of traditional architecture are primarily felt through the development of the plan and façade, allowing form to truly enter the conversation only during the Renaissance. An examination of these evolutionary lineages quickly illuminates the significance of precedent and typology. The impact and effect of precedent as an anterior examination allows for its extension and continuation. It is in this way that one can study the various reiterations through the chronology of architecture. Upon close examination of the lineage, elements emerge. Over time, these formal moves were developed to solve functional problems. Influenced by stylistic concerns or common practice these become aestheticized solutions. This is evident in the scrolls that were so common on church façades during the Renaissance. Originally this scroll was introduced by Alberti as a way of hiding the two angled roofs of the side naves of Santa Maria Novella in Florence. For the next century, numerous architects reworked this formula until Palladio offered the layering of classical temples as yet another way of solving this problem. Typically these lineages remain within a similar programmatic type; however there are instances when architects have hybridized other aspects of a building, like its geometry, as a starting point. The emergence of lineages has ensured through the centuries that architecture has remained a profession that is based on cultural evolution.

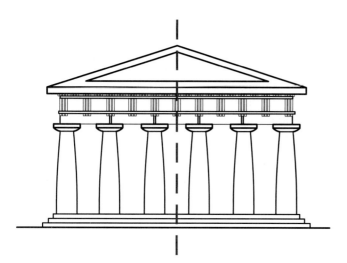
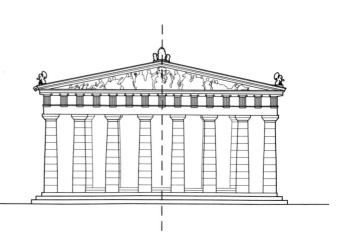
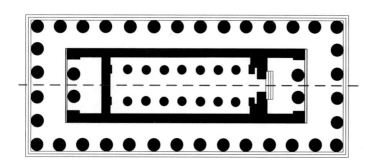
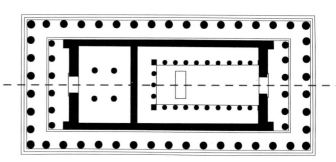

Temple of Hera II at Paestum
Paestum Italy 460 BCE
[Classical, pagan temple, masonry]

*The Temple of Hera II at Paestum is an early example of the temple type. Its façade has the typical six columns and the resulting central axis that leads to the porch and the main entrance. The columns that form the colonnade, or peristyle, sit on the stylobate, or base. The entrance leads to the larger room, the **cella**, which has two rows of smaller columns on either side of the axis. On the side opposite the entrance there is the **opisthodomos**, or treasury. This layout was common for the majority of Greek Temples.*

Parthenon, Phidias
Athens, Greece 432 BCE
[Classical, pagan temple, masonry]

*The Parthenon in Athens is a **Doric peripheral temple**. Here the number of columns was changed to eight on the façade and the general proportions were refined to reflect the size and importance of the building. The number of columns in antis, which is the space between the columns of the façade and the inner layer of the building, is dramatically increased from two in Paestum to six at the Parthenon. The cella, which housed the statue of Athena, had columns on three sides as compared to two, which was common in earlier temples, and the opisthodomos was considerably larger. The proportions, the frieze, and the statues that adorned the building, particularly in the pediment (the top triangular form created by the roof), were the main differences and improvements over earlier temples. These sculptures were essential to the communicative ability of the building.*

The temple plan is considered one of the initial architectural designs of civilized man. The following examples trace the development of the temple as an architectural type and historical precedent. The selected examples illustrate key thresholds in the evolution of the temple, highlighting significant changes and type deviations influenced by diverse localized factors including formal, political, religious, and site-based conditions.

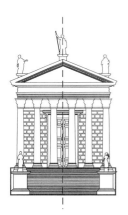

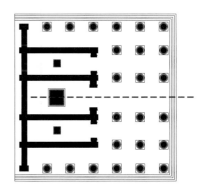

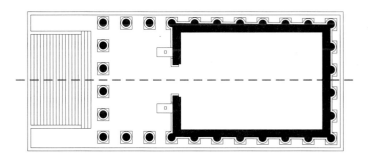

Temple of Jupiter
Rome, Italy 69 BCE
[Classical, pagan temple, masonry]

The Temple of Jupiter was different than other temples due to its site conditions. It was atop the Capitoline Hill overlooking the Forum. Due to this site condition it only prioritized three elevations instead of the typical four that Greek temples required. Like all Classical Temples, it had an even number of columns (six) on the main façade; however, the space between the center two columns was made larger as a way of imparting hierarchy. There were three layers of columns on the front porch and only one layer on the two sides. This plan form reflects the Romans' desire to limit the access to the front of the temple. There were three rooms in the temple that held sacred documents. This also marks a functional departure from the typical temple, which held the statue of the god that it represented.

Maison Carrée
Nimes, France 16 BCE
[Classical, frontal colonnaded rectilinear temple, masonry]

The Maison Carrée was built by Agrippa to commemorate his two sons. This transposition of function to honor mortals and not gods represents a significant programmatic evolution to the historical type. Unlike many earlier temples, it does not have the colonnade that encircled the building, but instead has columns that are engaged within the exterior walls. The freestanding aspect of the temple is still relevant; however, the Temple has now become completely dominated by the single entry axis. The single axis represents the power of the emperor rather than the democracy of Greece. The order of the Maison Carrée is Corinthian, reflecting a more precise and detail-oriented architecture. Additionally, the proportions of the façade were adjusted to give the temple a vertical emphasis, which was not seen in earlier examples.

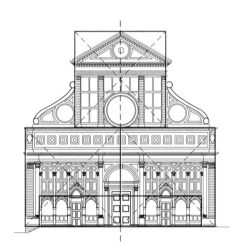
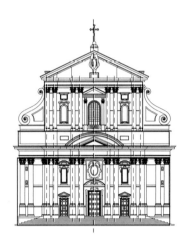

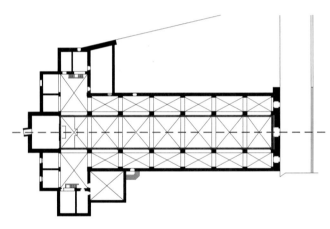
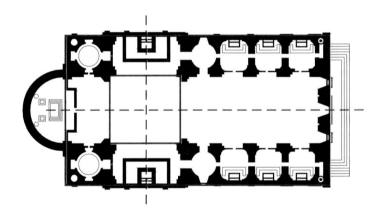

Santa Maria Novella, Leon Battista Alberti
Florence, Italy 1470
[Renaissance façade, rectangular nave with transept, masonry]

Almost two hundred years after this Gothic Church was completed, Alberti was called upon to complete its façade. He faced the problem of how an architect could incorporate the classical language of the Renaissance into the traditional shape of a church façade. Beyond this, Alberti was sensitive to the original language of the church that was firmly Gothic. Alberti combined the green and white marble of the traditional Florentine Gothic with classical elements like the temple front and proportioning systems to synthesize the two styles. The giant scrolls used to cover the side aisles were an original design feature that was adopted by others and used for centuries.

Il Gesu, Giacomo Barozzi da Vignola
Rome, Italy 1580
[Renaissance, Latin Cross, masonry]

Il Gesu represents the zenith of church design in the late Renaissance. Its façade builds on the ideas laid out by Alberti over a hundred years earlier. The **pilasters** (columns embedded in a wall) of the bottom section align directly with the temple elevation above. This accentuates the position of the nave, allowing the architectural elements used to reveal the organization of the interior of the church. The attic of the lower composition becomes the base of the upper composition thereby joining them together. There is a remarkable balance between the vertical and horizontal elements in the façade. Like Alberti, Vignola used the scrolls as a way of incorporating the roof of the side naves into the overall composition.

The development of the church façade is one of the most celebrated and clear lineages. Highlighted here is the evolution of the Renaissance church façade as a specific segment and moment in time. This lineage illustrates the reconciliation of the plan, which had essentially remained unchanged since the Gothic, with the new classical language of the Renaissance. Each project built upon the legacy of that which came before, using precedent to guide their design.

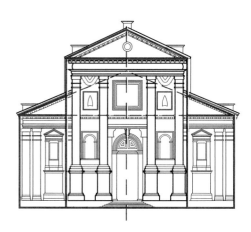

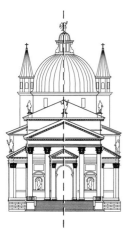

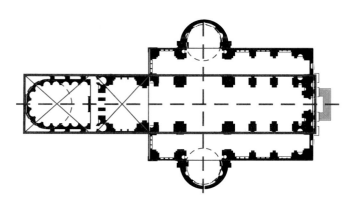

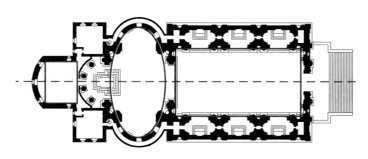

San Giorgio Maggiore, Andrea Palladio
Venice, Italy 1580
[Renaissance, transept and head additions, masonry]

The façade of San Giorgio Maggiore illustrates another significant transformation through the use of the pagan temple. Here, Palladio used two superimposed temples as he had done at San Francesco della Vigna. By using the natural triangle formed by the pediments, Palladio was able to fuse the elements to screen the roofs of the side aisles. This allowed the synthesis of classical elements into the form of the church typology. This solution proved much more holistic than the additive scrolls that were popular in southern Italy. Palladio took great liberties with the proportions and spacing of the elements but was able to use classical forms to solve the spatial problems of the typical Christian façade.

Il Redentore, Andrea Palladio
Venice, Italy 1591
[Renaissance, three layered plan, masonry]

At Il Redentore, as at San Giorgio Maggiore, Palladio used the concept of multiple temple fronts as a way to organize the façade of the church and to reflect its interior. The façade of Il Redentore used five temple fronts, which all spring from the same horizontal line. He employed a large **plinth** *(an elevated base), which allowed him to begin the temple fronts at the same level. This solved some of the difficulties of alignment and proportion seen in the earlier façades of San Francesco della Vigna and San Giorgio Maggiore. The frontal approach across the Grand Canal ensured that the façade was read in its entirety and on axis. A complex series of proportions (diverse geometric systems that govern the composition) ensured the unification of the collective composition. The multiple overlaid façades allowed Palladio to evolve his classical language to its highest complexity, representing a pinnacle of church façade during the Renaissance.*

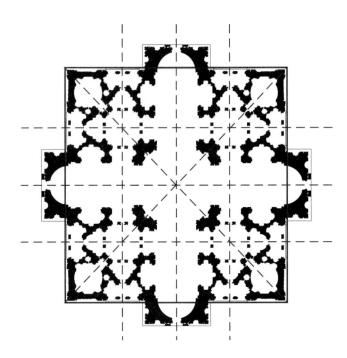
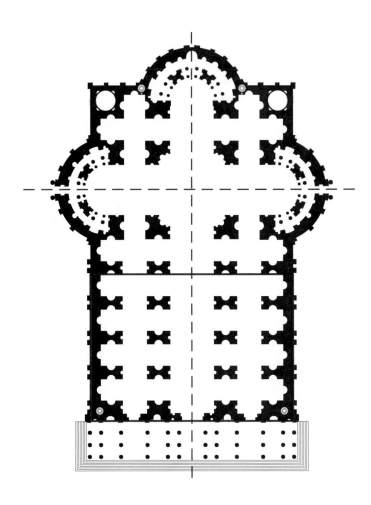

St. Peter's Basilica, Donato Bramante
Rome, Italy 1506
[Renaissance, Greek Cross, masonry]

Bramante used the Greek Cross when he designed the first plan of St. Peter's Basilica. This plan type was popular during the Renaissance as it embodied a purity of form. The plan was a large square that had two equal axes running through it at right angles. The axes represented the transept, the nave, and the chancel. They were all equal and each ended in an apse. The hierarchy that was naturally established in a Latin Cross plan was missing. At the intersection of the two main axes was the crossing, or high altar, which was to be covered by a dome based on the Pantheon. Placed diagonally from the main altar, in the four corners, were smaller domes representing separate spaces that were to house other chapels. This organization remained through the successive plans. Structurally, the four piers of the large dome were inadequate and subsequently were enlarged.

St. Peter's Basilica, Raphael
Rome, Italy 1513
[Renaissance, Latin Cross, masonry]

Raphael took over the design of St. Peter's Basilica after Pope Julius died and made a number of changes to the plan. Most notable were the reversion to the Latin Cross and the alteration of the nave into five bays with apsidal chapels off the aisles on either side. These changes, while significant, were instituted while keeping the same primary organization Bramante employed. The overall plan was made much larger and more rectangular. The apses of the transept and chancel were also made more prominent by the addition of ambulatories (covered procession ways). The overall effect was that the plan was not as dynamic as Bramante's original, but it solved the structural requirements and functionally provided room for the congregation.

St. Peter's Basilica presents a dramatic opportunity to illustrate lineage within one building. Due to its significance as a religious and cultural institution, a diverse collection of architects periodically operated on it using the ideals and principles of their disparate chronological eras. Each iteration built on the previous, simultaneously extending and reconfiguring the plan and elevation. The lineage documents the progression of the Cathedral from the initial plan by Bramante and continues through the successive plans by Raphael and Michelangelo and finally by Maderna.

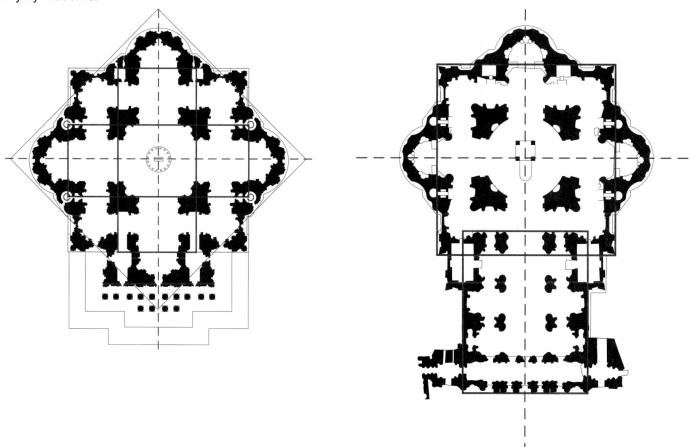

St. Peter's Basilica, Michelangelo Buonarroti
Rome, Italy 1547
[Renaissance, Greek Cross, masonry]

Michelangelo took over the project after a brief period during which Antonio da Sangallo the Younger oversaw the project. Michelangelo immediately reverted back to the Greek Cross plan, respecting Bramante's original vision. By making the walls and piers slightly thicker and more structural, he was able to achieve a cohesive, centralized plan that essentially still exists today. When he strengthened the walls, Michelangelo was able to modify them in such a way that Bramante's original projections of the apses were lessened. This allowed Michelangelo to create some of the most dynamic and undulating walls of the entire Renaissance. These revisions of the plan would also allow for the construction of the final piece of St. Peter's Basilica: Michelangelo's dome.

St. Peter's Basilica, Carlo Maderna
Rome, Italy 1607
[Renaissance, Latin Cross, masonry]

Maderna took over the project many years after Michelangelo died and originally wanted to alter it significantly by placing a ring of chapels around the building. Ultimately he would merely change St. Peter's Basilica back into a Latin Cross plan. During the Counter-Reformation it was believed that the Greek Cross was a pagan form and the Latin Cross was the symbol of Christianity. Maderna added three bays to Michelangelo's Greek Cross. He made the dimensions of the new bays slightly different from the original, thereby defining where the two projects met. Maderna also angled the central axis of the plan slightly so that the elevation would line up with the obelisk placed at the center of the piazza in front of the Basilica.

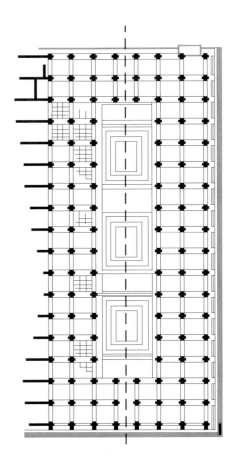

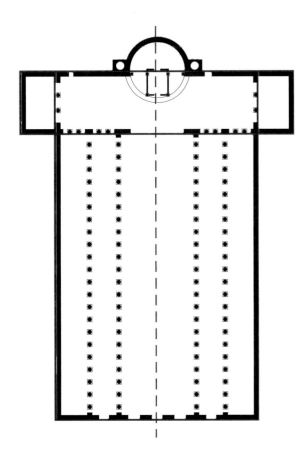

Basilica Giulia
Rome, Italy 46 BCE
[Classical, Basilica, masonry]

The Basilica Giulia was the largest building in the Roman forum. It was used by citizens for meetings and other official business. It was a large room, surrounded on all sides by a double colonnade. It is believed that this structure was a precursor to the five-aisle church. It could be entered at numerous points and hence had little hierarchy. The plan of the Roman Basilica served as the first model for the Christian Church. As a secular structure designed for the people, the Roman Basilica was clearly a more acceptable model than the pagan temple.

Basilica of Old St. Peter
Rome, Italy 326
[Medieval, Early Latin Cross, wood and masonry]

The original St. Peter's Basilica was a five-aisle church with a crossing transept at the altar end. Its plan and shape were reminiscent of the original Roman Basilicas, differentiated only by the fact that there was now an axial hierarchy and the crossing of the transept, which marked the position of the altar. A courtyard that marked the front of the church as well as the narthex further reinforced the primary axis.

The church has one of the most detailed, diverse, and complex lineages. It begins with a Roman prototype borrowed and deployed as the foundational building typology of the Christian Church. As the Christian culture developed, political and religious decisions were made that altered the design of each of the successive iterations. Structural and material abilities provided opportunity for forms to evolve, improving and developing both the aspirations and abilities of the church building type through precedent.

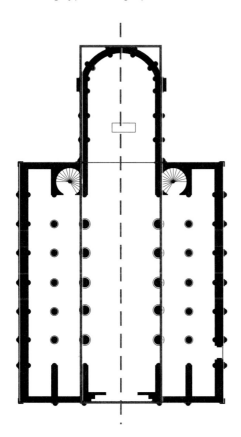

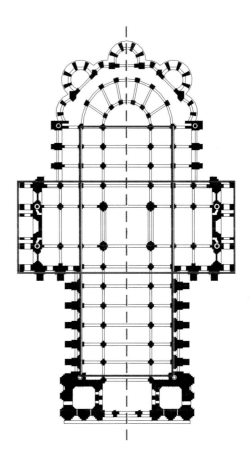

Sant Abbondio
Como, Italy 1095
[Romanesque, five-aisle church, wood and masonry]

Sant Abbondio in Como is one of the few remaining five-aisled churches. The transept is virtually unrecognizable in the plan, providing a more unified interior. There are two campaniles, or bell towers, that mark the transept's position on the exterior. The space around the altar, or chancel, is exaggerated and elongated. This form is then similarly carried through to the exterior. It is a clear representation of the architecture of the Romanesque, with its emphasis on the altar and chancel as governing figures of both the interior space and the exterior form.

Chartres Cathedral
Chartres, France 1260
[Gothic, Latin Cross, stone masonry]

Chartres Cathedral represents the pinnacle of the High Gothic Period. Using the Latin Cross in plan, the church is characterized by the extension of height and the dissolution of the wall to introduce light. The increased height of the nave reached a structural limitation with the associated wind loads. As a result, the secondary external structure of the flying buttress, which would become a hallmark of all Gothic cathedrals, was developed to provide the necessary bracing. With the wall now reduced to the delicate tracery of a structural skeleton, the infill of large glazing allowed for lighting effects to illuminate the spiritual spaces and the extensive stained glass through which biblical stories were told. The rose window was developed at the head of the nave both to resolve the vault of the roof and to give hierarchy to the tripartite façade.

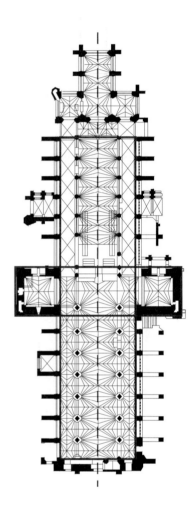
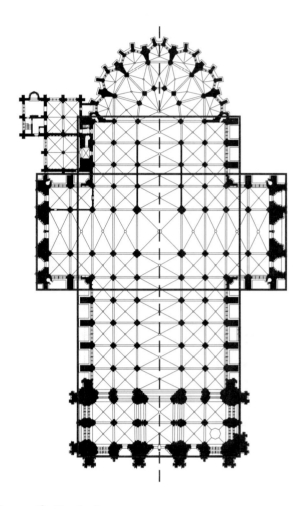

Exeter Cathedral
Exeter, England 1050-1342
[Romanesque to Gothic, Latin Cross, masonry]

At Exeter Cathedral, the evolution of the cathedral spanned the Romanesque through the Gothic eras. The front elevation resolves the center bay with a rose window, but leaves the vaults of the side chapels exposed. Extending the Gothic precedent, the front façade has a determinately large slope of both the nave and the side chapels. The inner ceiling of the central nave has highly ornamental tracery that flares out from the structural masts and transitions the vertical forces and geometries into the ceiling and roof. The intricate ornamentation reaches full capacity through the statuary, front façade, and highly articulated windows. The glazing uses the standard circle of the rose window, but then extends downward to capture the wall and dissolve the mass of the stone almost entirely. The ornament culminates in the highly formalized organ that has a multi-nodal presence in the nave.

Cologne Cathedral
Cologne, Germany 1248-1880
[Gothic, Latin Cross, stone masonry]

The Cologne Cathedral reiterates with Germanic refinement the principles of High Gothic ecclesiastical architecture. Developing the front towers with positionally symmetrical, steeply pitched, slate shingled roofs, the Latin Cross plan is reiterated with an encircling ring of highly detailed and formally dynamic flying buttresses. The skeletal expression of the structural forces of lateral thrust for both the lower and upper wall creates a density of elements. The collective perception of this structural filigree dominates the composition.

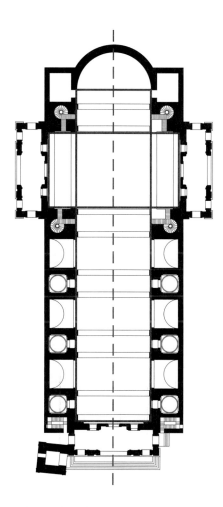

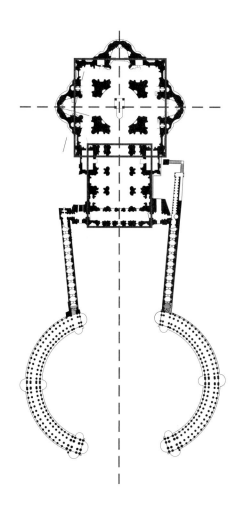

Sant Andrea, Leon Battista Alberti
Mantua, Italy 1476
[Renaissance, Latin Cross, masonry]

The plan of Sant Andrea reflects Alberti's obsession with the wall as an architectural device. The Latin Cross plan is conceived of as one massive space and is not divided by the columns that were so prevalent in Brunelleschi's churches. There are no side aisles, but instead one continuous space of the nave. Alberti incorporated a series of private chapels that replace what were traditionally the side aisles. The span of the nave, which is much larger than in earlier churches, allowed Alberti to take this monumental step forward in church design. Additionally, the proportional systems of the plan are carefully related and reiterated in the main façade.

St. Peter's Basilica, various
Rome, Italy 1506-1667
[Renaissance, Latin Cross, masonry]

St. Peter's Basilica in the Vatican is interesting as the evolutionary culmination of four separate architects over more than a one-hundred-year span. The original design by Bramante was for a Greek Cross. Raphael then altered the plan into a Latin Cross. Michelangelo later reverted the plan back to Bramante's original idea. Ultimately Maderna added three bays to Michelangelo's design to solve both religious and functional conditions. Today the plan expresses both the centralized concept of Michelangelo and the addition by Maderna to create a centralized, albeit Latin Cross, plan. Bernini's arms extended the spatial control of the Basilica into a baroque piazza.

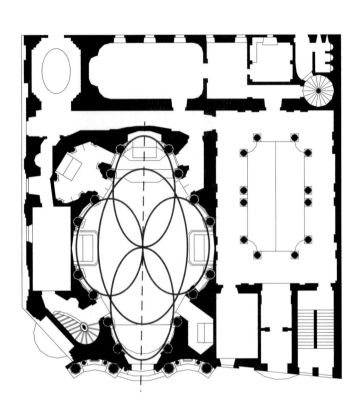
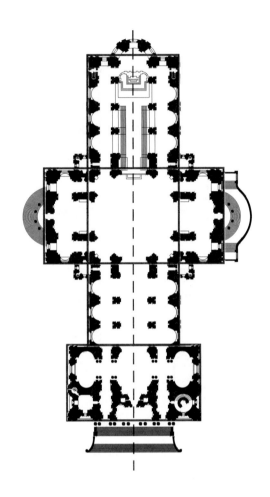

San Carlo alle Quatro Fontane, Francesco Borromini
Rome, Italy 1641
[Baroque, oval, masonry]

The plan of San Carlo alle Quatro Fontane reflects Borromini's fascination with the oval as a spatial type. Inspired by Galileo's recent discovery that the planets orbited in oval paths (as opposed to the previously believed circular path), the plan is a series of overlapping ovals that combine to form an intricate and intimate space. The walls of the church undulate and respond to the pressures of the various oval geometries. The altar and two chapels are now participants in the main space, no longer hidden in aisles, transepts or chancels. The dome of the church takes the shape of a large oval, which controls the complexities of the plan below. The intricacy of the geometry is fully manifested in all aspects of the dynamic form.

St. Paul's Cathedral, Sir Christopher Wren
London, England 1668-1710
[Late English Renaissance, Latin Cross, masonry]

Built on one of the oldest church sites in London, St. Paul's Cathedral was redesigned multiple times. Its final iteration was built by Sir Christopher Wren following the destruction of the earlier church by the London fire of 1666. Designed in a Late English Renaissance style, the plan takes the form of a lengthened and thus modified Greek Cross. The nave and the choir are equal, differentiated by an apse at the east end and a narthex with two large chapels at the west end. The transept is equal in width to the nave and choir, but shorter in length. This allows for the correct hierarchical relationship between the parts of the cathedral, yet still obtains a centralized reading. The high altar is placed not at the crossing, but at the extreme east end of the choir. To avoid the need for flying buttresses, the walls were thickened, providing a massiveness and presence while minimizing the articulation of detail on the side elevations. The final composition merges the Gothic traditions of the cathedral with the Late Renaissance language of the era in which it was built.

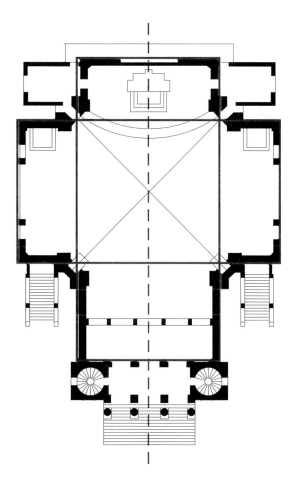
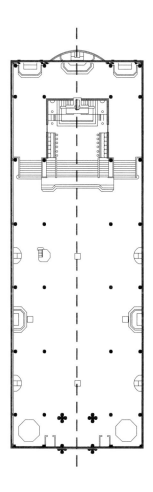

Steinhof, Otto Wagner
Vienna, Austria 1907
[Art Nouveau, Greek Cross, steel and masonry]

The Steinhof Church by Otto Wagner sits at the top a large hill and is part of the Steinhof Psychiatric Hospital. Due to the small size of the congregation, it has a simple Greek Cross plan that is extended towards the entrance to form a narthex. It is a symmetrical plan that borrows heavily from the ideal plans that were proposed during the Renaissance, such as the Greek Cross plan of Michelangelo at St. Peter's Basilica. The remarkable aspect of the Steinhof Church is the materiality and ornament that are evident on both the interior and exterior of the building. The evolution to new techniques of construction allowed the material technologies of steel and masonry to be applied to the traditional historical precedent of the church.

Notre-Dame du Raincy, Auguste Perret
Raincy, France 1922
[Modern, rectangular plan, concrete]

The plan of Notre-Dame du Raincy is a rectangle that has four rows of freestanding columns that support the large concrete canopy. It is an extremely simple plan that allows the visitor to concentrate on the materiality of the building. The chapels, narthex, and altar all participate in one large space. There is a sophisticated series of proportions that govern the plan and interior space. The building's significant evolution in the historical precedent of the church comes from the translation of traditional forms through the new materiality of cast-in-place concrete.

| 82 | PRECEDENT — PRINCIPLE | LINEAGES — ORGANIZATION | CHURCH — TYPE | PLAN — READING | ARCHITECTURE — SCALE |

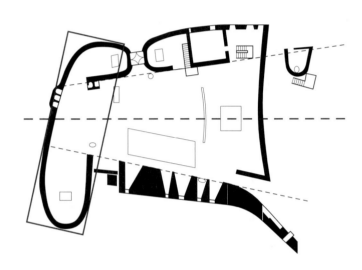

Notre Dame du Haut, Le Corbusier
Ronchamp, France 1955
[Expressionist Modern, free plan church, masonry and concrete]

As an application of modernist principles executed with expressionist, curvilinear forms while maintaining the traditional rituals and ceremonies of the church, Ronchamp hybridizes traditional church design with modern sensibilities. Containing the essential components of altar, bell tower, stain glassed window, and sacristy (a room for keeping the vestments), the organic and experientially based forms provide a highly composed experience. Developing the use of light, the undulating geometries produce a spirituality through the uniqueness of form.

Thorncrown Chapel, Fay Jones
Eureka Springs, Arkansas 1980
[Late Modern, lattice framed sanctuary, wood and glass]

Thorncrown Chapel is a sanctuary that emerges out a celebration of the specific beauty of the natural site. Its architecture is fully in service of a tectonic expressionism that celebrates and reiterates the material, repetition, verticality, and connection to the surrounding woods. Laminated wood elements, sized by the weight that two men could carry through the forest, overlap and repeat to create a dense structural field. The simple rectangular plan subscribes to a traditional organization of central aisle connecting to frontal altar. The transparency of the glass perimeter dissolves the boundary of the wall. The repetitive yet delicate columns allow for a further evaporation of enclosure. This permits a connection to the dappled light, verticality, and beauty of the surrounding trees and forest. Even the floor grows formally and materially out of the site, organically blending into the surrounds. The power of the space comes from its connection and association, blending with the beauty of its natural surrounds.

PRECEDENT

PRINCIPLE | 83

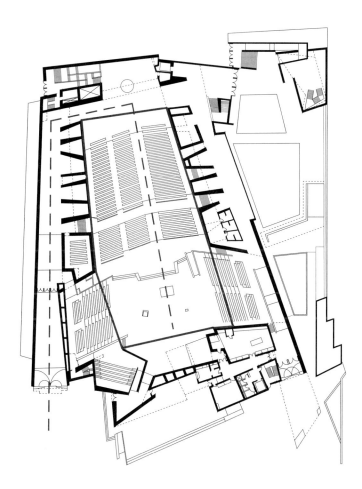

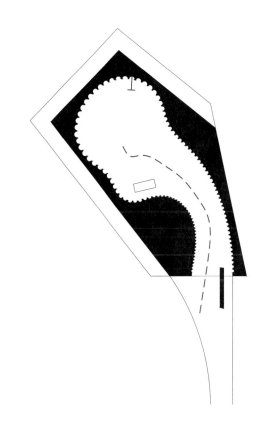

Cathedral of Our Lady of the Angels, Raphael Moneo
Los Angeles, California 2002
[Postmodern, folding axis, concrete and stone]

Built as a contemporary iteration of the traditions of Catholicism, the Los Angeles Cathedral uses a highly postmodernist approach to the design. Maintaining the prominence of the central worship space, the circulation, materiality, form, and spatial subdivision are all rethought. The pedestrian approach ascends through an urban plaza, past massive and highly ornamental doors located to the side, drawing the viewer to the back of the sanctuary and then folding them back into the main congregation hall. This is an inversion of the traditional sequence. The materiality transposes the traditional stone body to an exposed cast-in-place, reinforced concrete structure. Stained glass windows are reiterated as thin sheets of alabaster stone that allow for a naturally composed and majestic transmission of light. The form and organization is fissured, porous, and non-orthogonal. Using diverse angles, the spaces openly flow into one another, allowing the open side chapels to be nested in the pillars lining the sides of the central space. The collective composition is a referential yet spatially innovative iteration of the traditional cathedral configuration.

Brother Claus Field House, Peter Zumthor
Wachendorf, Eifel, Germany 2007
[Postmodern, material process formed chapel, concrete and lead]

The Brother Claus Field House is a contemporary worship space that emerged out of the limited resources and the specific construction skills of the remote community it was to serve. Founded in the material process, the chapel is an organic plan formed by the construction logic. The interior form is derived from a series of lean-to logs that create an organic perimeter. The outside is formed conventionally with smooth, faceted surfaces. The space between the two is then filled with sequential layers of concrete. The inner logs were then burned out of the figure to reveal the void of the sanctuary space and simultaneously produce a scalloped wall surface with a distinct charred patina. The blunt honesty and informal legibility of the material process of construction produces a simultaneously abstract yet primal form and experience. The hierarchy and organization of the historical precedent of the church are fully informalized to transition the power of spirituality to the monumental experience of the space.

| 84 | PRECEDENT
PRINCIPLE | LINEAGES
ORGANIZATION | ROTUNDA
TYPE | PLAN
READING | ARCHITECTURE
SCALE |

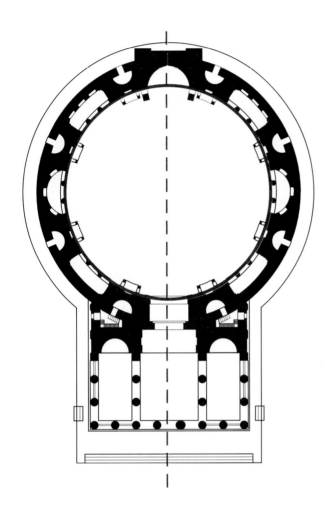

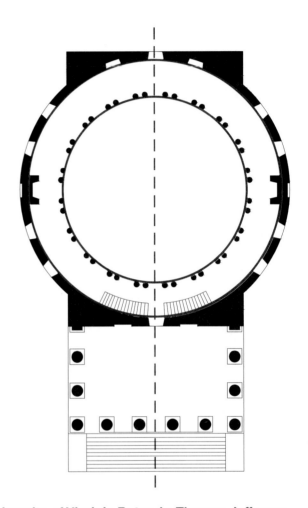

Pantheon
Rome, Italy 126
[Classical, temple, masonry]

University of Virginia Rotunda, Thomas Jefferson
Charlottesville, Virginia 1826
[Neoclassical, library, masonry]

The dominance of the Pantheon's circular architectural drum is most clearly seen on the Nolli Plan of Rome. The centralized plan hearkens back to the ancestry of the Greek Temple of Apollo at Delphi, but employs a heroic scale to make the circular space more spiritual and effective than functional. The open circular **oculus** *of the dome reestablishes the geometry in the ceiling and calibrates the passage of time through the movement of light.*

The Rotunda at The University of Virginia deploys the planning and figure of the Pantheon. The figure is clearly recognized through its striking formal similarities. The formal geometry and classical language of these two structures relate them in many ways. Like the Pantheon, the Rotunda occupies a significant position within its particular context. The Rotunda utilizes a half sphere, rather than an entire sphere, as its spatial context typology. Instead of entering the main space, as in the Pantheon, Jefferson instead designed the entry to access a lower floor that contains two ovaloid rooms that accept and acknowledge the all-important axis of the Lawn. The beginning of this axis, in collaboration with the entry portico, begins the production of the primary axis of the building, lawn, and entire campus. The ovaloid geometry begins the campus plan that leaves open the view, focus, and extension to the expansive manifest destiny of westward expansion, establishing both figuratively and philosophically the future and opportunity of America.

	PRECEDENT	
	PRINCIPLE	85

The rotunda is a fundamental typology founded in form, function, and geometry. It is an element that can be traced from its foundations in Roman architecture through contemporary application. As a readil y identifiable figure, it has been a symbol of the earth or the heavens and has remained a powerful type across diverse religions and cultures.

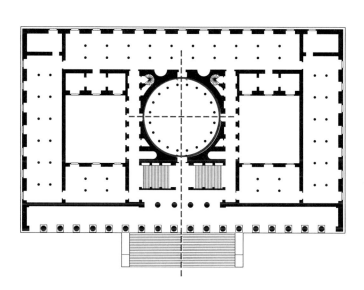

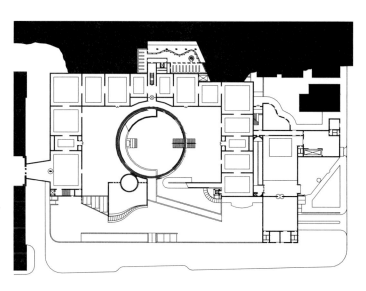

Altes Museum, Karl Friedrich Schinkel
Berlin, Germany 1830
[Neoclassical, symmetrical rectangle with rotunda, masonry]

The central rotunda of the Altes Museum again hearkens back to the form, figure, and effect of the Pantheon. Employing the rotunda in a classical manner, it is central to the building and axially positioned. Dominating the rest of the building in height, its multistory chamber has a statuary recessed between the twenty-one columns ringing the perimeter of its walls. As an organizational chamber for the museum as a whole, the room draws the viewer through the portico, past the stairwell and into the majesty of its scale and experience. The rotunda directs the visitor out the cross-axial circulation pathways into the subsequent hallways and galleries of the museum.

Neue Staatsgalerie, James Stirling
Stuttgart, Germany 1983
[Postmodern, open rotunda, masonry]

In the Neue Staatsgalerie, Stirling uses the circle at the center of the museum as a void. Deploying the rotunda as an exterior room, this primary spatial node is engaged by the circulation pathways, providing a continual return. Touching both interior and exterior circulation paths, the space bridges the varied interior galleries and mediates the dramatic change in section between the neighboring streets. The simplicity of the space and the power of the void are entirely dependent upon the geometry, reference, and effect of the reinterpreted rotunda. The precedent of the rotunda remains centralized in location, but is spatially inverted to become an exterior space.

| 86 | PRECEDENT / PRINCIPLE | LINEAGES / ORGANIZATION | VILLA GIULIA / TYPE | ELEVATION / READING SECTION | ARCHITECTURE / SCALE |

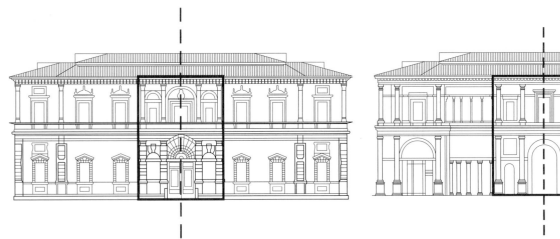
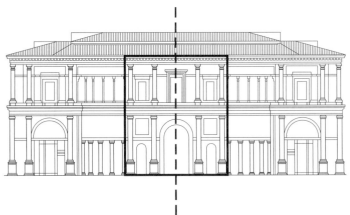

Villa Giulia, Giacomo Barozzi da Vignola
Rome, Italy 1555
[Late Renaissance, layered courtyard house, masonry]

The entry façade of the Villa Giulia is dominated by the central bay, which is heavily rusticated with bold stonework. It represents primitive ideas of architecture and is a metaphor for the cave or primary concepts of architecture. The proportions of this central bay are vertically extenuated, drawing attention to the hierarchy of the central axis. The rustication acts not only as a symbol of the cave, but also as an image of protection. Ultimately the repetitive **facades** of Villa Giulia depict the historical lineage of architecture itself. This entry façade begins that sequence.

Villa Giulia, Giacomo Barozzi da Vignola
Rome, Italy 1555
[Late Renaissance, layered courtyard house, masonry]

The second gate of the Villa Giulia is the triumphal arch gate that is seen as part of the semi-circular façade in the main courtyard of the villa. Here the proportions are similar to the entry gate, however the architecture has become more refined and subtle. Niches that appeared in the entry are now openings and the wall is considerably less thick. This begins a trend throughout the villa whereby each gate becomes more refined and more transparent.

In this lineage, the examination focuses on a series of parallel evolutionary façades that occur in one building: the Villa Giulia by Vignola. In the Villa Giulia, there is a series of elevations, or gates, that begin with the rusticated entry portal and end with the Serlio window in the garden. Essentially, this sequence is an architectural attempt to record and illustrate the history of architecture (cave, triumphal arch, frame and Serlio window) through the chronological sequencing of elevations within one building.

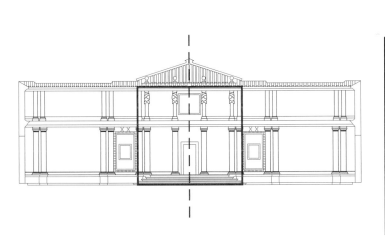

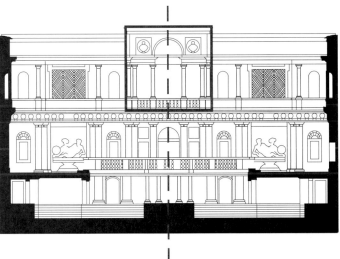

Villa Giulia, Giacomo Barozzi da Vignola
Rome, Italy 1555
[Late Renaissance, layered courtyard house, masonry]

The third gate leads to the casino and the nymphaeum, or grotto, below. It is formed by three large openings that no longer rely on the arch as the structural element. The spans of the openings are equal and are all spanned by using a **lintel** *(a horizontal spanning beam). Above the openings there are attic panels with caryatids (columnar stone figures), which emphasize the structure of this gate. Their position is removed from the wall and exposes the idea of a frame-like structure rather than the ancient concept of wall as structure.*

Villa Giulia, Giacomo Barozzi da Vignola
Rome, Italy 1555
[Late Renaissance, layered courtyard house, masonry]

The final gate in this sequence is seen across the nymphaeum and remains purely a visual part of this sequence as the gate that leads to the final garden of the villa. Here the Serlian or Palladian motif is utilized as the most architecturally sophisticated gate. This façade element represented, at that time, the highest and most refined element of architecture. These gates, in effect, record the history of architecture through a controlled linear, axial sequence.

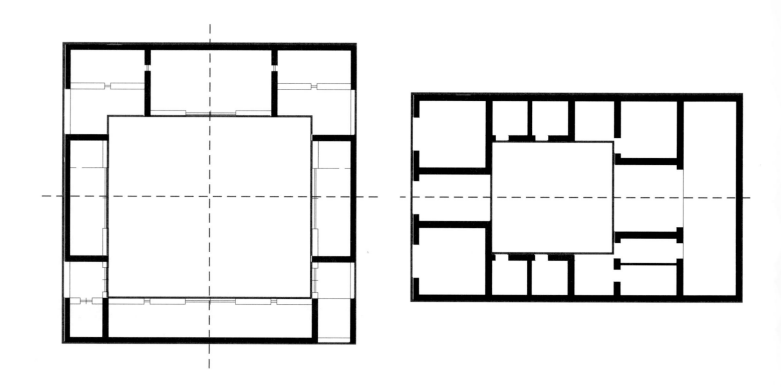

Chinese Courtyard House
China 1122 BCE
[Chinese, centralized courtyard, wood and masonry]

The Chinese Courtyard House, or siheyuan, focuses upon the geometric purity of the inner courtyard as a square and establishes a distinct object-like quality to the surrounding pavilions, each laid out along a north-south and east-west axis. Thus, unlike the subtractive and additive relationships of the European and Middle Eastern traditions, the courtyard is formed through the aggregation of units around a space. Made of repetitive and unitized pieces of wood and masonry, the additive, elemental, assembly-based construction system allows for distinct compositional segmentation that carries through from part to whole. Each house around the courtyard maintains independence of ownership often belonging to a different member of the family. The courtyard itself is for privacy and contemplation as opposed to community, often with multiple courts receding into the site, offering more privacy.

Etruscan House
Pompeii, Italy 80 BCE
[Roman, sequential courtyard house, masonry]

The typical Roman House in Pompeii has a rectangular plan that is almost entirely devoid of exterior windows. The house has a series of atriums and courtyards around which the interior rooms were organized. Movement begins in the first courtyard that has a ceiling conpluvium (a roof aperture) that funnels water into the impluvium (a depression in the floor) around which the sleeping rooms are organized. This is followed by the tablinium (a meeting, reception, or dining room). The sequence ends in the back courtyard that contains the garden and is surrounded by a peristylium or colonnade. Over the evolution of the type, the size and number of courtyards increased, eventually having a single house cover almost an entire Roman city block, or insula.

The courtyard, found in numerous countries and cultures, is one of the most ubiquitous architectural elements. At its most basic level, it is a way to bring the exterior into the center of the house. It provides the heart of the house through a space that relates to many of the functions and rooms. It is a type that has crossed centuries and known few boundaries in terms of materiality or construction. It is difficult to imagine an urban context that does not have courtyards. The courtyard allows the building to become a series of thin layers that permit light and air to enter into the most interior spaces.

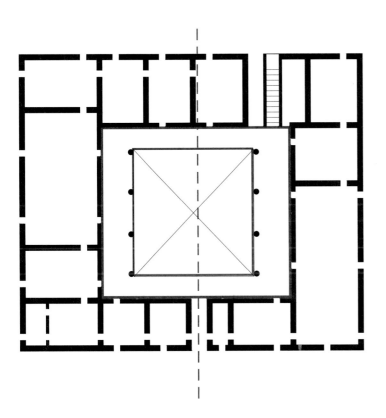

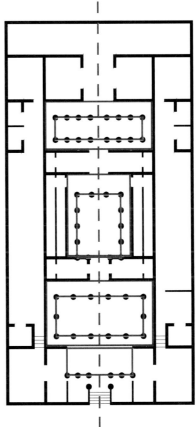

Florentine House
Florence, Italy 14th century
[Renaissance, centralized courtyard house, masonry]

The Florentine House, the basis of the Italian Palazzo type, is situated around the central courtyard. The internalized courtyard produced an new exterior that translated the architectural issues and considerations of the façade to the interior. This introduced the problem of how to turn a corner in three-dimensional space. The history of Renaissance Architecture can be witnessed through the decisions made on how to resolve the corner. The perimeter colonnade of the central courtyard served as the circulation for the one room deep perimeter band allowing for light and ventilation to penetrate the diminished cross section. Here the courtyard type also deals with the translation to a multi-story configuration.

Persian Courtyard House
Iran 16th century
[Islamic, symmetrical layered courtyards, masonry]

In the Persian Courtyard House, the majesty of Iranian architectural ability is fully displayed through the layered ornamental courtyards. A full perimeter band of various functions brackets three interior courtyards; each court is uniquely dimensioned and variably encircled with a ring of columns that produce a shaded ambulatory, or perimeter walkway. The courtyard itself contains water elements and plantings that create an inner garden and a passive cooling system for the space. In the Persian Courtyard House, the rooms do not open directly onto the court, but rather preserve the garden as a destination space, not simply a circulation space. Here the rooms are not given dedicated functions, but used as appropriate through the seasons.

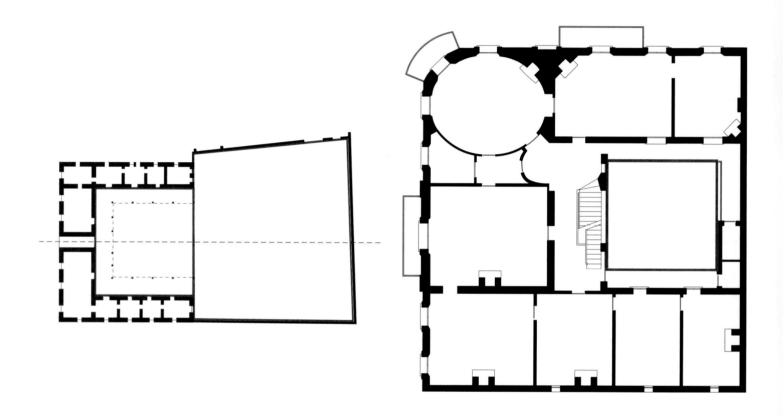

Spanish Courtyard House, Casa de Estudillo
San Diego, California 1827
[Spanish Colonial, U-shaped with central courtyard, wood and masonry]

The Spanish Courtyard House as typified by the San Diego old town courtyard house is a simple centralized court. Rooms line the edges of the courtyard, relying on the exterior circulation of a covered veranda to connect them. The front edge of the house has the larger formal spaces, including the living room and chapel, and guest and master bedrooms; the wings are smaller secondary workrooms and bedrooms. The intense symmetry of the plan provides a formal hierarchy and positions the courtyard as the connective space of the house. Thick adobe walls provide both structural and thermal barriers, whereas the narrow one room deep configuration allows for light and air to penetrate the interior.

New Orleans House
New Orleans, Louisiana 19th century
[French Colonial, central courtyard house, wood and masonry]

There are a number of types of courtyard houses in New Orleans. They are most common in the denser urban fabric of the French Quarter. These houses are typically multi-storied and the primary circulation runs along the edge of the courtyard, either along a balcony on the upper floors or under an overhang on the ground floor. Often the major rooms of the living floor open out into this courtyard. Typically a fountain is placed at the center of the courtyard along with many plantings that offer shade and passive cooling to combat the oppressive heat. Unlike other courtyards, the entrance into the New Orleans courtyard is secondary and located along the edge.

PRECEDENT	
PRINCIPLE	91

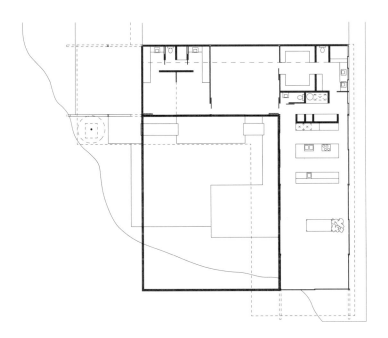
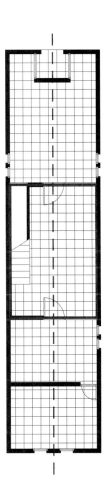

Case Study Houses [#22]
Los Angeles, California 1945-1966
[Modern, free plan indoor-outdoor, wood, steel and glass]

The Case Study Houses were experiments in modern American residential construction. Embracing the new materials and technologies of the post-war era and bringing them to the moderate climate of Southern California, the series produced a collection of exquisitely experimental yet simple modern homes. Sponsored by Arts & Architecture magazine, the list of participants included, but was not limited to, Pierre Koenig, Ray and Charles Eames, Eero Saarinen, Craig Ellwood and Richard Neutra. Each architect deployed differing forms and materials, but the goal was the same: houses that had modern open plans and connected the interior and exterior of the home. Courtyards, pools, and large planes of operable glass all worked to highlight the beauty of the surroundings and dissolve and streamline the composition of the home. The collection became emblematic examples of modern domesticity.

Azuma House, Tadao Ando
Osaka, Japan 1976
[Postmodern, courtyard houses, concrete]

The contemporary Japanese Courtyard House has embraced the confines of the extreme expense and small spatial allocations of Tokyo and adapted the traditions of the courtyard house to the dense infill sites. As a cultural precedent, the Azuma House evolves the dedication to the spatial modularity of the tatami mat. Extending the frame of the teahouse to the domestic realm, the house combines the geometry with the philosophy of life. In balance with nature, accepting of its transitions, the courtyard itself becomes not just a space but a physical and formal conversational element, framing the seasons and illustrating the pensive and chronological nature of time. As a void, the perceptual openness of the courtyard, despite its small scale, allows for the flow of the house and the continuation of the natural into the interior. The courtyard serves not simply a functional or practical purpose, but also an aesthetic and philosophical one.

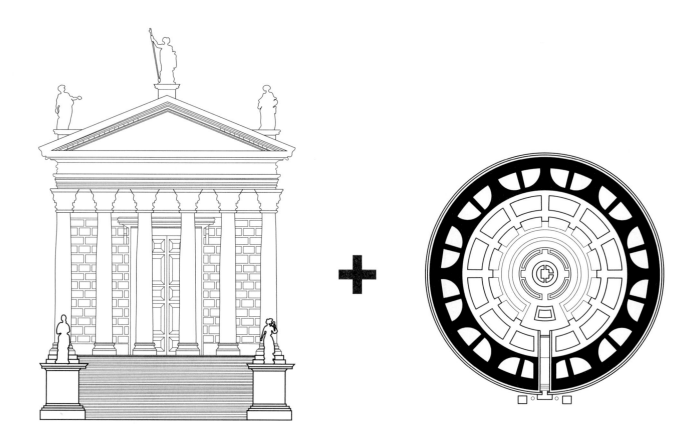

Temple Front + Circular Mausoleum = Pantheon

This equation might be considered one of the initial assemblages. The equation took a circular structure such as the Mausoleum of Augustus (28 BCE), which was actually a solid building, and attached to it a temple facade, resulting in the Pantheon (126 CE). The temple front, represented by Maison Carrée (16 BCE), added an entry and axis into the building, and, by virtue of its increased importance and monumentality, extended the language of the temple and classicism. As a result, the Romans now had to conceive of the large interior space of the Pantheon and subsequently how to span that distance.

The concept of assemblages comes about when an architect takes two buildings or building types as precedents and combines them into one cohesive whole to define a new parti. Throughout history this has been a very effective practice that often allows an architect to develop new concepts and types. As collagist assemblages or synthetic hybrids, the interrelationship of the two parts results in fully unique third.

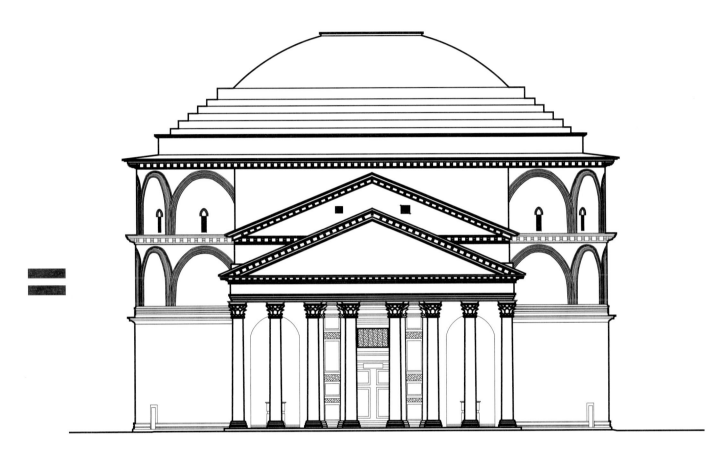

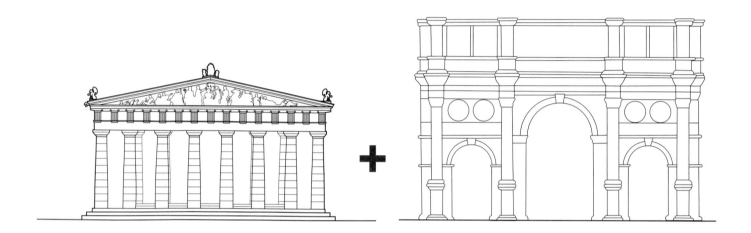

Temple Front + Roman Triumphal Arch = Sant Andrea

Alberti formed this equation when designing Sant Andrea in Mantua (1476). He layered two pagan elements to form the façade of this church. Alberti used the Triumphal Arch and the Temple front; collaging them together reinterpreted not only the façade, but the spatial aspect of this church. Using pagan elements and layering them compositionally was to become an obsession with Renaissance architects for years to come. Left out of the equation, however, was the rose window, which Alberti struggled with; he concluded by using it as an awkward element in the façade, which was a hood atop the pediment.

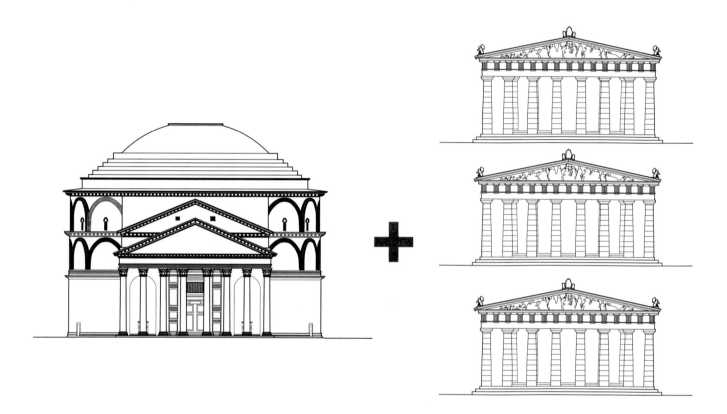

4 Temple Fronts + Pantheon = Villa Rotunda

To form the Villa Rotunda (1571), Palladio took a centralized building like the Pantheon and added three more temple fronts so that the two main crossing axes would be recognized. The Villa Rotunda, which sits on hill overlooking Vicenza, enjoys magnificent views in all directions; hence this multi-focal directionality worked perfectly. This is an illustration of how an earlier assemblage (the Pantheon) can be further modified to create a new assemblage. The layering of precedent illustrates the depth of architectural lineage and historical reference.

PRECEDENT
PRINCIPLE | 97

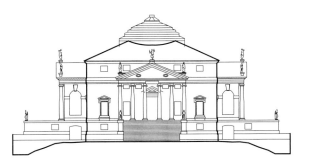

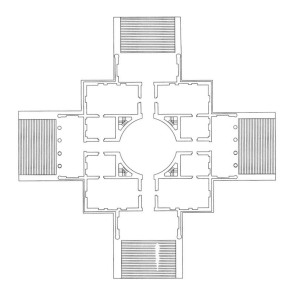

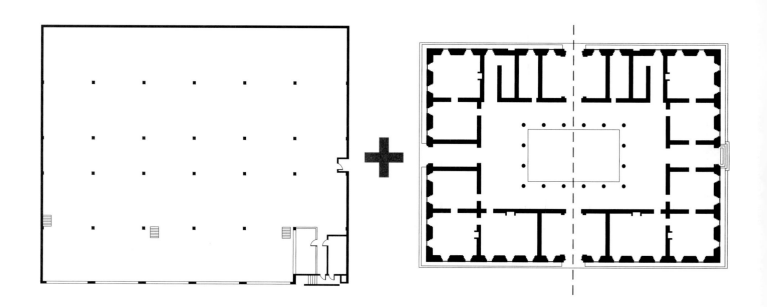

Warehouse + Palace = Museum

Throughout history, many museums have been housed in former palaces. The Louvre in Paris and the Hermitage in St. Petersburg are prime examples of this particular equation. Represented here, we combine a typical free plan warehouse with the formality of the Palazzo Strozzi (1538), which results in a new museum typology, as illustrated by the British Museum (1850). These buildings, which served as lavish residences for royalty are now primarily storage facilities for priceless works of art. This conception equates the combination of a warehouse and a palace to produce a museum. This equation served for generations as museums continued to be designed to resemble palaces, but typically with no or few windows, hence the warehouse typology was continually present.

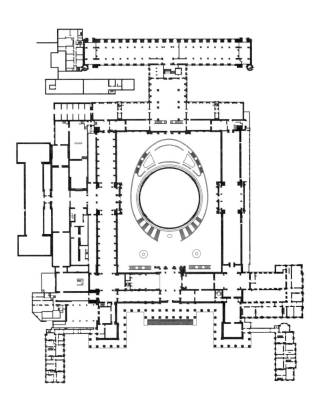

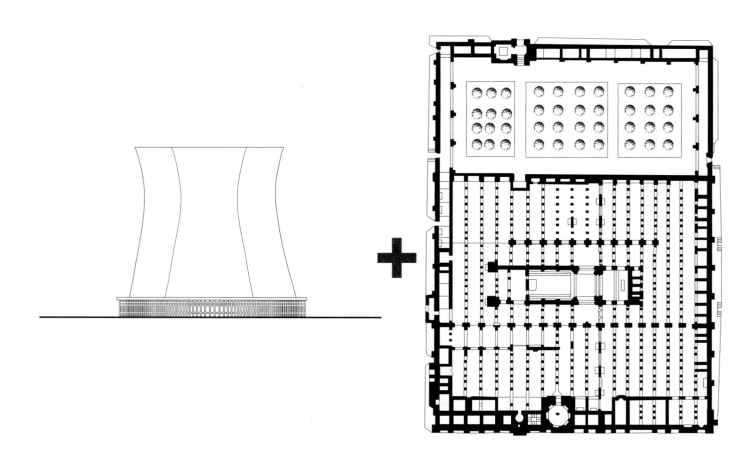

Nuclear Reactor + Mosque = Palace of Assembly, Chandigarh, Le Corbusier

The Palace of Assembly (1963) by Le Corbusier is the central building of the Chandigarh complex. Housing the main assembly hall, the building is designed to facilitate the congregation and interaction of the governmental body. The form derives from the earlier five points developed by Le Corbusier. The ribbon window evolves into a deepened **brise-soleil** *to mediate the light and blur the boundary between interior and exterior. Pilotis are translated into an exaggerated column grid that blankets the composition and allows the extended field of the mosque colonnade (originally intended for overflow exterior seating), to be deployed for structural and compositional functions. The roof garden is translated from an occupiable green space into a sculpted and formed roof accentuated through the skylight chimney. The free form façade and free plan dominate the multivalent field bounded by the perimeter figure bars. The main meeting hall is top-lit through a massive light cannon with a parabolic shape almost identical to the cooling towers of the nuclear reactor—an otherwise purely functionally derived form, here deployed for its experiential and effectual qualities. Placed within a grid of columns in a mosque-like configuration, the combination is a highly abstract, and yet distinctly referential, formal composition.*

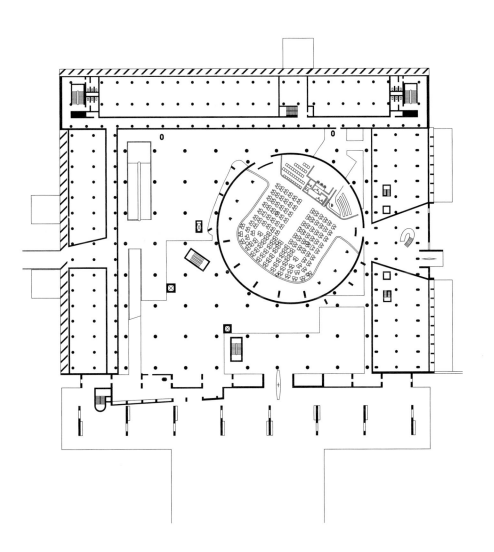

PRECEDENT
PRINCIPLE 101

102	PRECEDENT	ASSEMBLAGES		FUNCTION	ARCHITECTURE
	PRINCIPLE	ORGANIZATION		READING	SCALE

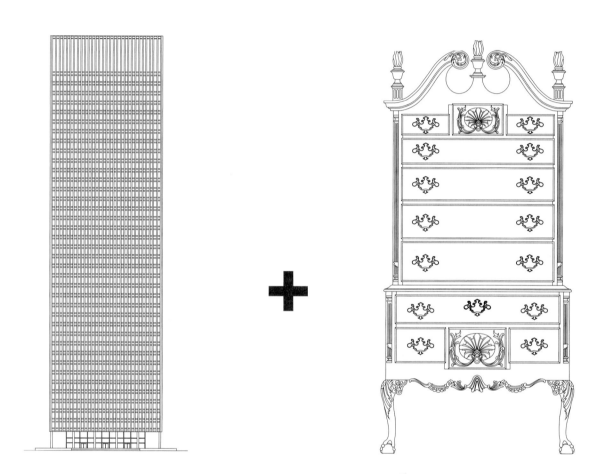

Seagram Building + Chippendale Furniture = AT&T Building, Philip Johnson

As a play on the purity and ubiquitous nature of the Modernist high rise (as represented by the Seagram Building by Mies van der Rohe, 1958), the AT&T Building (1984) hybridizes the streamlined nature of the modernist composition as a "middle" by adding in the classical notion of base, middle, and top. Here Johnson merges the Seagram building with a signature piece of Chippendale furniture (1754), crowning the building with a decorative top that plays with history and scale. The design was a well-timed application of a historical reference as a hood ornament for corporate identity. The iconic visual association and recognizability served the booming corporate clientele well in producing the beginnings of a branded architecture. Identifiable by form and associated with a discrete corporate entity, architecture transferred to logo.

PRECEDENT
PRINCIPLE | **103**

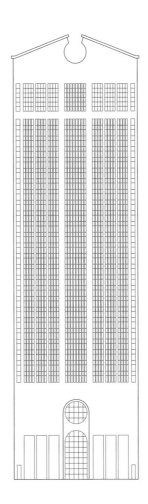

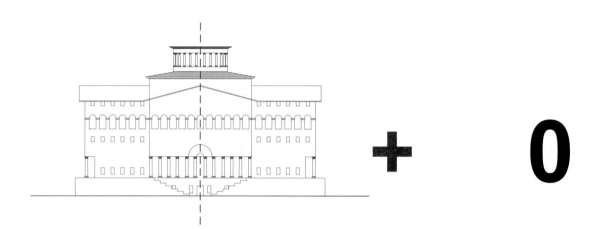

House of Education, Ledoux + 0 = University of Houston, Philip Johnson

At the University of Houston School of Architecture (1985), Philip Johnson reached the pinnacle of postmodernism. In his desire to look towards history as a collagist opportunity of reference and form, Johnson literally transposed an unbuilt project to a new site and program. Ledoux's House of Education, from the late eighteenth century, was intended as a support structure for the Royal Saltworks at Arc-et-Senans. Johnson adopted and transposed the form of the building, changing only its functional planning, material, and scale. The new organization positions four stories of design studios around a centralized square atrium. The square, open-roofed tempietto (a small circular temple) atop the building has a glass floor that serves as the skylight of the main interior space. The materiality of the design was switched from Ledoux's originally intended stone to a brown brick to match the rest of the material palette of the University of Houston campus. The outright appropriation of the form illustrates an indifference to originality and instead adopts the postmodernist sensibility with a cavalier appropriation of historical precedent. Here the issue of the "copy" clearly surrounds the intention and execution.

PRECEDENT | **105**
PRINCIPLE

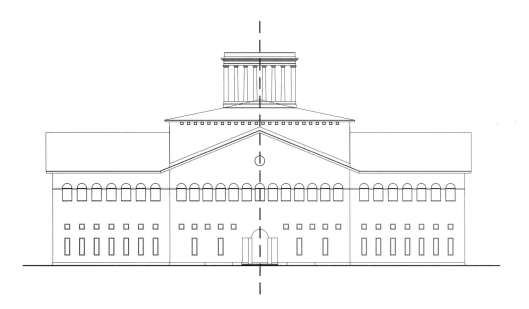

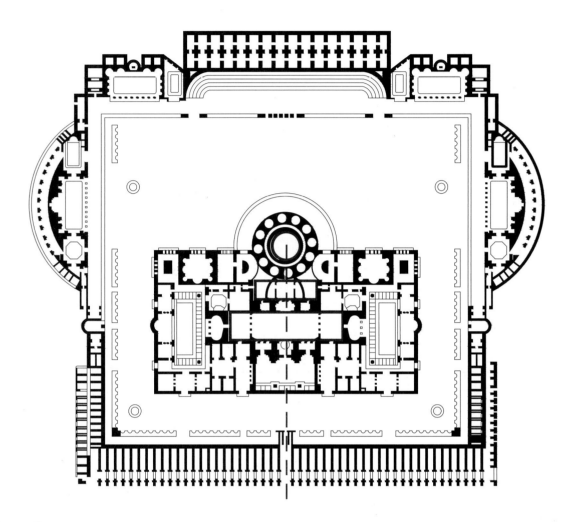

Baths of Caracalla
Rome, Italy 216
[Roman, baths, masonry]

*Like most Roman baths, the Baths of Caracalla, consisted of three main spaces: the **frigidarium**, the **tepidarium** and the **caldarium**. These three space were separated from each other by a series of thresholds that allowed an axis to connect them but simultaneously ensured their separation. These rooms dominated the baths spatially and functionally and also expressed the highest sense of Roman architecture and engineering. Palladio used his knowledge of Roman baths as a way to address the problem of this church. Like the baths, Il Redentore consists of three main spaces: the nave, (with its side aisles), the sanctuary (at the crossing), and the monastic choir (beyond the choir screen). These three spaces recall the three spaces of the Roman bath. Beyond this, Palladio used the typical Roman bath window type as clerestories (high windows above eye level) for the church.*

There is a series of buildings that are very closely related and are often discussed together as pairs. These morphological analyses examine how a number of these projects are related and how one has led to the other. The use of precedent is so strong, the two projects become innately associated.

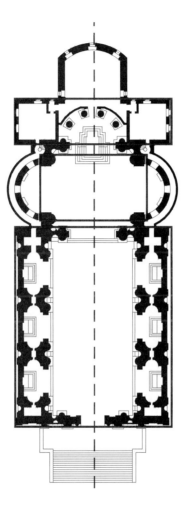

Il Redentore, Andrea Palladio
Venice, Italy 1591
[Renaissance, church, masonry]

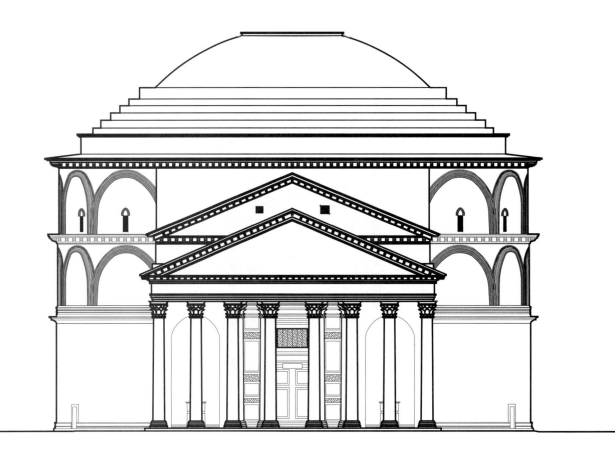

Pantheon
Rome, Italy 126
[Classical, temple, masonry]

The Pantheon and the Rotunda at the University of Virginia present just one pairing with the Pantheon, with myriad ensuing examples of buildings influenced by the power of its presence, clarity of form, and geometric purity. The combination of the circular rotunda with the classical temple front created a hybrid form that has been reiterated throughout history. The Rotunda at the University of Virginia is an excellent example of its enduring influence. Built eighteen hundred years after the Pantheon, the form adopted by Thomas Jefferson is nearly identical in composition. The precedent is adopted for both its physical dominance and axial hierarchy. Equally significant is the historical reference to Roman Architecture, representative of a society founded in law, intellect, and democracy. The translation comes through the hierarchy that the building establishes and the repercussions in the surrounding context and landscape. At the Pantheon in Rome, the building is simply set within the fabric. It is a product of its context. In The UVA Rotunda in Charlottesville, the building becomes the main building for the entire campus. Creating the westward axis of the campus, establishing the lawn and beginning the flanking colonnade of dorms and classrooms, the hierarchy and dominance of position is established through the form.

PRECEDENT 109
PRINCIPLE

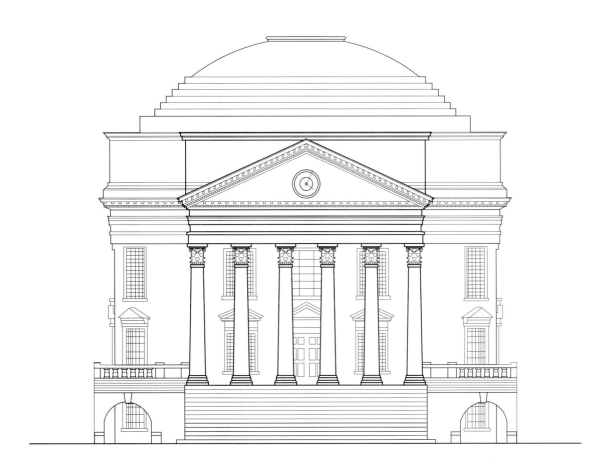

UVA Rotunda, Thomas Jefferson
Charlottesville, Virginia 1826
[Neoclassical, rotunda, masonry]

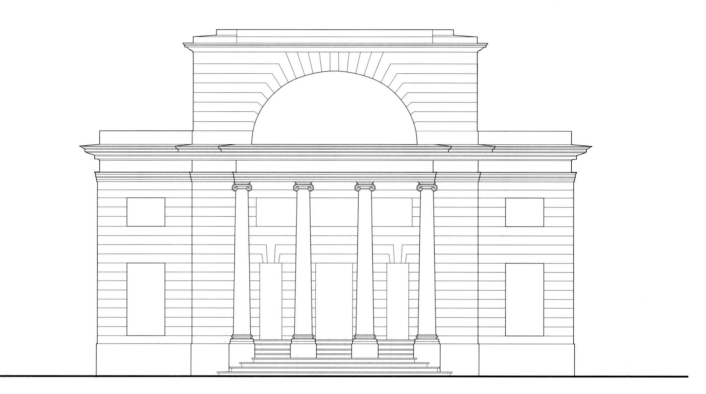

Hotel Guimard, Claude Nicolas Ledoux
Paris, France 1770
[Neoclassical, residence and theater, masonry]

The Hotel Guimard contained a courtyard between the small theater at the entrance and the hotel proper. The entrance to the Hotel from the courtyard was an apse-shaped semicircle that extended up into a half dome. Cutting across this element in the elevation was a cornice line supported by two freestanding columns and two engaged pilasters. Jefferson, having held the position of the American Ambassador to France, would have undoubtedly been familiar with the Hotel Guimard. On the Lawn at the University of Virginia, Jefferson essentially reconstructs the elevation of the hotel in Pavilion 9, the residence of the Dean of the Architecture School. The main difference between the two structures is that the entirety of the composition at the Pavilion is not as easily understood because of the deep colonnade that passes in front of the entrance.

| PRECEDENT | 111 |
| PRINCIPLE | |

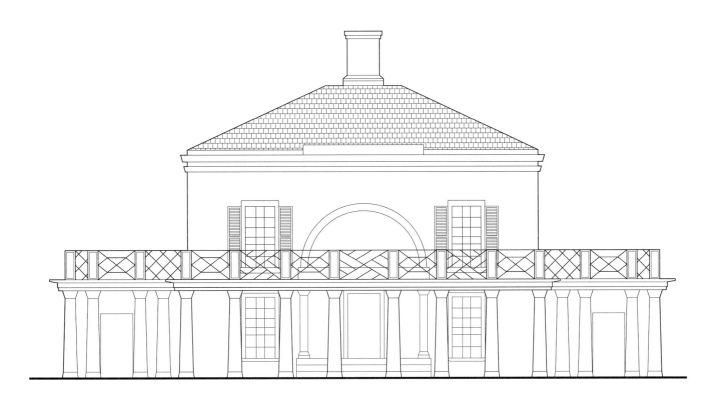

Pavilion 9 at UVA, Thomas Jefferson
Charlottesville, Virginia 1826
[Neoclassical, residence, wood and masonry]

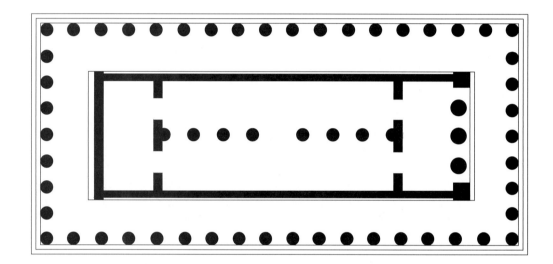

Basilica at Paestum
Paestum, Italy 550 BCE
[Classical, basilica, masonry]

The Basilica at Paestum is singular in that it contains nine columns in the front elevation as compared to the more traditional even numbers of columns. This results in a column occupying the center axis of the building. It was believed that this was an attempt to illustrate that this building was not for the pagan gods, but was instead a building for the citizens' use. Research has shown, however, that this was actually a temple to the Goddess Hera. What is critical is the fact that during the eighteenth century, it was believed to be a Basilica and not a temple. Labrouste was aware of the readings of the Basilica at Paestum and used the typology of the center row of columns as a model for the Bibliotheque St. Genevieve. Labrouste believed that the center row of columns identified the basilica as a civic building for the people rather than a religious one for the gods. He felt that this library should also be for the citizens of Paris and thus utilized this design parti.

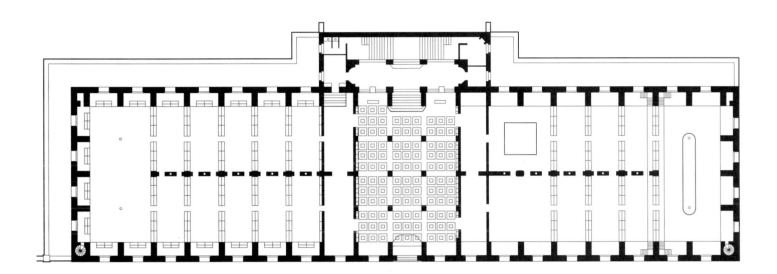

Bibliotheque St. Genevieve, Henri Labrouste
Paris, France 1851
[Neoclassical, library, steel and masonry]

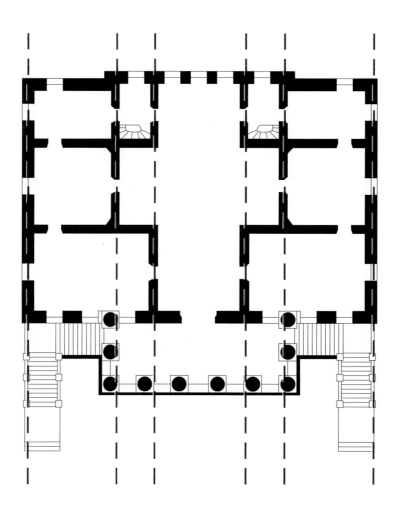

Villa Malcontenta, Andrea Palladio
Mira, Italy 1560
[Renaissance, symmetrical bayed plan, masonry]

The plan of the Villa Malcontenta includes a bay system that dictates the structure and organization of the building. The rhythm of the system is developed around an A-B-A-B-A arrangement of alternating larger then smaller bays. The structural walls follow this organization and determine the size and shape of the rooms. The stairs, for example, are found in the smaller "B" portions of the plan, whereas the more important public rooms are located in the larger "A" portions. When Corbusier developed the plan of the Villa Stein at Garches, he used an identical structural bay layout. Rather than using structural walls, Corbusier used the free plan as a way to spatially develop the plan of the building. This connection between these two buildings was originally noticed by Colin Rowe and significantly changed how architects and historians viewed the work of Corbusier.

PRECEDENT
PRINCIPLE 115

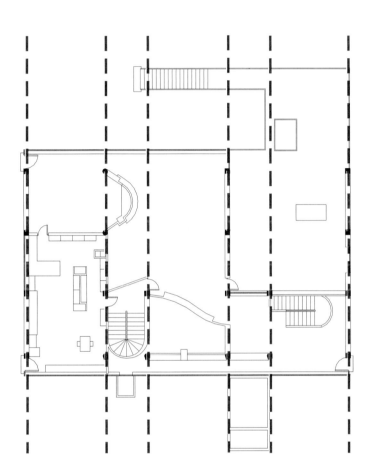

Villa Stein at Garches, Le Corbusier
Garches, France 1927
[Modern, free plan, concrete and masonry]

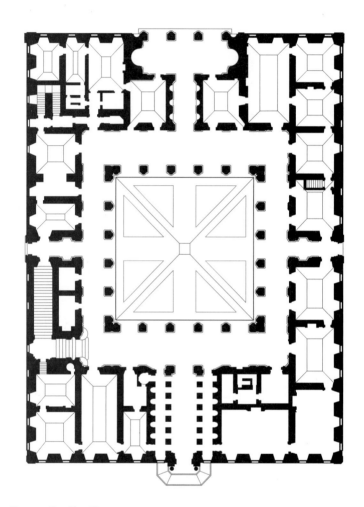

Palazzo Farnese, Antonio da Sangallo the Younger
Rome, Italy 1534
[Renaissance, Palace, masonry]

The Palazzo Farnese is one of the most grand of all Renaissance Palazzi in the sixteenth century. Its extensive plan occupies an entire urban block. It is the quintessential palazzo plan, following the typology set by earlier Florentine Palazzi. It is a three-storied building with a square central courtyard. Most of the rooms on the ground floor are accessed from the courtyard, but many were connected enfilade (an axial arrangement of doorways) in the interior. In the Palazzo Farnese there are two axes that cross in the center at right angles. The major axis runs the long distance of the palazzo and joins the entrance to the gardens behind the building; the minor axis runs in the short direction. There are a number of connections between the Palazzo Farnese and the Casa del Fascio. Both are rectangular buildings with square courtyards, both are multi-storied, both have a similar proportional system, and both occupy full urban blocks. The language of the plan is where the similarities are most notable. Terragni used the Renaissance palace as a model for the Casa del Fascio as a way of building on the greatness of Italy. Terragni's design differs from the Palazzo Farnese in that the courtyard is now an interior space, and the strong axial relationships are not present. However, the surrounding rooms and their relationship to the courtyard as an organizational device are continued.

PRECEDENT 117
PRINCIPLE

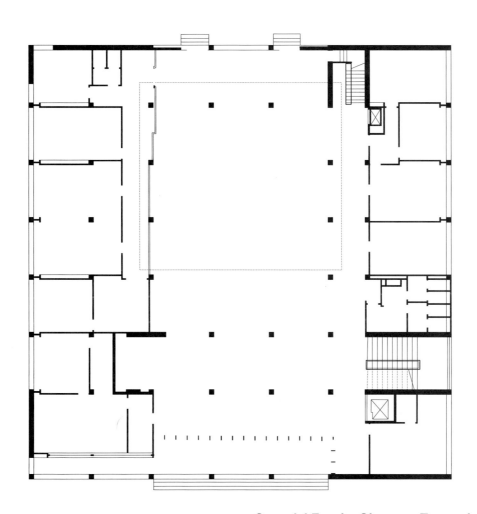

Casa del Fascio, Giuseppe Terragni
Como, Italy 1937
[Italian Rationalist, centralized atrium, concrete and masonry]

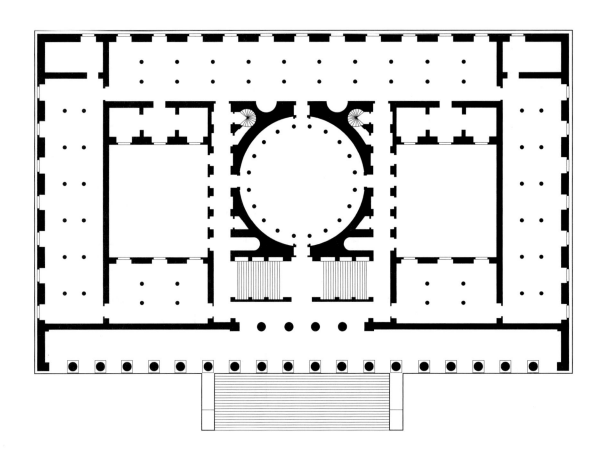

Altes Museum, Karl Friedrich Schinkel
Berlin, Germany 1830 [Neoclassical, museum, masonry]

The plan of the Altes Museum is a simple rectangle with a circular element placed at the center. This circular element is further expressed on the exterior as a square figure that rises above the roof of the museum. Schinkel has essentially placed a pantheon into the center of a palazzo. At the front of the museum is a large, monumental, open colonnade that addresses the large square in front of the museum and also organizes the entry sequence. When Corbusier designed the Assembly Hall at Chandigarh, he used the model set out by Schinkel at the Altes Museum. One can clearly see the U-shaped building that surrounds the giant column filled room in the center. The pantheon in Schinkel's museum has been converted into the main assembly space. The large colonnade has, for the most part, remained intact. The major differences are that the rotunda has been shifted off-center and the overall building has gone from a rectangle to a square.

PRECEDENT | 119
PRINCIPLE

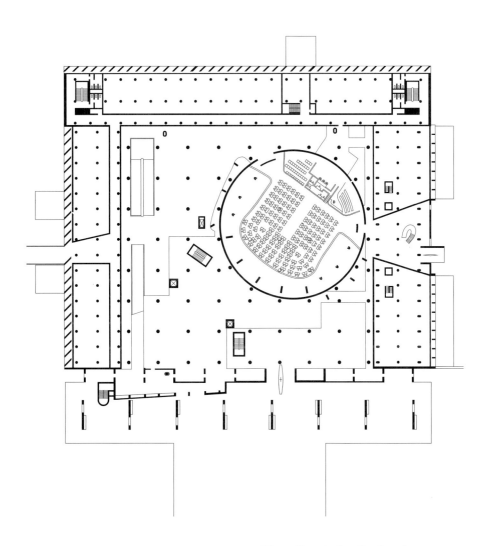

Chandigarh, Le Corbusier
Chandigarh, India 1963
[Modern, parliament house, concrete]

120	**TYPOLOGY**			
	PRINCIPLE			

Typology

03 - Typology

Typology is the taxonomic classification of characteristics commonly found in buildings and urban places according to their association with different categories. It is a study of types. Architecturally this has a far-reaching impact on the pedagogy and methodology of architecture. The history of architecture has innately established relationships across time. The impact of that which has occurred in the past architecturally and that which an architect is attempting at the present time is interrelated. It is fundamental to the design profession to understand the history of a building type and to work with and against it to build, evolve, and respond to the lineage. The implications of the precedent of building type on the design process and the resulting iterative evolution of that type is the basis of architecture. Within the scope of typology, there are a number of subcategories that concentrate more specifically on a distinct classification topic. Four of the most prevalent are geometric or form, programmatic or functional, organizational, and material.

Geometric or Form Typology Form is one of the most powerful and recognizable attributes of any design. Buildings are more often identified and described by their form than by any other characteristic. There are examples where the function of a building has changed yet the form remains the same. Dominant geometries and formal figures can maintain significance across functional types. For example, a cubic building can function as a tomb, or a house, or a variety of diverse building types. The formal classifications can be governed by either two-dimensional or three-dimensional systems, allowing readings of either the overall form or the discrete reading of plan, section, or elevation. If one considers the myriad possibilities within the simplicity of a square plan or the diversity of a geometrically composed façade, one quickly comprehends the power and value of historical and associative compositional precedents.

Programmatic or Functional Typology Functional typology is based on the programmatic requirements of the building. The functional requirements of a building typically change slowly over time. This slow evolution provides clear patterns through the repeated production of a building type. The study of these precedent buildings of similar function can assist in any attempt at creating another model. Some of the overtly dominant building types include housing, churches, schools, and hospitals, for instance. There are evolutionary trends within each of these functional categories, but also lateral relationships across categories that provide opportunity for cross-pollination and hybridization.

Organizational Typology Organizational typology refers to planning systems deployed as governing techniques. The history of organizational systems is the history of architecture (see Chapter 01), rooted in the fundamental plan systems: centralized, single-loaded corridor, double-loaded corridor, free plan, pods, and raumplan. The organizational typology allows for a diversity of forms and functions. A centralized building can be a church, a tomb, a house, or a museum, unified through the common bond of its geometric organization. Buildings that similarly deploy a free-plan organizational typology can have completely different forms and functions, yet they are related due to their organization. Historically these systems were often extremely simple. Early linear or central organizations evolved with the introduction of other elements such as the courtyard, the façade, ideas on sequence, perspective, and ultimately the hybridization of them all. The connections of organizational typologies are intrinsic. Whether intended or not, their referential framework is an innate basis for the reading of architecture.

Material Typology The material of a building, like its form, is a powerful and recognizable trait that can be identified and classified. The material and construction of a building will always relate it to the formal and methodological traditions of what has gone before. Consequently, all wood-frame buildings (or any other material and its associative principles) are related by their materiality. Material typology is founded in the knowledge of how that material is made, used, and applied as a design consideration. This typology is in many ways the most fundamental as its recognition is immediate and requires no overt knowledge of the function, form, or organizational system to identify.

One of the key issues associated with typology is the knowledge of, and belief in, history as a value system. Fundamentally, in every design project there must be an understanding and valuation of the knowledge that has been created before. The examination and analysis of typology is critical to clearly understand the complexity and legacy of associated problems and opportunities. As a result, the designer has an active imperative to study buildings of the past as a way of creating buildings that can add to and extend their lineage. Typology can be rooted in a commonality of use or simply a material system. Regardless, it establishes the foundation and datum that allow invention and the relationship to new forms.

Geometric Typology / Form Typology

Geometric or form typology examines the interrelationship of formal types through the primary geometries of square, circle, oval, triangle, polygon, and star. Requiring a parallel description at both the architectural and the urban scale, the reading of the figures equally requires analysis both as positive geometric forms and negative subtractive spaces. Taken across the variations of shape, scale, and definition as space or object, these cross-referential systems allow for the development of a series of subclassifications within each type.

Square One of the most common and recognizable shapes, the centralized square can be found throughout the world in many different cultures. The predictability of the square allows a person to understand a space without seeing the plan or its entire space. This shape carries with it a sense of stability and formality that provides a formal independence. Simultaneously it has an intrinsic subdividing geometry and organization, providing the opportunity to relate to other elements through its internal axes. It is readily deployed on both the architectural and urban scales as buildings, courtyards, blocks, and piazzas.

Circle The centralized circle is the most ancient of all organization systems. Like the square, it is a geometry that has existed in diverse cultures and eras. Examples such as Stonehenge give testament to the primal power of this system. Sites such as Hadrian's Tomb or Aalto's Woodland Chapel reiterate the evolutionary description and timeless nature of its form. Nearly every religious building uses the square or circle (or both) in their design. Urbanistically, the circle made its most dominant appearance in cities during the Georgian Era of eighteenth century where its purity and uniqueness was deployed to produce identity and locality within the rapidly expanding urban fabric.

Oval The Romans typically used the oval as a prevalent type for an amphitheater or stadium, a required element in the city planning deployed throughout the Roman Empire. During the Baroque period the complexity and multi-nodal directionality of its geometric shape proved extremely popular. Utilized extensively as both an architectural and urban geometry, the oval found its dominant application in Baroque Rome. It continues to be used by designers of modern and contemporary buildings who seek geometric continuity with spatial complexity. It gives a direction in plan not provided by the neutrality of the circle.

Triangle The triangle, while not as common as the square or circle, has typically been employed as either a courtyard on the architectural scale or as a piazza, resolving a number of circulation paths, on an urban scale. The triangle naturally gives a strong sense of direction through both the sides and vertices, focusing sight lines across and movement through the funnel space. The triangle is typically reliant upon external forces such as property lines or converging axes to employ its geometry.

Polygon The polygon has been used throughout history to indicate nodal importance. It is more complex than the circle or square and was typically used to announce a singular room or event as hierarchical. The Baptistery in Florence and the Tribune Room at the Uffizi are examples of spatial types that employ the polygon to differentiate and prioritize. The polygon frequently occurs as an urban void that allows for a complex center that accommodates numerous axes.

Spiral The use of the spiral originated with the Tower of Babel. The spiral naturally recognizes a center, however it is in a very different way than other centralized organization systems. The spiral typically has a visual reference to the center when used as an interior space, and conversely has a visual reference to the surrounding context (landscape or urban fabric) when used as an exterior typology. The experiential quality unique to the spiral is that with each revolution, the observer's relationship to the center changes.

Star The star shape was used almost exclusively as a fortress. The pointed ramparts of this shape acted as deflectors of projectiles such as cannonballs and established firing sight lines. The star quickly became obsolete as a defensive structure with rapid advancements in modern warfare. The significance of its highly centralized form and the artifactual presence of its historical deployment in contemporary urban fabrics make it a significant geometric type to consider and understand.

Square as Figure / Form—Architecture

The square as a formal and figurative generator has typological consequence in plan, section, and elevation. The purity of the form, the legibility of the geometry, and the associated predictability make the square a fundamental form in architecture. Providing the ability to cognitively map a space despite the inability to fully perceive its entirety, the square provides a great clarity and power through its primal geometry. In plan, the square has evolved from an ordering method of providing organizational and spatial control (as in Andrea Palladio's Villa Rotunda), to a compositional non-directional figure that allows for abstraction (as in Mies van der Rohe's Neue Nationalgalerie Museum). In elevation, the square similarly developed from a primary hierarchical framework to a non-directional boundary (neither landscape nor portrait) and used as either a universal pattern (as a grid module) or a specific composition (as a focused composition such as SANAA's Zollverein School of Management and Design).

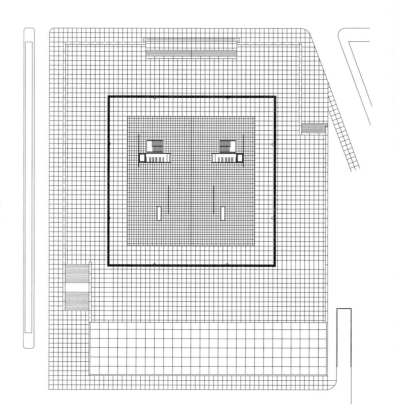

Neue Nationalgalerie Museum, Mies van der Rohe
Berlin, Germany 1968
[Modern, free plan / universal space, steel and glass]

*The Neue Nationalgalerie by Mies van der Rohe deploys the square firmly in its plan. With a glass perimeter, the weight, module, and overt presence of the roof defines the form. The square in plan gives a non-directionality that, when coupled with the larger **plinth** and recessed columnar structure, allows for a perceptually infinite spatial extension. The result is the legibility of the artwork in the context of a larger horizontal spatial field, which dissolves at its perimeter, allowing the exterior to become part of the gallery through a perceptual extension of the boundary of enclosure.*

TYPOLOGY
PRINCIPLE 127

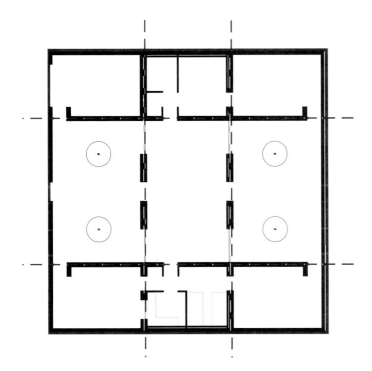

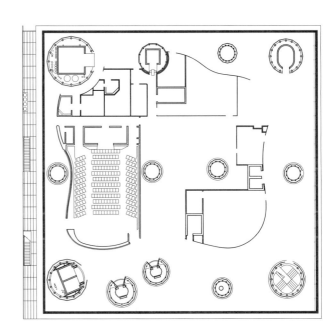

Casa Gaspar, Alberto Campo Baeza
Zahora, Spain 1992
[Minimalist, centralized nine-square courtyard, masonry]

Casa Gaspar subdivides the square plan into nine segments. The middle band is open and transparent, whereas the edge bands close to provide discrete spaces. The geometry in combination with the uniform whiteness of the masonry presents an abstract and artificial space where the natural objects (trees) are highlighted and objectified through their visual complexity and contrast. Exploiting the natural symmetry of the square, the plan bifurcates the house in one direction and striates the house into bands of courtyard / interior / courtyard. The further segmentation of the courtyard bands allows for four privatized courts associated with their adjacent programs and two larger public courts that flank the transparent central living space. The continuous adjacency of interior and exterior spaces, their equal methodology for spatial subdivision, and the continuity of materials from inside to out all collaborate with the simplicity of the geometry to produce a powerful spatial abstraction.

Sendai Mediatheque, Toyo Ito
Sendai, Japan 2000
[Structural Postmodern, free plan, steel and glass]

In the Sendai Mediatheque, Ito uses the strong geometry of the square to establish a plainness to the form of the glass perimeter, which acts as a transparent boundary. This allows for contrast in form and figure through deformed tubular cores that sectionally migrate through the building, shifting laterally in variable increments from floor to floor. The square enclosure allows for organicism and structural filigree of the internal towers. Platonic base geometries of circle and square are carefully juxtaposed with delicate and dynamically calibrated structure. The pairing allows for recession of the muted singular form of the perimeter glass enclosure in favor of the objectification of the figurally dynamic structure and service cores.

Square as Figure / Form—Urban

The square as a form and figure on the urban scale defines both the independent object building and the collective unit block. Deployed as a base geometry that establishes the module of the repetitive field, its proportion and scale establish the spatial quality of the urban fabric. It is the most common underpinning figure for the urban grid, but is actually rarely perfectly deployed.

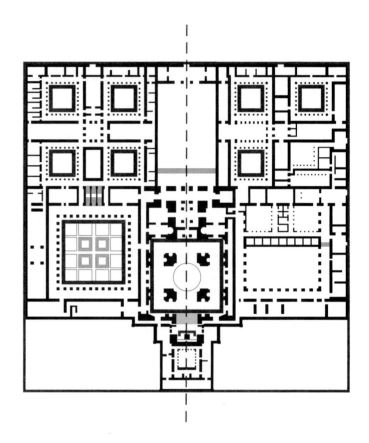

El Escorial, Juan Bautista de Toledo
San Lorenzo de Escorial, Spain 1584
[Renaissance, central axis palace with multiple courtyards, masonry]

The massive urban figure of El Escorial is underpinned by the repetitive use of the square module. As a positive figure defining the outermost frame, the square establishes the boundary. Reoccurring through the fabric of the building, the square is used through a multiplicity of spaces carved out of its massive figure. Repetitive courtyard spaces, rooms, and even the chapel are defined by the square.

	TYPOLOGY
	PRINCIPLE **129**

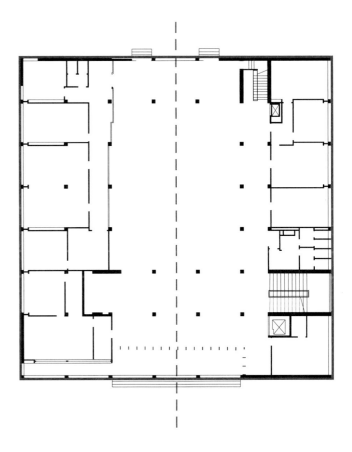

Barcelona Block, Spain
1850s
[Neoclassical, repetitive field with chamfered corner, masonry]

Casa del Fascio, Giuseppe Terragni
Como, Italy 1936
[Italian Rationalist, urban object, masonry]

In the mid-nineteenth century, the expansion of Barcelona allowed for the city to develop from its medieval foundations to a gridded block. The square is the base figure for the gridded field; chamfered corners create voids that become urban courtyards. The square thus exists as both a positive and negative field. The Barcelona Block begins with a conventional square block uniformly determined by carefully proportioned scale and height. The hollowing out of a secondary square courtyard within each block creates a centralized courtyard that optimizes the depth of floor plate for light and ventilation. The final chamfering of the four corners (achieved through the removal of a rotated square, clipping the four corners of a quadrant of blocks) creates broader faces for addressing the street and produces a distinct and discretely figured urban void at every grid node, allowing the linear experience of the street to open up on a regular increment from block to block.

The Casa del Fascio was executed as an authoritative governmental project under Benito Mussolini for the Italian Fascist Party. It establishes its hierarchy and importance through its urban geometry. Set at a significant crossroads in the city of Como, the square figure of its plan is foiled against the medieval wall and the slightly irregular geometry of the existing city fabric. The reductivism of its form and the anonymity of its material and scale assist in the power and uniqueness of its presence.

Square as Void—Architecture

In addition to being used as a positive formal figure, the square is equally legible as a negative or architectural void. As a geometric figure used to define a space, the removal of a volume and the remaining void are equally dependent upon geometry of form. The void of a negative space creates a bounded room that is affected by the perimeter as a figure and surface. The square geometry has associative midpoints and axes that provide opportunities for localized articulations of the bounding surfaces that govern the figuration and reading of the space and form.

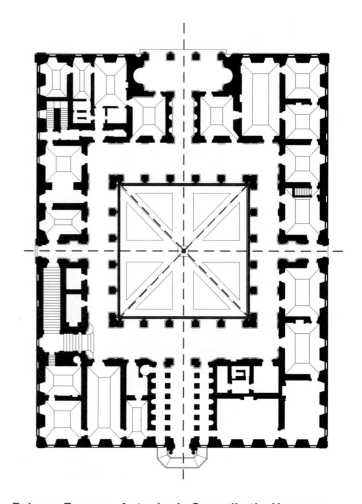

Palazzo Farnese, Antonio da Sangallo the Younger
Rome, Italy 1541
[Renaissance, courtyard building, masonry]

Adhering to the typical courtyard typology, the Palazzo Farnese is defined by a central square court. Ringed by a colonnade, which is not equal in dimension, the edges of the pure inner space are given varied hierarchies. Furthering the importance of the central courtyard, two axes, established by dominant ceremonial circulation pathways, run in a cross-configuration through its center. This central void was considered one of the paramount spaces or rooms of the Palazzo.

TYPOLOGY
PRINCIPLE | **131**

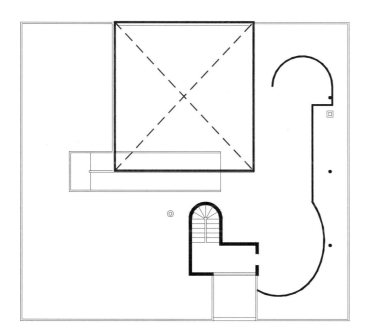

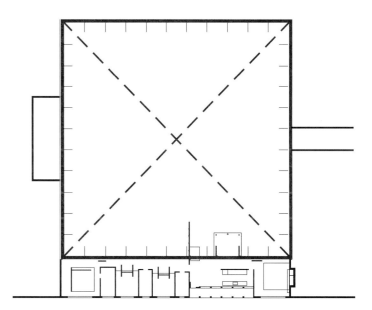

Villa Savoye, Le Corbusier
Poissy, France 1929
[Modern, International Style, free plan, concrete and masonry]

The complete removal of historical reference in both form and material requires the Villa Savoye to rely upon the geometric abstraction and systemization of the square in the organization of its structural grid. This figure becomes fully revealed as a spatial void on the second level where the building's square composition crescendos in the roof-level courtyard. A perfect square in plan, the void is defined by the third floor plate as a canopy. The free plan allows the space of the living room, ramp, and adjacent covered outdoor area to flow independently across and through this geometry.

Marshall House, Denton, Corker, Marshall
Phillip Island, Victoria, Australia 1990
[Minimalist, single-loaded courtyard, concrete]

The Marshall House is a linear interior space edging a massive, walled square courtyard. The house rides as a datum against the architectural void. The square courtyard is engaged with the rooms of the house by the thickened black concrete edge, which objectifies and contrasts the solidity and mass of the wall against the void of the adjacent space. The hierarchy of the geometry of the square is furthered in importance by the axial entry to the courtyard prior to the asymmetrical entry to the house.

| 132 | TYPOLOGY
PRINCIPLE | FIGURE / FORM
ORGANIZATION | SQUARE: VOID
GEOMETRY | PLAN
READING | URBAN
SCALE |

Square as Void—Urban

The square as a spatial void in the urban fabric operates similarly to the square as an architectural void, but adapts to the scale of the city. Dependent on the perimeter buildings surrounding the space to define the void, the importance of the space is formally determined by the fullness and continuity of the contextual fabric, while the civic hierarchy of the space is determined by the programmatic functions of the contextual fabric.

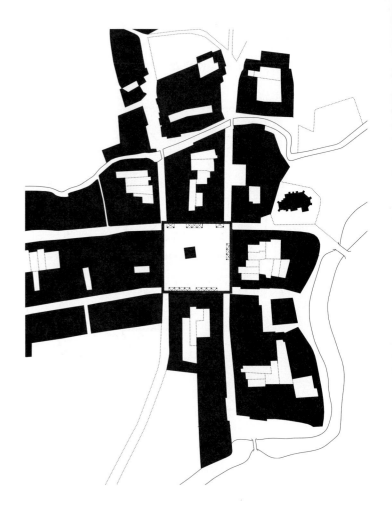

Bruges, Belgium
1128
[Medieval, fabric with square, geometric void in irregular field]

As a medieval new town, Bruges represents the contrasting geometries of the figured and irregular cityscape with the purity of the square's geometric subtraction. The void carved out of the dense pedestrian fabric creates an urban room where aggregation of activity and congregation of people can occur. The four corners of the square radiate the four main streets into the surrounding urban fabric.

	TYPOLOGY	
	PRINCIPLE	**133**

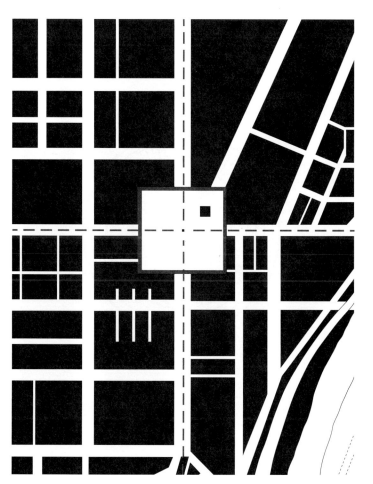

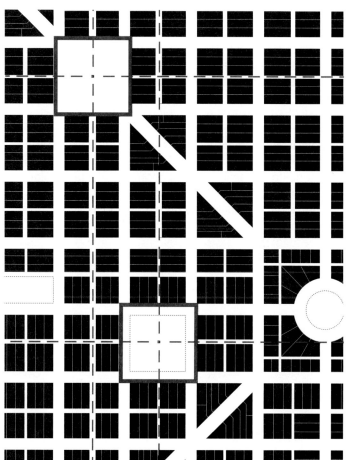

Cleveland, Ohio
1796
[Jeffersonian Grid, grid with nodal void]

Cleveland is a typical Midwestern city that arises out of the Jeffersonian Grid. Here the geometry of the grid is dominant. Not responding to natural features, the grid does not recognize or respond to the presence or organicism of its river, instead it simply ends at its banks. The urban void is carved from the gridded fabric with a center on a grid node. As a result of this disruption in the field, the geometry of the square is given a spatial hierarchy. Similarly the public square is given programmatic privilege as the civic center of the city. The location of the square directly on the crossroads of vehicular circulation subdivides the single urban room into four figured islands established by the extension of the urban edge of the eroded block.

Indianapolis, Indiana
1821
[Jeffersonian Grid, quadrant square grid]

The Indianapolis urban fabric is dominated by the square grid. Each square defined by the street grid is further subdivided. Similar to Cleveland, the grid unyieldingly marches across the local geography of the river. Within the fabric there are a series of removals. These urban voids are remnants of city blocks and produce spatial squares.

Circle as Figure / Form—Architecture

The circle as an architectural form and figure establishes a natural centric organization. It is a universal geometric figure that occurs in every culture. Creating a point in space, it is a self-resolving geometry that produces an intrinsic hierarchy that is difficult to break, deny, or alter. Non-directional (with the exception of the **axis mundi** as a vertical extension), the circle is an equally radial form that references only itself. The difficulty in modifying or extending the self-resolving geometry of the circular form creates inherently object buildings. Historically it is deployed for this formal hierarchy and its visual panoramas.

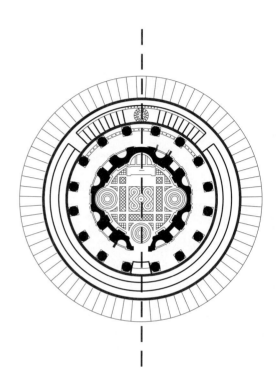

Tempietto, Donato Bramante
Rome, Italy 1502
[Renaissance, roman temple, masonry]

The Tempietto uses the circle to mark the spot of the martyrdom of St. Peter. Its geometry is highlighted by the juxtaposition of the circular plan with the orthogonal rectangularity of the courtyard in which it is located. Formulated of four concentric circles, the layers each take a different form, function, and density. The stair and plinth create the outermost circle; the structural columns form the second ring; the inner masonry wall establishes the third circle; and the interior chamber completes the fourth circle.

	TYPOLOGY	
	PRINCIPLE	**135**

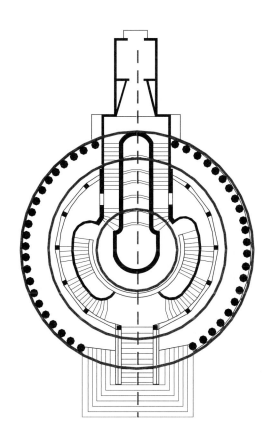

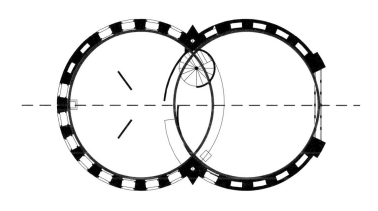

Glass Pavilion, Bruno Taut
Cologne, Germany 1914
[Modern Expressionist, axial symmetry, glass and metal]

Taut deploys the circle to create a monumental figure. Emphasizing a singular directional grain (as opposed to the typical cross axis) the pavilion uses the circulation and waterfall to create an axis. Using concentric layers of plinth, enclosure, and secondary plinth for the pool, the repetitive circular shapes consistently emphasize the formality of the central geometry.

Melnikov House, Constantine Melnikov
Moscow, Russia 1927
[Constructivist, free plan, plaster masonry]

The Melnikov House uses two intersecting circles to define its organization. The axis established by connecting the centers of the two circles and the entry door produces a linear grain. The ovaloid shape created by the overlapping circles is used for circulation. The rooms in the house are not completely affected by the central quality of the shape and instead exist more as figures within a neutral condition.

	TYPOLOGY	FIGURE / FORM	CIRCLE	PLAN	URBAN
136	PRINCIPLE	ORGANIZATION	GEOMETRY	READING	SCALE

Circle as Figure / Form—Urban

Similar to the principles established by the geometry of the circle on the architectural scale, the form continues its centric dominance when transposed to the urban scale. The singular form is inflated to the scale of a city block. The geometric self-resolution of the circular form allows for an objectification of the geometry through the contrast of curvilinear to rectilinear within the density and orthogonality typical of its surrounding urban context. The contrasting shape denotes place, hierarchy, and dominance of a moment within the broader field.

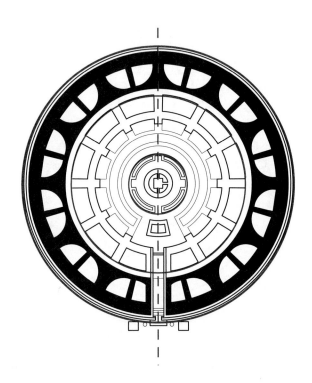

Tomb of Augustus
Rome, Italy 28 BCE
[Classical, centralized, masonry]

The Tomb of Augustus is defined by its isolationist, circular figure. As the mausoleum for the Roman Emperor, it housed his sacred remains. As a place of great importance, the cypress hill (a constructed, conical, urban berm covered in cypress trees) had a secondary circular chamber nested at its core. This drum held the actual remains. The scale and form denote the hierarchy of the function within the surrounding orthogonality of the urban fabric.

TYPOLOGY | PRINCIPLE | **137**

La Bourse de Commerce
Paris, France 1783
[Neoclassical, centralized, masonry]

Designed as an agrarian products exchange, the circular shape underwent a series of modifications to its roofs, skins, and levels of enclosure. Despite these various reconfigurations, the circular form and its dominating central hall always remained primary. Its current configuration, with a neoclassical skin of circular form, nestles it as an object that bridges the density of the Parisian urban fabric on one edge and the gardens of Les Halles on the opposing face.

The Hirschhorn Museum, Gordon Bunshaft
Washington, D.C. 1974
[Late Modern, concentric rings with central void, concrete]

The Hirschhorn Museum uses the form of the circle for the optimization of its circulation. The outer circulation ring is rendered entirely opaque, housing paintings and protecting them from ultraviolet light. The inner ring is entirely glazed and houses the sculpture collection bathed in natural light. The circular void at the center and the expressive elevated donut shape of the modernist composition use the geometry to produce an abstract and minimalist formal statement. Its geometric identity is further exaggerated by its curvilinear contrast with it surrounding rectilinear neighbors.

Circle as Void—Architecture

The removal of figured space to create an architectural void within a project places even more emphasis on the space when done with a circular geometry. The consistency of the surrounding façades prevents any localized individuation within the architectural surface. Relying on uniformity of elevational design, the emphasis is on the space created as opposed to the object by which it is defined. The circular void has a non-directional hierarchy and suggests an importance of center over edge.

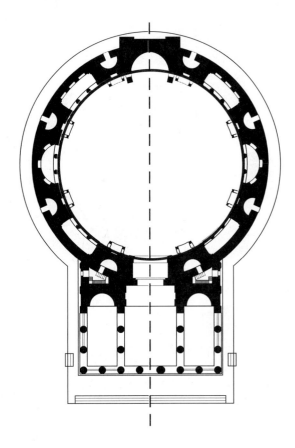

Pantheon
Rome, Italy 126
[Classical, temple, masonry]

The Pantheon's circular dominance in its architectural drum is most clearly seen on the Nolli Plan of Rome. The centralized plan hearkens back to the ancestry of the Greek Temple of Apollo at Delphi, but employs a heroic scale to make the circular space more spiritual than functional. The oculus (an open circular aperture at the top of the dome) reestablishes the geometry in the ceiling and marks the passage of time through the movement of light.

TYPOLOGY PRINCIPLE **139**

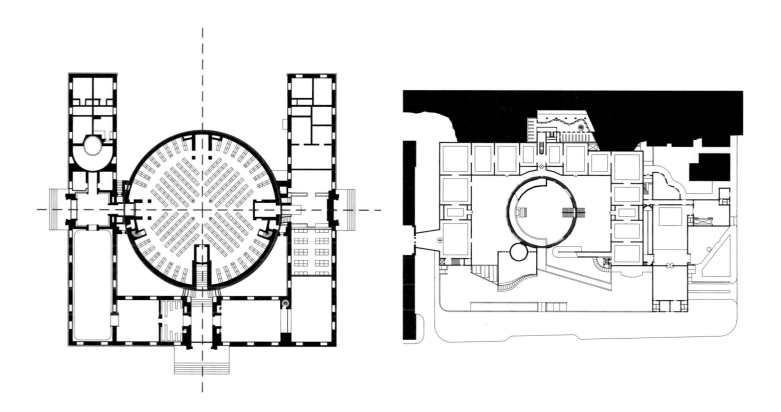

Stockholm Library, Gunnar Asplund
Stockholm, Sweden 1927
[Early Modern, centralized drum, masonry]

The Stockholm Public Library is dominated by the circular central drum of the reading room. Bracketed in plan with a U-shaped support building, the drum as the central space in both location and functional hierarchy uses the circle to emphasize that hierarchy. From the exterior, the height of the drum illustrates its significance, whereas from the interior, the walls of the drum act as the bookcases. Housing the knowledge of the accumulated texts, the circular drum is reinforced through the marriage of geometry and program.

Neue Staatsgalerie, James Stirling
Stuttgart, Germany 1983
[Postmodern, linear enfilade, stone cladding]

Stirling uses the circle at the center of the Staatsgalerie as a void. As the primary spatial node along the circulation path, it is a space that is continually returned to throughout the architectural promenade. The courtyard overlaps both interior and exterior circulation paths and bridges the section between the neighboring streets. The simplicity of the space and the power of the void are entirely dependent upon the geometry of the circle.

140	TYPOLOGY	FIGURE / FORM	CIRLCE: VOID	PLAN	URBAN
	PRINCIPLE	ORGANIZATION	GEOMETRY	READING	SCALE

Circle as Void—Urban

The circle as an urban void is typically used to create a circulation hub or roundabout. Pierced by radial entry points and typically bounded by uniform architectural façades, the form of the void is given constant reinforcement. The integrity of the shape allows for an equality and anonymity to the edges that emphasize space over object.

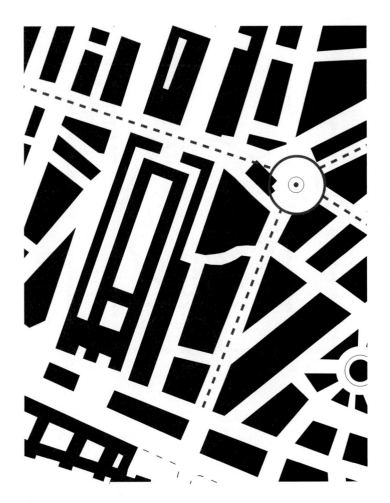

Place des Victoires, Jules Hardouin Mansart
Paris, France 1685
[Baroque, circular node]

The Place des Victoires is a circular void pierced by six boulevards. At the center of the circle and in a position of greatest hierarchy is an equestrian monument honoring King Louis XIV. The circular room is emphasized by the standardization of the façades and the signature stepping back of the Mansard roof.

The Circus, John Wood the Younger
Bath, England 1766
[Georgian, circular ring with central green]

As a pure geometric form, the Circus in Bath establishes an arrayed residential perimeter defining a circular urban room. Typical of the Georgian Style, the center is filled with a landscaped park. The circular perimeter is broken by three evenly spaced axes. The result is a nodal room within the urban fabric.

The Royal Circus
Edinburgh, Scotland 1820s
[Georgian, circular ring with park center]

As one in a series of circular voids developed in the Georgian Style, the Royal Circus emphasizes the continuity of the architectural edge bounding a picturesque central park. The circulation edges the perimeter sandwiched between the architecture and the landscape. The resulting architectural room is filled with natural content.

Oval as Figure / Form—Architecture

The elliptical form of the oval introduces a multicentric geometry that both increases the complexity of the form and evolves the perceptual and directional qualities of the resulting figure. The evolution from the circle to oval marks a similar accentuation of detail and complexity that occurred during the transition from the Renaissance to the baroque periods. With an elongated cross axis, the intrinsic geometric nature of the oval introduces a grain and the potential for either a longitudinal or transverse approach. As a result the figure has intrinsic hierarchy within its basic geometry.

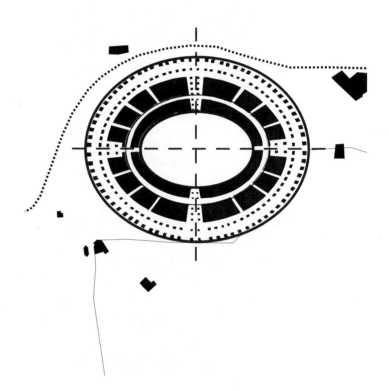

Roman Coliseum
Rome, Italy 80
[Roman, classical plan, masonry]

The Roman Coliseum uses the ovaloid configuration for visual efficiency. Attempting to provide and optimize the sight lines of the onlooking audience, the configuration provides spaces for a perimeter audience encircling a central arena base for gladiator fighting, sporting events, performances, and even naval battles. The sectional slope of the terraced seating, in collaboration with the base geometry of the oval plan, establishes a primary plan type for sports arenas that proves optimal even today.

	TYPOLOGY	
	PRINCIPLE	143

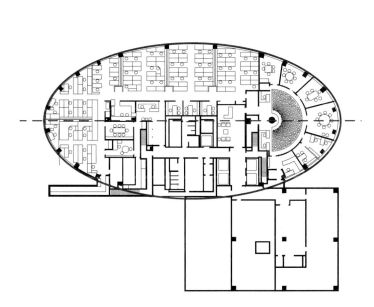 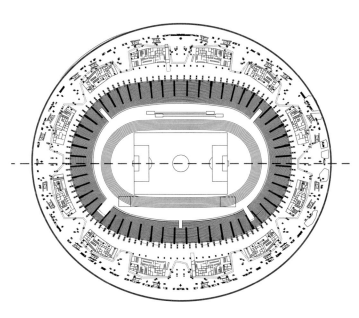

885 Third Avenue, Lipstick Building, Philip Johnson
New York, New York 1986
[Postmodern, ribbon-windows, steel, glass and masonry]

In the high rise tower at 885 Third Avenue, Philip Johnson creates a postmodern combination of a functional plan configuration (allowing for an optimization of perimeter transparency and a narrowing of the depth of the floor plate) with a sleek, streamlined shape. The building has been nicknamed the Lipstick Building for its figure and the associated form and flair of the cosmetic applicator. The combination of the banded layering of ribbon windows with the elliptical plan creates a universally directional view engaging a full 360 degrees. The form translates the conventional rectangular high-rise-building plan as an extruded condition of the orthogonal urban block into an aerodynamically formed, geometric figure.

The Bird's Nest Stadium, Herzog and de Meuron
Beijing, China 2008
[Postmodern, irregular lattice frame stadium, concrete and steel]

The Bird's Nest, built for the 2008 Olympics, adopts the plan typology of the Roman Coliseum. Designed to host the track and field events, the elliptical running track establishes the primary inner form reiterated through the surrounding ring of seating. The adoption of a functional and historically predetermined planning typology fits well with the agenda and methodology of Herzog and de Meuron. Their architecture is defined through the development of skin and the collaborative ingenuity of material, tectonic, surface, and pattern to produce identity on a traditional and simplistic form. In the case of the Bird's Nest, densely lapping and intersecting bands non-orthogonally intersect and aggregate to define a porous and homogenous yet dynamically individuated shell.

Oval as Void / Space—Architecture

The oval as void uses the same geometric distinctness of the shape's formal elongation with a smooth yet distinctly grained perimeter and a multi-nodal center, and deploys this unique figuration as a negative space—a subtraction from the solidity of the architectural or urban fabric. When deployed as a nodal condition within baroque urban planning, the resulting room allows for hierarchical multi-linear nodes of approach either axially or radially. Either urban or architectural in scale, the result is a space that creates an evolutionary perspective continually transitioning in effect as one moves through the figure. The result is a dynamic shift over time of the scale and perception of the void's form. The dynamism and complexity permit and encourage the experiential exploitation innate to the geometry.

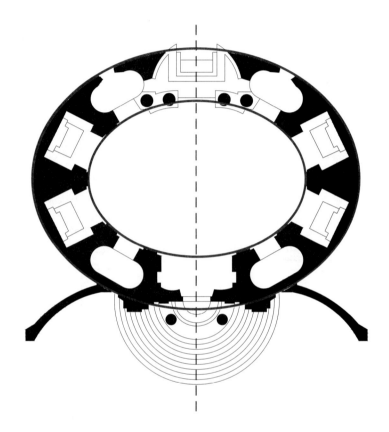

Sant'Andrea al Qirinale, Gian Lorenzo Bernini
Rome, Italy 1678
[Baroque, short-axis elliptical plan, masonry]

Sant'Andrea al Qirinale represents the pinnacle of baroque form and geometry. Deploying the oval as the figure of the sanctuary, the compressed spatial geometry is a result of the contextual condition that constrained its perimeter. The geometry and its perception is further complicated by the engagement of the multi-foci figure and its associated compression of form. The narrow cross axis establishes the lateral entry in connection to the altar. The perimeter is ringed by a series of discrete subtractive chapels.

TYPOLOGY	
PRINCIPLE	**145**

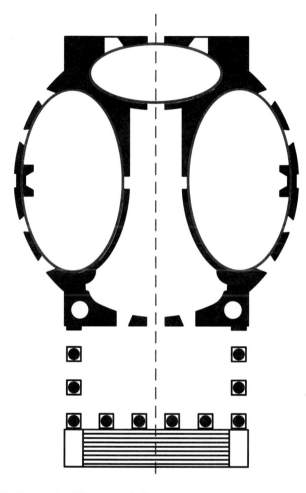

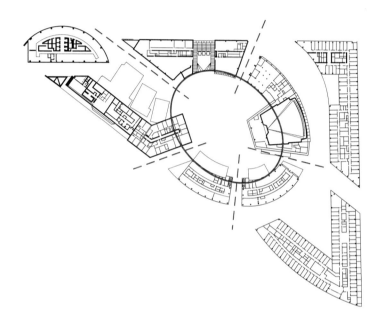

UVA Rotunda, Thomas Jefferson
Charlottesville, Virginia 1826
[Neoclassical, Neo-Palladian rotunda, masonry]

The Rotunda at the University of Virginia deploys the planning and figuration of the Pantheon. The uniqueness of the lower plan comes from the iterative transition of the central Rotunda to parallel ovaloid-shaped twin chambers, with a secondary rear lateral oval on the lower level. The resulting warping of the circular and even fully spherical definition of the rotunda (as at the Pantheon), in collaboration with the directionality of the entry portico, begins the production of the primary axis of the building, lawn, and entire campus. Establishing the orientation of the central lawn to the west, the ovaloid geometry begins the reiterated campus plan that leaves open the view, focus, and extension to the manifest destiny of westward expansion establishing both figuratively and philosophically the future and opportunity of America through the architecture of the University.

Sony Center, Helmut Jahn
Berlin, Germany 2000
[Postmodern, elliptical center court, glass and steel]

The Sony Center, is part of the development of the Potzdamer Platz zone, which was left as an urban vacancy after World War II, scarring the central urban fabric of Berlin. The Center extruded the newly established block perimeter to create a multi-building conglomerated mass. The discrete component buildings re-establish their identity and unification through the central elliptical subtraction. Capped in an intricately structured glazed roof, the transparency allows for the legibility of the elliptical subtraction both from within and above.

Oval as Void / Space—Urban

Like the circle, the oval as an urban space is typically a circulation hub that is penetrated by numerous axes. The oval became popular during the baroque period when science identified and proved that the planets rotate around the sun in an oval rather than a circle. This celestial complexity translated to an increased desire to reflect such configurations in architecture. The oval was a more complex shape that allowed a more dynamic reading, which was a natural progression from the more ordered ideas of the Renaissance.

Piazza dell'Anfiteatro
Lucca, Italy circa 100
[Roman, remnants of amphitheater]

The Piazza dell'Anfiteatro in Lucca is formed by a series of buildings constructed on the foundations of an earlier Roman amphitheater. The shape of the urban space is made by the optimization of sight lines required by its previous functions. A series of axes penetrate the continuous surface of the bounding wall. This complete enclosure of the space by the surrounding walls makes the oval reading of the space more powerful and compelling.

TYPOLOGY
PRINCIPLE 147

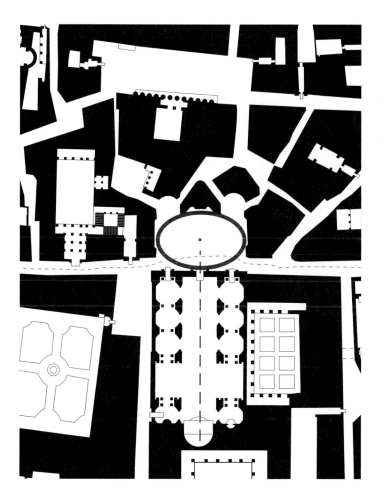

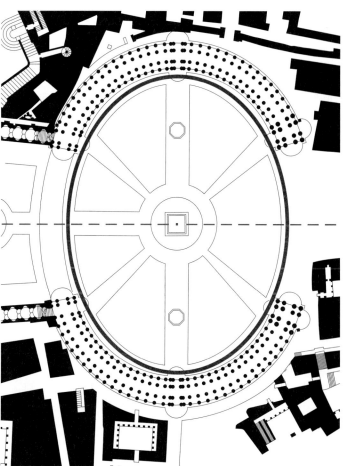

Piazza at Sant Ignazio di Loyola
Rome, Italy 1626
[Baroque, elliptical piazza]

This piazza, which sits in front of the Church of Sant Ignazio in Rome, is a series of ovals and circles that were constructed during the baroque era. It represents one of the initial attempts at using the baroque ideas at an urban scale. It necessitated the destruction of a number of existing buildings and hence is constructed in a very uniform fashion. There are a number of axes that enter the space, however unlike the Renaissance spaces that gave preference to the axes, the baroque geometry emphasized the corners.

St. Peter's Basilica, Gian Lorenzo Bernini
Vatican City, Italy 1667
[Baroque, elliptical colonnade]

Unlike the Piazza dell'Anfiteatro and the Piazza Sant Ignazio, the Piazza at St. Peter's Basilica is completely transparent as the surrounding edges are defined by the repetitive columns of Bernini's massive colonnade. Of the numerous architectural and urban experiments of the baroque era, the Piazza at St. Peter's Basilica is undoubtedly the largest and grandest. The axis from the cathedral bisects the space and aligns with the obelisk at the center. There are two large fountains that mark the foci that determine the size and shape of the space itself.

Triangle as Figure / Form—Architecture

The triangle is the simplest geometry able to establish a planar form out of the least number of elements. Its shape intrinsically creates visually accelerated spaces formed by the shallow angle of two planes converging. The resulting form produces a narrowing of the space which increases the perception of distance in perspective. Often defined by the figure of a contextual site, the form is typically generated by streets positioned obliquely across an otherwise regular orthogonally gridded grain of an urban fabric. The triangular figure of a building defined by this urban condition extends and accentuates the form.

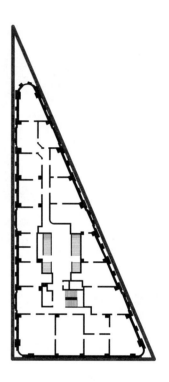

Flatiron Building, Daniel Burnham
New York, New York 1902
[Early Modern, early skyscraper, steel and masonry]

The Flatiron Building is an exemplary precedent for the triangulation of form based upon the urban condition. Located at a forking crossroad, the urban form is pronounced and reiterated through the height and thinness of the architecture. Having a shallow, textured façade, the sleekness of the design is read through the vantage of an urban opening that is produced from the angled intersection of crossing roads. Extending a form created urbanistically at street level, the triangulated figure extends the shape of the fabric and accentuates the **focal cone** of perspectival recession to further the drama of the urban condition.

TYPOLOGY	
PRINCIPLE	149

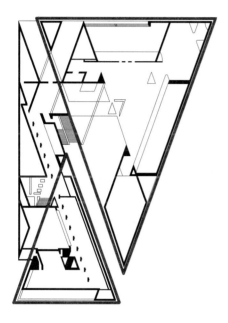 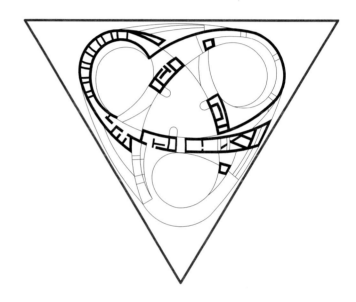

East Wing of the National Gallery, I.M. Pei
Washington, D.C. 1978
[Late Modern, central triangular atrium, concrete and stone]

The East Wing of the National Gallery derives its form from the radial streets of L'Enfant's plan for Washington D.C. It is located at the head of the National Mall, which is flanked by the varied museums of the Smithsonian Institution. The East Wing of the National Gallery is an addition that both spatially and curatorially expands the collection to include the artworks of the modern movement. The triangulated footprint is a result of the rectangular mall set on axis with the U.S. Capitol Building, juxtaposed with the radial boulevard originating at the Capitol and extending northwestwardly. Due to its awkward shape, the site was left vacant for many years prior to its development. The triangular shape of the site is further iterated through the formal bifurcation into two more triangles. The shape is reiterated through the entire building, evident in the triangulated waffle-slab structural system and pyramidal module of the central atrium's skylights. The geometry fully pervades the conception and perception of the building at all scales.

Mercedes Benz Museum, UN Studio
Stuttgart, Germany 2006
[Postmodern, double helix circulation, steel and glass]

In the Mercedes Benz Museum, the triangular form becomes a boundary condition that is defined by the curvilinear, planametric organization of the building's circulation. Defined as three connected loops, the floor plates incrementally ascend across the building's section and spiral around the central atrium. The architectural promenade ascends through the atrium to the top of the building and then descends along the two paths of automobile and pedestrian movement. The resulting looping defines the overarching three-pointed triangular figure of the building.

Triangle as Void / Space—Architecture

Triangular spaces as architectural voids are not as common as squares or circles, yet they have been used effectively throughout history. The triangle is typically used in response to the constraints and influences of a localized site condition that requires a different geometric approach not possible with a conventional rectangular courtyard. The triangular form, despite its atypical geometry, still remains responsive to the constraints and capabilities of typical architectural language. Similar to other geometric forms, the triangle's intrinsic geometric characteristics and proportional qualities remain subject to architectural principles. Rarely is the triangle utilized when the architect is not forced to do so by site conditions.

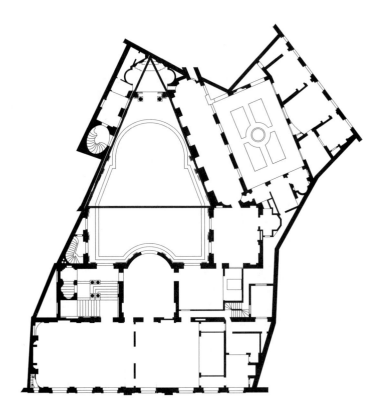

Hotel de Beauvais, Antoine le Pautre
Paris, France 1656
[Baroque, triangular courtyard, masonry]

The Hotel de Beauvais is set on an irregular site that would not allow the typical square or rectangular courtyard. Instead, le Pautre, having established the entry at a central location on the main façade, continued the axis into the site utilizing a triangular courtyard as a way to negotiate between the site and the geometric language. Like many palazzi, the rooms that surround the courtyard are allowed to morph to the shape of the site in order to maintain the clarity of the courtyard's figure.

TYPOLOGY PRINCIPLE 151

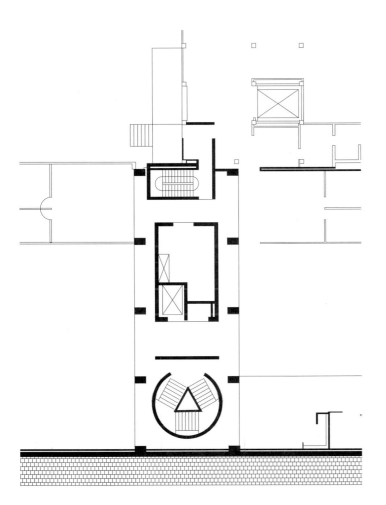

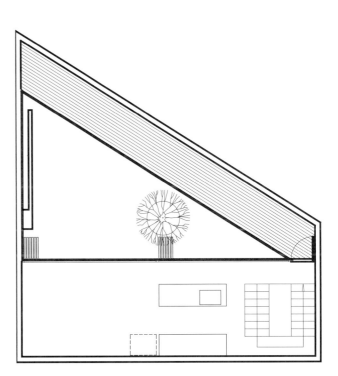

Yale Fine Arts Gallery, Stair, Louis Kahn
New Haven, Connecticut 1953
[Modern, Platonic forms - triangle stair, concrete]

At the Yale Fine Arts Gallery, Kahn uses the triangular form as a special element within the building's formal organization. The building uses a rectangular grid to organize its structure and spaces. Set into this orthogonal matrix, Kahn places a concrete, circular stair tower that houses a triangular stair within it. This gesture, composed of two variant Platonic geometries, separates this formal element from the others in the gallery and speaks of its importance and individual characteristics. This stairway is the primary vertical circulation of the building, requiring patrons to continually move from the grid of the museum to the Platonic form of the stair.

Garden House, Takeshi Hosaka Architects
Yokohama, Japan 2007
[Postmodern, operable façade, concrete and glass]

The Garden House is a simple rectangular figure set next to a triangular courtyard. All the aspects of the house are focused on this bounded courtyard. Collectively, the building and adjacent courtyard resolve the irregular geometry of the site. While maintaining a pure, positive architectural form, the variable remainder is given as a negative-space side courtyard. The triangular courtyard forms a green oasis within the heavily urban area of Yokahama.

Triangle as Void / Space—Urban

The triangle as an urban void similarly emerges from the resultant geometry of the urban fabric. Established by the positive figured field of buildings, it tends to occur more often in conditions where the space is either too small to occupy with an additional block or where the spatial and visual vantage of approach is desired to establish an urban prelude to the approach. The widening of the triangular void, when approached through the axis of the triangle's tip, establishes a position of innate hierarchy and focal perception based on approach and projected sight lines.

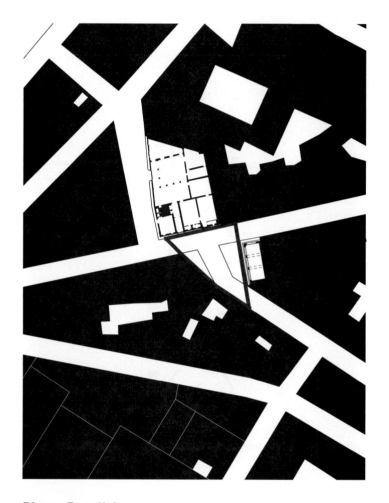

Piazza Rucellai
Florence, Italy 1451
[Italian Renaissance, Piazza as focal cone for adjacent palazzo, masonry]

Piazza Rucellai is the triangular forecourt in front of the Palazzo Rucellai. Formed by the intersection of two urban street axes, the Piazza becomes a webbing of approach lines. As a result of the urban condition, the site is established as a point of hierarchy in both position and approach. In the otherwise dense urban fabric of Florence, the Piazza is a rare moment that allows for a spatial expansion that permits the viewer distance from the surface of the street edge. This spatial release allows for a head-on reading of façade as opposed to the typically oblique vantage permitted by the narrow road width of the urban condition. The architect takes advantage of the expanded visual cone to develop the highly composed façade treatment of the Palazzo Rucellai. The geometry of the urban space allows for an architectural composition to emerge directly from its adjacent form.

TYPOLOGY
PRINCIPLE **153**

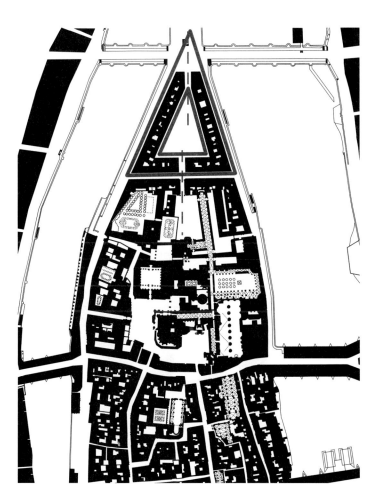

Place Dauphine, Ile de la Cite
Paris, France 1607
[French Classical, axial symmetry, masonry]

The Place Dauphine is located on the tip of the Ile de la Cite in Paris. It is a product of the urban fabric meeting the natural geography of the converging rivers. Formed by the erosion caused by the movement of the Seine River, the island shape reiterates in the urban form. The triangular courtyard creates a hierarchical figure that links the Central Courts of the French Government with the equestrian statuary located at the tip of the island. The form funnels from the planar surface of the courts (defining the base side) to the statuary and axis of the river beyond (establishing the point that completes the triangle). The triangular figure is articulated by the homogenous and repetitive façade that produces a continuous perimeter and creates a continuity of figure. The resulting triangular void is a park space, symmetrically flanked and axially skewered.

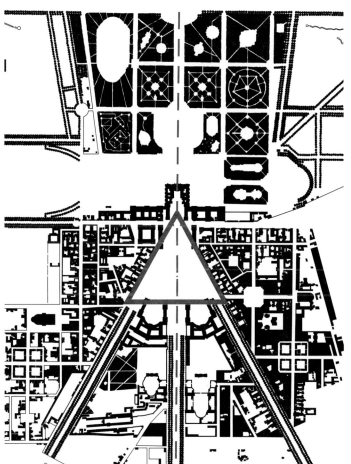

Versailles Entry Court
Versailles, France 1774
[Baroque, royal—axial symmetry, masonry]

The triangular entry court at the Palace at Versailles is a fragment of an extended geometry focused on the dominance of King Louis XIV. Originating at the king's bed, the axial geometry of the house, garden, and adjacent urbanism, is a production of radial hierarchy. Near the tip of this radial condition is the triangular entry court to the palace compound. Here the transition from the urban fabric to the palace fabric occurs. The triangular figure amplifies the majesty, dominance, and hierarchy through the funneled focal cone. The shape directs both movement and views toward the palace and ultimately the power of the institution and the king.

Polygon as Figure / Form—Architecture

The polygon as a multi-sided geometrical figure is described here to include shapes beyond the three- and four-sided geometries and introduces the complexity of forms that arise through a multiplicity of surfaces. These geometries, typically emblematic of numerological or referential figures, are largely centralized. Determined by a singular point with an arrayed perimeter of linear facets, the number of surface edges, or faces, determines the polygon type and shape. For example, the pentagon has five faces, the hexagon has six, the octagon has eight, and so on. The number of faces determines the resolution of the perimeter shape and the perception of the individual facet, or the form of collected facets, as a singular surface.

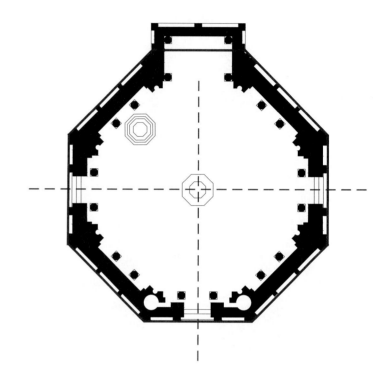

Florence Baptistery
Florence, Italy 1059
[Romanesque, centralized biaxial symmetry, masonry]

The Florence Baptistery occupies a unique position as one of three objects (including the Duomo and the Campanile) within the Duomo complex and the city proper. The baptistery is an object in the round. Rare in the tight solidity of the Florence urban fabric, the baptistery adopts an octagonal geometry that both denies a primary "front" while equally preventing a "back." The octagonal form creates an equality of directionality. The eight-sided nature allows for the development of a series of multivalent axes that in their collection cancel the hierarchy of one another and produce an equally distributed field.

TYPOLOGY
PRINCIPLE **155**

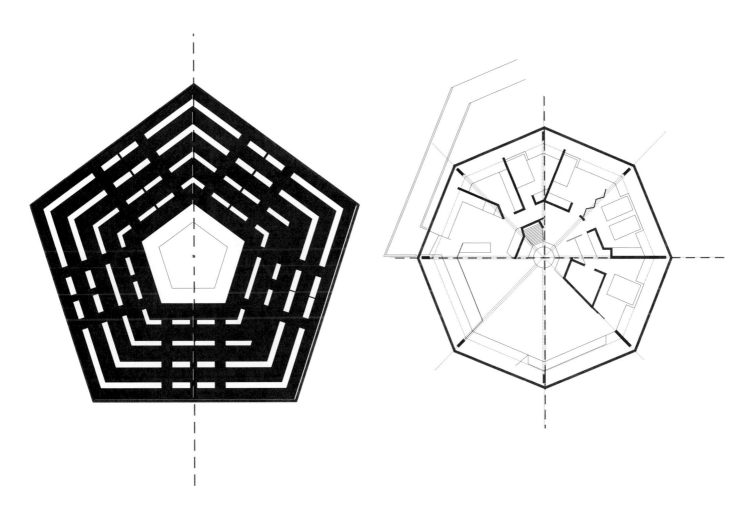

The Pentagon, George Bergstrom
Washington, D.C. 1943
[Neoclassical, concentric rings, steel and masonry]

The Pentagon houses the headquarters of the United States Military. It adopts the five-sided geometry of the pentagon to give equal preference and privilege to each of the five branches of the military services: Army, Navy, Air Force, Marine Corps, and Coast Guard. The pentagonal form is reiterated with a series of concentric layers in plan. Each ring is a double-loaded corridor that alternates layers of building with courtyards between to maximize light and ventilation. The five faces represent the individual military branches with their custom expertise and specialties, but combines them in a single shape to express their unity and collaboration. The geometry establishes the form, function, and symbolism of the architecture through the pentagonal shape.

Malin Residence, John Lautner
Los Angeles, California 1960
[Heroic Modern, Chemosphere, concrete, steel and glass]

The octagonal plan of the Malin Residence (often referred to as the Chemosphere) is derived from the constraints of a precarious site condition and a limited budget. Set on a steeply sloped site in the Los Angeles hills, the house uses a central columnar foundation, which reduced the cost of developing the steep hillside site. Deploying an octagonal geometry, the shape cantilevers from the central mast with struts connecting the outer octagon at its nodal points. The plan organization divides the house in half; the north half is a large single living space, whereas the south half is divided into four wedge-shaped rooms that complete the geometry. The uniqueness of the octagonal shape derives the structure, planning, and form through its innate geometric organization.

Polygon as Figure / Form—Urban

On the urban scale, the polygon is deployed for function reasons: optimization of defensive sight lines, differentiation of form from the rigidity of the orthogonal grid, and creation of secondary gaps in the fabric. In each scenario, the multifaceted geometry employs the straight line of the surface for ease of construction, but bends at the joint to create forms that can calibrate and orchestrate the form and resulting internal or external spaces.

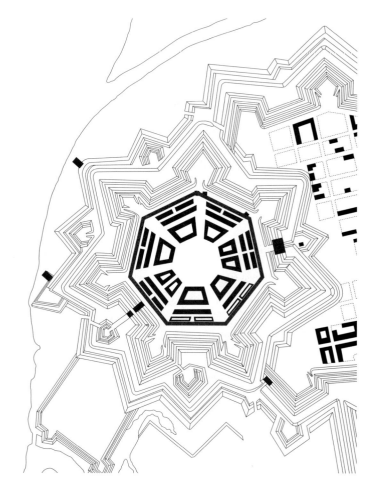

Fort Mannheim, Germany
1645
[Military Fortress, defensive fortification, masonry]

Mannheim, Germany is a polygon-formed city built entirely for defensive purposes. Defined by sight lines and ultimately projectile trajectories, the shape is one that provides visual and militaristic dominance in all directions. Subdivided by a traditional urban grid, the city streets and blocks are laid out in an orthogonal organization. This fabric is encircled with a triangulated perimeter that maintains a continuity of edge and creates the multifaceted geometry of the encircling surface.

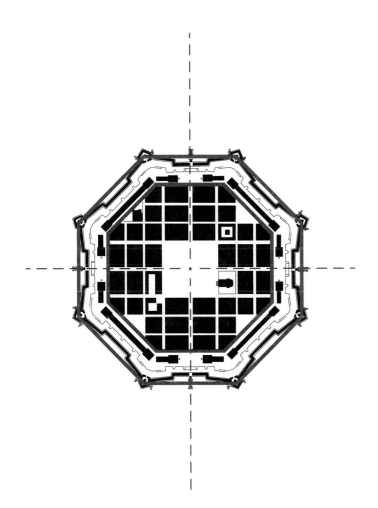

Neuf-Brisach, France, Marquis de Vauban
1696
[Military Fortress, defensive fortification, masonry]

Neuf-Brisach is a French town derived from concentric geometries. Beginning with a central square, the grid of the city extends outward with axially organized blocks extending from the sides of the originating figure. The Greek Cross of the urban-fabric grid is then bounded by four extended planes that are angularly connected to define a perimeter octagon. This polygonal shape is then further encircled with a series of layers of triangular defensive **earthworks***. The purity of the initial geometry is evolutionarily iterated in form, while rigorously maintained with each successive encircling layer.*

Barcelona Block, Spain
1850s
[Neoclassical, repetitive field with chamfered corner, masonry]

The Barcelona Block uniquely deploys the polygonal shape as an urban unit within a larger gridded field as opposed to either a centralized organizing figure or an overarching boundary form. The Barcelona Block begins with a conventional square block that is hollowed out to form a centralized courtyard and chamfered on the four corners to address the street intersections. In Barcelona, the polygonal figure is responsible for the exquisite fabric that balances light, density, movement, and infrastructure with a powerful square grid and a polygonal block.

158	TYPOLOGY	FIGURE / FORM	POLYGON: VOID	PLAN	ARCHITECTURE
	PRINCIPLE	ORGANIZATION	GEOMETRY	READING	SCALE

Polygon as Void / Space—Architecture

The polygon is typically a centered space that permits a number of forces to be resolved in a unified and synthesized whole. The polygon allows for numerous axes to penetrate a space and remain ordered. Often the sides of a polygon are equal and balanced, however they can be more complex and dynamic.

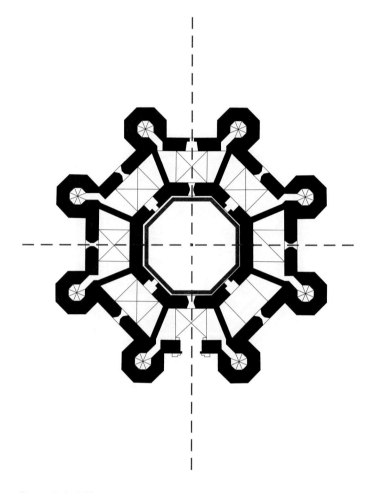

Castel del Monte
Bari, Italy 1250
[Medieval, fortress with central octagonal court, masonry]

Castel del Monte is a study in the geometry of the octagon. It takes the form of an octagonal building with an octagonal courtyard, surrounded by eight octagonal towers. The exterior towers contain either stairs or lookout positions. The shape of the main castle is formed by intersecting lines drawn from these towers to form a Greek Cross plan, which leads to the octagonal courtyard. There are three entrances into the courtyard from the castle. The entire width of the castle is one-room thick. These rooms have a centered window to the exterior and interior and are connected to each in an enfilade, or repetitive layered, method. The interior structural walls all radiate from the center and align with the eight towers.

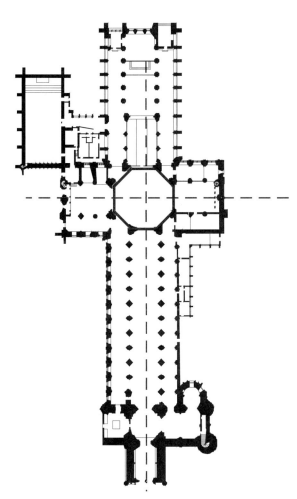

Ely Cathedral Dome
Cambridgeshire, England 1300s
[Medieval, polygonal central dome, masonry]

The central crossing of the nave and transept of Ely Cathedral is octagonal. This shape allows the larger center aisles of both the nave and transept to align with this central space. The four angled sides then correspond to the smaller side aisles. The octagon is primarily formed by the large supports that are comprised of numerous grouped columns. This allows the centralized space to exist within a Gothic interior and not break with the transparency that is experienced in the rest of the cathedral.

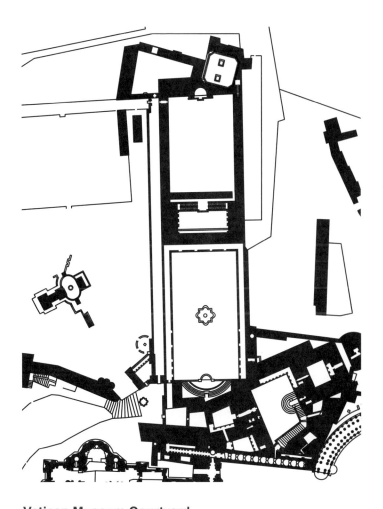

Vatican Museum Courtyard
Vatican City, Italy 1701
[Classical, octagonal courtyard, masonry]

The octagonal courtyard in the Vatican Museum acts as a point of reference that joins elements gathered at the far end of the Pigna Courtyard. This octagon negotiates the various angles that connect the Circular Hall, the Hall of the Muses and the Animal Room with the Round Vestibule and the Vestibule of the Torso. There are two doors into the courtyard and each corner is occupied by important sculptures in what are referred to as cabinets. This is the major exterior space within the museum and acts as a mediator and a resting place.

| 160 | TYPOLOGY
PRINCIPLE | FIGURE / FORM
ORGANIZATION | POLYGON: VOID
GEOMETRY | PLAN
READING | URBAN
SCALE |

Polygon as Void / Space—Urban

The primary advantage of a polygonal void in an urban condition stems from the shape's ability to receive multiple axes from different directions. The majority of these polygonally formed spaces are vital to a city's urban life. Unlike a conventional square (which can typically only resolve two axes), a polygon can exploit numerous axes, views, and field conditions.

Campo, Sienna
1349
[Medieval Hill town, fan-shaped city center]

The medieval city of Sienna has a non-symmetrical, polygonal central space. Referred to as "the Campo," this central plaza forms the major political space of the city and is famously utilized as a horseracing track during the Palio. There are eleven entries of various sizes that access the Campo. Unlike Renaissance spaces, there is little hierarchy and there are no axes in the space. This free-flowing plaza is similarly illustrated in section. The horizontal surface of the Campo is not even or flat but instead is undulating and ramped and divided into nine fan-shaped sections.

Palmanova, Italy
1845
[Military Fortress, concentric rings around center void]

In the center of Palmanova there exists a public space that takes the shape of a hexagon. There are six major roads that lead into this space. Three of the roads are directly connected to the three gates of the city. The overall shape of the city is a nine-point star. It was conceived as a fortress city, where the shapes of the exterior walls were governed by the functional concerns of defense. Through a series of geometrically ordered moves, the street patterns and central polygonal space were formed with rational geometric planning. Their rigorous formation results in a tightly figured urban organization.

Canberra, Australia
1911
[Early Modern, concentric plan with radial connectors]

The urban plan of Canberra is a poly-centered plan that uses the circle and the polygon as its primary geometries. The densest area of the city is centered on a hexagonal space that has a circular park at its center. There are six separate roads that enter at the midpoint of each side. The scale of the space, primarily related to the automobile, lacks the intimacy that is found in the more traditional polygon spaces of Europe. Unlike the Campo in Sienna, the walls of the polygon are constructed of numerous buildings that sit as individual objects, thus ensuring a discontinuous wall that does not encompass the space effectively.

Spiral as Figure / Form—Architecture

The spiral as a dually directional shape serves as a concentric yet continuous linear pathway. As either centripetal (inwardly spiraling towards the center) or centrifugal (outwardly spiraling away from the center), the directionality of this reading implies either an expansion or a contraction. The centripetal spiral implies a hierarchy towards center where the approach and destination of the middle demands the importance. The centrifugal spiral is about a hierarchy of edge, end, or perimeter where the outward layering of the spiral becomes the focal extension.

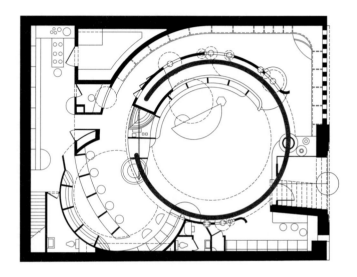

V.C. Morris Shop, Frank Lloyd Wright
San Francisco, California 1948
[Modern, circular spiral ramp, concrete]

At the V.C. Morris Shop, Wright deploys the spiral as circulation. Ramping the perimeter of the shop floor with an inclined plane, the scheme is a diagrammatic precursor to his Guggenheim Museum in New York City. In both schemes, the sectional spiral of circulation defines the primary formal reading (here an interior, while at the Guggenheim Museum both interior and exterior) and establishes the architectural promenade. The slow, ascending, perimeter ambulatory movement establishes a highly choreographed cinematic experience. The spiral becomes both the sequence and the form.

TYPOLOGY PRINCIPLE 163

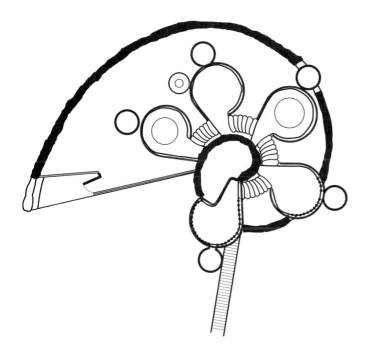

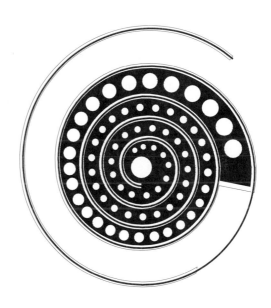

Bavinger House, Bruce Goff
Norman, Oklahoma 1955
[Expressionist Modern, spiral-terraced house, masonry and steel]

The Bavinger House uses the spiral as a terracing sectional organization (Raumplan) rotating around a single central point. The suspended pods each assume a discrete programmatic function. They are arranged in a sectional and radial spiral that simultaneously establishes the sequence and hierarchy of each individuated component and the collection as a whole. Suspended from a central mast, the trays are structurally and experientially dependent upon their geometrically defined position.

The Chapel of Thanksgiving, Philip Johnson
Dallas, Texas 1976
[Postmodern, spiral chapel, concrete]

The Chapel of Thanksgiving, stemming from a postmodern approach to geometry, uses the spiral form to create an iconic figure. Like the Tower of Babel, or a spiraling ziggurat, the concrete chapel uses the spiral form to create the shape and space of the sanctuary. The spiral as an icon establishes a unique perception through the complexity of its geometry.

Star as Figure / Form—Architecture

The star is the most complicated and specific geometric typology. Unyielding in its form, the possible variations are limited. It derives its functionality from the defensive figure that it offers. As a highly faceted perimeter, the form originates in the sight lines and firing trajectories. The optimization of these factors, combined with a centralized organization, a desire for a geometric and organizational purity, and a finite space of defensible territory makes the star a closed and efficient geometric figure.

Villa Farnese, Giacomo Barozzi da Vignola
Vitterbo, Italy 1560
[Renaissance, pentagonal plan circular courtyard, masonry]

The Villa Farnese deploys a pentagonal plan with curving inner façades. The articulation of the nodes defined by the connection of the five sides establishes points in space that devise the five-pointed star. The outer geometry is balanced by the inner circle, carved out of the mass of the figure. Vertical stairways and antechambers fill the interstitial spaces between the two geometries. The site plan superimposes a hierarchical axis extending from the town, through the oval stairway, and branching into the two projected square gardens. This sequence breaks the evenness of the five sides. Designed for defensive purposes, the shape and the resulting projecting bastions provide an ideal vantage for raking fire. The combination of the fortified planning with the fineness of the Renaissance order and ornament create a complimentary composition to one another and the site.

TYPOLOGY PRINCIPLE 165

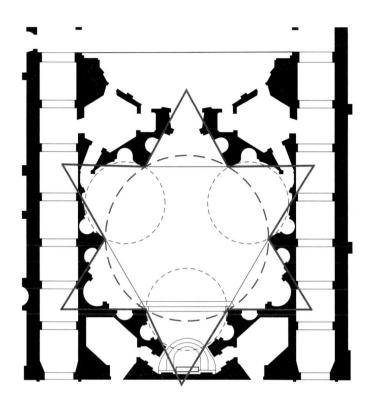

Saint Ivo Della Sapienza, Francesco Borromini
Rome, Italy 1650
[Baroque, star-shaped sanctuary, masonry]

*Saint Ivo Della Sapienza is set in the cloister at the University of Rome. It deploys a six-pointed star to define the interior figure of the plan. As a centralized circle with six semi-circular spaces located evenly along the circumference, the side spaces establish the entry and altar (and create an extension to the axis of the courtyard) with the other four as side apses. The complexity and figuration of the dome is determined through the extension of the hemispherical side spaces that conically taper and intersect to resolve at the circular **cupola**. The resulting ribs of the intersecting roof geometries reiterate the six-sided nature of the star form, introducing movement and undulation to the surface.*

Fort McHenry
Maryland 1798
[Military Fortress, pentagonal court with central courtyard]

*Fort McHenry is a five-sided pentagonal military complex surrounded by a five-pointed defensive perimeter wall. A sixth protrusion happens on one face to establish an entry that is reiterated in a **sallyport** through the entry building. Defined by the sight lines and trajectories of cannon fire, the fort is a defensive outpost surveilling the grounds and adjacent waterways of Baltimore Harbor. It gained its historical significance during the War of 1812 as the site where Francis Scott Key composed the "Star-Spangled Banner" while defending the harbor from the British Navy. The concentric rings of building, fortification, and dry moat allow for iterative transitions from polygon to star, as well as from built free-standing forms to earth-sheltered building to earthwork.*

Star as Figure / Form—Urban

The star shape, when used as an urban figure, exclusively represents a fortress or a citadel. The points of the star were constructed in a way so that projectiles that were fired toward the fortress would be deflected. This particular typology was found primarily during the nineteenth century, but was soon obsolete as the machines of war advanced.

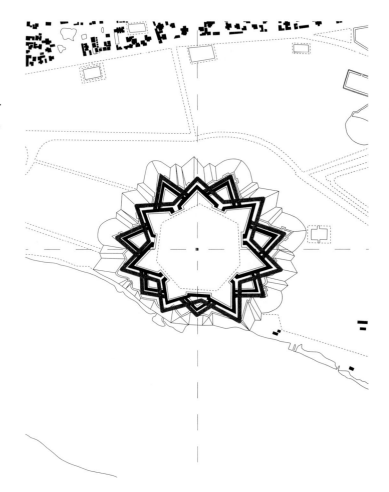

Fort William, Calcutta
1842
[Military Fortress, two overlain seven-pointed stars]

Fort William, which sits in the center of the Eden Garden in Calcutta, takes the form of a multi-point star figure. It has three series of walls that surround a central area that is loosely based on a grid structure. The multi-faceted star figure is particularly powerful in this setting, where it is separated from the rest of the city by the large gardens. The outer wall has thirteen points; the middle wall has seven points and the inner wall has nine points. All of these points were conceived as fortifications and shaped to deflect projectiles.

TYPOLOGY PRINCIPLE 167

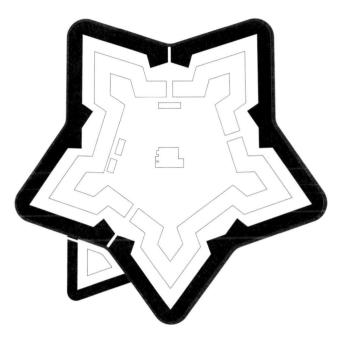

Palmanova, Italy
1845
[Military Fortress, nine-pointed star atop nine pointed star]

The city of Palmanova exists as a fort or citadel as much as it does as a city. There are a number of minor squares that are organized in a radial pattern around a central hexagonal space. The form of the city is based on nine projections that are set within another nine-point fortress. As in most of these fortress cities, the points are there for defensive purposes. Palmanova is singular in that it has survived in its original form and is certainly one of the best examples of this typology.

Goryokaku Fort, Hakodate, Japan
1866
[Military Fortress, five-pointed star, masonry]

This fort, placed in the city of Hakodate in southern Japan, was one of the largest western forts constructed in Japan. It is in the shape of a five-pointed star, which allowed for more gun displacements than the traditional Japanese fortress. As in the majority of these forts, the shape and formal characteristics were dictated by engineering calculations that were concerned with cannon warfare. As were many of its European counterparts, the fort was completely surrounded by a moat of the same shape.

Programmatic / Functional Typology

The most common form of typology is programmatic or form typology. This refers to a series of buildings that share the most basic elements of program. The associative formal evolutions are typically developed through a series of incremental changes in technology and construction. For every building type there is a lineage that is formed from the initial models and then continually manipulated and transformed. Often these transformations occur almost without notice. However, there are moments when, due to either technology or cultural events, a building type can change radically. The primary classifications of programmatic or functional typology include the following:

Temple/Church The temple and church typology has been one of the most widely used and most manipulated of all the typologies. This stems from the strong and functional form of the temple or church, which permits slight modifications and iterations to resonate boldly. The basic idea of a large room that is focused in one direction is fundamental to all peoples. As the programmatic type of one of the oldest surviving institutions, the church typology has continued uninterrupted for centuries, allowing and requiring repetitive use hybridized by cultural transitions.

Palazzo The palazzo typology was developed primarily during the Renaissance. It was an extension of the typical Roman house, complete with the interior courtyard. The palazzo originated to serve the powerful families of Florence and formally manipulated to accommodate increased scale, diverse programs, concepts of defense, and urban responses. Typically these buildings were three-stories tall and contained a regular-shaped courtyard in the center that allowed light and air to penetrate into each of the rooms. This courtyard emerged as one of the most important spaces in the building and indeed the city.

House The house is undoubtedly the most widely used and ubiquitous of all the typologies. Every culture has developed its own version of this typology. Environmental necessities, cultural hierarchies, and local laws have dictated many alterations and variations. The basic need to shelter one's family has been a constant since the first man first employed a cave or tree to serve in such a capacity. The house was the first building type and remains an experimental mechanism (due to scale and quantity) for exploring ideas of architecture.

Museum The first museums were typically buildings originally designed for other functions, such as palaces or offices, emerging from the residence of the king, emperor, or czar. These buildings were easy to convert to museums as their room size and circulation patterns worked well with the idea of exhibition. Over time, architects have examined the function and character of this building type. It is a typology that continues to develop as the nature of art changes.

Library The library is an ancient building typology with its roots originating in Egypt at about 2000 BCE. Originally these buildings contained clay tablets and scrolls, usually of a governmental nature. During the Hellenic and Roman times, the library developed into a depository of not only records, but also fiction. During the Middle Ages, the Church was responsible for much of the continuation of the library as a typology. Libraries continue to change as the material that is stored continues to change.

School Linked inexorably to the library, the school typology has been in existence for thousands of years. Originally it was formed as a residence associated with a single person. Later, the Greeks developed the school typology as an institution with a foundational model. The Romans continued to develop the school until it was ultimately adopted and hybridized by the Church, resulting in the models that are typically used today.

Prison It is unclear exactly when this typology began, though the need to incarcerate criminals emerged early in human history and has continued through time. The Greeks and Romans had distinct forms of prisons that establish a model remarkably similar to what is used today. During the Middle Ages, people were typically housed in the dungeon of the city castle or government building. Prisons form a strict typology, firmly rooted in its standardized function with only minor architectural changes over time.

Theater The theater as we know it was originally developed by the Greeks and has undergone various alterations as it has progressed throughout the ages. Architecturally, this progression has witnessed the move from the amphitheater of the Greeks and Romans, through the development of the proscenium theater of the Renaissance, and finally arriving at the architectural experiments that have accompanied modern theater, dance, and film.

Office / High-Rise / Commercial The office/high-rise/commercial typology is concerned with commercial architecture in general and hence it covers a wide range of building types. Certainly the first offices and commercial spaces were originally part of residences. Interestingly this function did not actually require a discrete typology as commerce could literally occur anywhere. The markets of Greece and Rome, the souks of the Middle East, and the various structures that have served other civilizations have all contributed to this rich and varied typology, which has developed into the skyscrapers, boutiques, and office parks of today.

Parking Garages The parking-garage typology has only recently emerged with the advent of the automobile. Though new, it is dependent upon the functional engineering of the automobile's mechanized constraints. Vehicle size and turning radius determine the base needs, establishing constant forms and optimized arrangements. The changes that have emerged over the years are typically in stylistic language, material tectonic, and ornament.

Campus The word "campus" was first used to describe the grounds of Princeton University, coming from the Latin term to describe a field. Today, the typology typically describes the conglomeration of buildings and lands that make up an educational or cooperative institution. The campus includes multiple buildings, their functional arrangement, and relative organization, along with stylistic and formal characteristics.

Temple / Church

Throughout history, the temple or church typology is one of the most recognizable programmatic or functional typologies. The sacred space of the temple or church has typically followed the rather simple formula of a linear space that has a hierarchical element (often an altar) placed at the end opposite the entrance. This fundamental typology, though undergoing numerous iterations, adaptations, and manifestations through history, has essentially remained constant. The Egyptian temple used this model and later the Greeks continued with the numerous temples that were critical to their culture. The difference being that the Egyptian temple was seemingly carved out as a negative space whereas the Greek temple was always a freestanding object. The Romans used this model as their own and manipulated it further. The Roman temple was visually very similar to the Greek, however the Romans typically had a stronger axial relationship with the surrounding environs. The Romans mostly constructed their temples in urban conditions where the building's relationship with built context was critical. During the dark ages, the temples were replaced by the Christian basilica, a type that generally hybridized the modern Christian church and the Roman basilica. This model became the five-aisle church that later developed into the Gothic cathedral. The Gothic cathedral changed the formula slightly by recognizing the hierarchical position of the altar, resulting in the transept. This Latin Cross church form then dominated for centuries through the Gothic and Romanesque periods. With the arrival of the Renaissance, there was a desire to develop either the centralized church or Greek Cross. Due to the functional requirement of seating the faithful, the Greek Cross (though favored by the Renaissance architects) was rarely constructed. Other concerns during the Renaissance dealt with the conversion of the multi-aisle church into a single space. This typically occurred by manipulating a series of chapels on the sides and by increasing the span of the roof over the nave. Another formal concern that appeared was the idea of forcing the space of the transept into a more regular form. Since the Renaissance there have been numerous challenges to this typology; however, the basic formula remains largely unchanged.

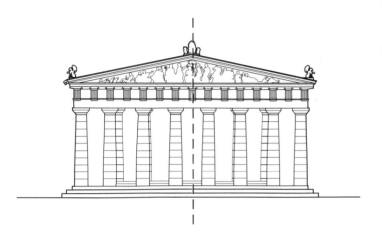

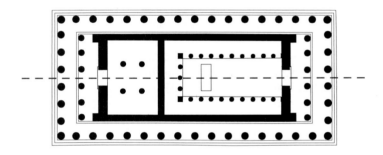

Parthenon, Phidias
Athens, Greece 432 BCE
[Classical, circular rotunda temple, masonry]

The Parthenon in Athens is a Doric peripheral temple. It has a symmetrical plan that is surrounded on all four sides by Doric fluted columns. There is a single large room (the cella) that dominates the plan and contains the large statue of Athena. This is the primary space for religious ceremonies. At the back of the Parthenon there is the opisthodomos which is essentially a treasury. The colonnade that circulates around these rooms is also critical to the building and its function as a temple.

TYPOGRAPHY | **171**
PRINCIPLE

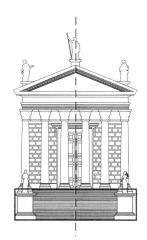

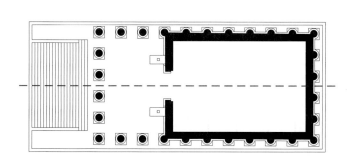

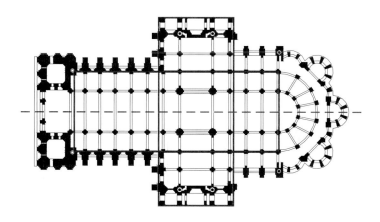

Maison Carrée
Nimes, France 16 BCE
[Classical, frontal colonnaded rectilinear temple, masonry]

The Maison Carrée represents a key shift in the theory of the temple. Unlike the Greek temple which could be approached from all sides, the Roman temple instead focused on the primary entry axis. The Maison Carrée is similar to the earlier Greek temples, however the columns along the sides are now engaged into the wall, negating any entry other than along the primary axis. The building is also placed on a higher pediment to increase its monumentality.

Chartres Cathedral
Chartres, France 1260
[Gothic, Latin Cross cathedral, masonry]

Chartres Cathedral has a Latin Cross plan that begins with two towers and a small narthex. The nave of the cathedral is three-bays wide. The nave transforms as it extends past the transept into a five-bay system, ultimately terminating in a large, complex, semi-circular ambulatory that contains numerous chapels. The increase in the number of aisles in the rear of the cathedral reflects the importance of that portion of the plan over the frontal nave.

| 172 | TYPOLOGY
PRINCIPLE | FUNCTION
ORGANIZATION | TEMPLE /
CHURCH
TYPE | PLAN
ELEVATION
READING | ARCHITECTURE
SCALE |

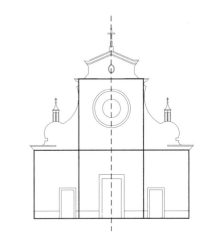

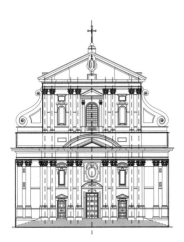

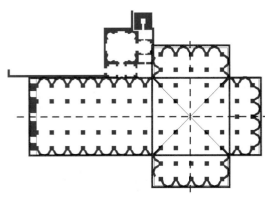

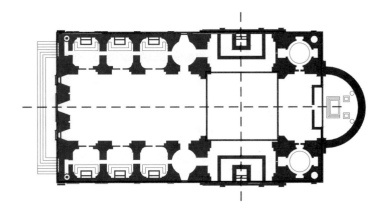

San Spirito, Filippo Brunelleschi
Florence, Italy 1482
[Renaissance, centralized Latin Cross church, masonry]

San Spirito represents many of the ideas of church design during the Renaissance. The bays of the interior colonnade relate to the larger square bays of the nave. In fact, the entire plan is executed using the square as the primary building block. The plan forms a perfect geometric Latin Cross that allows the observer to recognize the purity of the geometry. One of the most original aspects of San Spirito is the continual colonnade that occupies the edges of the nave and transept and literally runs around the entire interior of the church.

Il Gesu, Giacomo Barozzi da Vignola
Rome, Italy 1584
[Renaissance, centralized Latin Cross church, masonry]

Il Gesu represents the zenith of church design in the late Renaissance. The plan reflects many of the aspects formed by the Council of Trent. There is no narthex, instead the faithful were thrust directly into the church. Vignola also designed the church so that the nave was one singular space rather than the more traditional multi-aisled church. Replacing the side aisles are a series of chapels. This large space allows everyone to see the high altar at all times. The crossing of the transept is more compact, thus focusing attention on the high altar.

TYPOLOGY	
PRINCIPLE	173

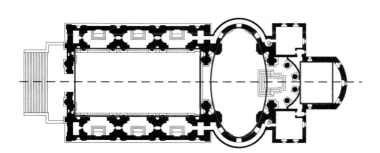
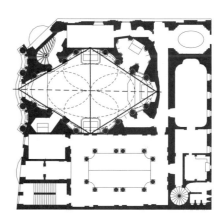

Il Redentore, Andrea Palladio
Venice, Italy 1591
[Renaissance, centralized Latin Cross church, masonry]

The plan of Il Redentore reflects Palladio's interest in history and the architecture of the ancients. Like Il Gesu, it also reflects changes in church design brought on by the Council of Trent. The plan uses a spatial organization that is reminiscent of the Roman Baths. The building is divided into three distinct parts that correspond to the frigidarium, tepidarium, and caldarium, all components of the bath's organization. Palladio uses the nave, transept, and chancel respectively, as a way to tie the architecture to the classical past. In the same way as Vignola, Palladio replaced the side aisles with chapels and reduced the transept, thereby focusing more attention on the high altar.

San Carlo alle Quatro Fontane, Francesco Borromini
Rome, Italy 1646
[Baroque, centralized oval church, masonry]

San Carlo alle Quatro Fontane, represents for many the prime example of baroque church design. Due to the small site, Borromini was unable to utilize either the Latin or Greek Cross plans. Instead he used two triangles that form a diamond shape. The main altar is placed at the long end of the axis and two minor altars are placed on the central cross axis. The fact that the altars are exposed, and not hidden from view, marks a dynamic shift in church design. Overlaid on this plan are a series of ovals (most evident in the ovaloid dome) that synthesize the diamond-shaped plan into a complex and singular space that has influenced architects ever since.

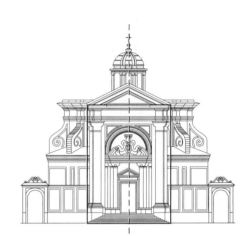
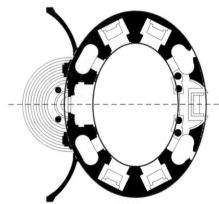
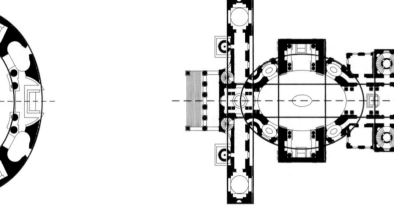

Sant'Andrea al Quirinale, Gian Lorenzo Bernini
Rome, Italy 1670
[Baroque, centralized oval church, masonry]

Karlskirche, Johann Fischer von Erlach
Vienna, Austria 1737
[Rococo, temple portico with axial layering, masonry]

Sant'Andrea Quirinale is one of the seminal churches of the baroque period. The plan takes the shape of a large oval with chapels buried in the thickened poche of the wall. The main axis of the church runs in the short direction, rather than the more typical long direction. The side chapels all point to their respective foci rather than the geometric center. However, the floor pattern reinforces the centrality of the space, allowing the incongruency of these elements to be unified with the overall effect being one of harmony.

The Karlskirche church is an eclectic design composed of a number of architectural elements that seemed to be plucked from completely different eras. What is remarkable is how Fischer von Erlach was able to synthesize these apparent disparate elements into a cohesive whole. This church is to be seen in the round; the façade-oriented sensibility of the Renaissance has given way to a completely three-dimensional extravaganza. The entire complex tends to represent a small city rather than one building. The plan is composed of an elevated classical entrance portico, which leads into a vestibule that is penetrated by a cross axis that connects two towers. Beyond this is the main sanctuary, which is a large oval that is bisected by two axes. The cross axis connects two large chapels that are symmetrically placed on either side of the church. The altar is expressed as yet a third element that extends beyond the body of the church.

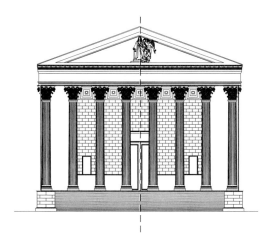
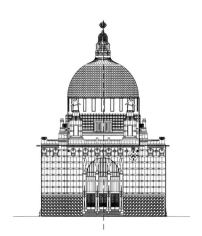
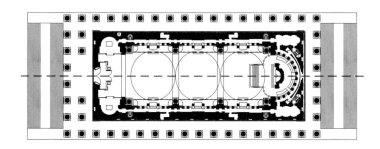
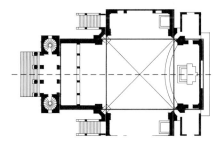

La Madeleine, Pierre-Alexandre Vignon
Paris, France 1842
[Neoclassical, colonnaded rectilinear church, masonry]

This church began as a monument and was later consecrated as a church. The architecture was a reaction against the baroque and rococo styles that had flourished in the 18th century. This building was based on the Maison Carrée at Nimes and reflected the desire to return to a simplicity in architecture. The plan is a simple rectangle that has a second building, or colonnade, that sits within the larger building. There are three domes that correspond to the three bays. On each side of the bays are located large chapels that are connected to each other through passageways that run parallel to the nave. The altar is situated under a half dome at the appropriate end. The interior of the church borrows heavily from the Roman baths in both planning and decoration.

Kirche am Steinhof, Otto Wagner
Vienna, Austria 1907
[Art Nouveau, Greek Cross church, masonry]

The Kirche am Steinhof uses the ideal Renaissance church plan as its organization device. The Greek Cross plan, considered harmonious and completely centralized, worked well for Wagner in this singular church. Wagner was able to execute this plan due to the fact that this was not a public church, but was instead situated as part of a hospital. Therefore the number of people in the church was relatively small and could be controlled. As a result, Wagner did not have to concern himself with allowing large numbers of the faithful into the church, which typically resulted in the more common Latin Cross plan. The ornament of the church was devised with more invention and did not follow any particular typology.

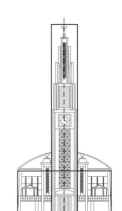
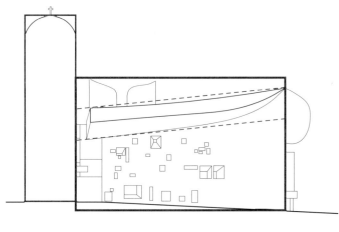
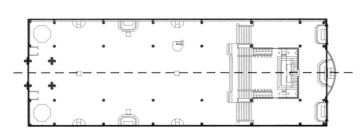
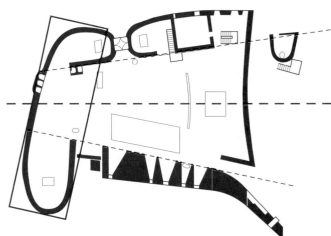

Notre Dame du Raincy, Auguste Perret
Raincy, France 1923
[Early Modern, axial church, concrete]

This church, by Perret, was a modern interpretation of the traditional cathedral. It is essentially a three-aisle church with chapels along the sides. At the entrance there is an abstracted notion of a narthex; the altar occupies its traditional location on axis from the entrance. The construction of the church, which is cast-in-place concrete, allowed for a new material language to be introduced. The columns on the interior are very thin and express the abilities of the new material, as do the shallow concrete arches that span the nave. The verticality of the structure recalls the Gothic. The concrete is left unfinished and is expressive of its construction. This is a good example of how materiality can add to the evolution of a type, whereas the planning has remained relatively unchanged.

Notre Dame du Haut, Le Corbusier
Ronchamp, France 1955
[Expressionist Modern, free plan church, masonry and concrete]

Notre Dame du Haut, by Le Corbusier, marked a quantum shift in his work. Le Corbusier had relied essentially on types and models when designing his earlier work. In this building he attempts instead a new language of metaphor. Though the imagery of the building is dominantly different, its planning remains grounded within the typical church typology. The building is essentially a space that is dominated by a linear axis that passes through a central altar. The pews are placed to one side, which attempts to deny this symmetry. Like the traditional church, the chapels are placed along the edges. One of the more remarkable inventions of the building is the large, thick entry wall, which was constructed using the stones of the original church that had previously occupied the site. Also the large sweeping concrete roof attempts to negate the straightforward axial alignment. Despite these radical moves, the building remains within the tradition and evolution of the church.

TYPOLOGY PRINCIPLE **177**

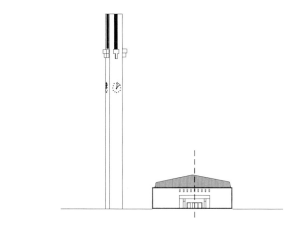

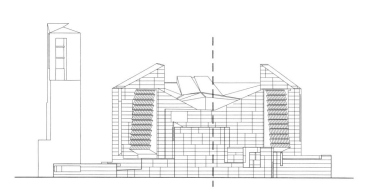

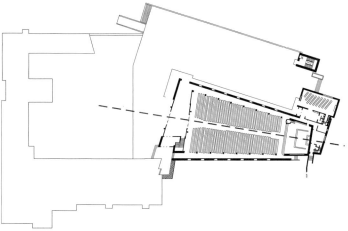

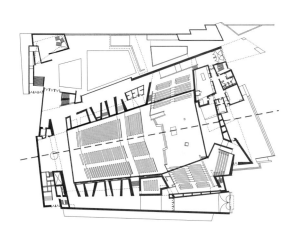

Lakeuden Risti Church, Alvar Aalto
Seinajoki, Finland 1960
[Modern, free plan church, wood, masonry, concrete]

This church follows many of the same patterns that have been set down by the rituals of the Christian church. This building differs in that it allows a series of entrances rather than the typical hierarchical central entrance, which, however, does exist within the complex. Formally the building has taken the overall shape of a symmetrical trapezoid. The building is comprised of two separate parts: the nave and the altar. The altar appears as a smaller building inserted into the larger nave building. The signature Aalto curved surface is used in the ceiling to amplify light and acoustics. This church questioned the church type from a formal and acoustic standpoint.

Cathedral of Our Lady of the Angels, Rafael Moneo
Los Angeles, California 2002
[Postmodern, axial cathedral, concrete]

The Cathedral of Our Lady of the Angels is perched on the edge of the 101 Freeway in Los Angeles and is another manifestation of the traditional church. Morphed into a building that takes into consideration its extraordinary site, the entrance to the cathedral complex is through a large courtyard that occupies the center of the site. The entrance into the cathedral proper is along the left side of the building, in what can be described as a sort of narthex that leads to the baptistery, which is on axis with the altar located at the opposite end of the church. All the traditional ecclesiastical elements are present; they are merely set up in a way that recognizes the odd angles and conditions of the site.

Palazzo

During the Renaissance there was a great deal of attention given the development of the palazzo typology. Most of this development occurred in Florence during the early Renaissance. The typical layout of the palazzo was a large, three-storied building with an interior courtyard. The main rooms of the palazzo were located on the second floor or **piano nobile**. Proportions were very important to these architects and typically the palazzo plans were comprised of squares and golden sections. The elevation treatment was often comprised of three different types of stone divided by a **stringcourse**. The stones typically differed in that the most rusticated were on the ground level, becoming more refined in the second and third floors. This same hierarchy of refinement was applied to the use of the orders in the pilasters moving from Doric to Ionic to Corinthian. This model essentially remained unchanged throughout the Renaissance. When an architect was confronted with an irregular site, the purity of the courtyard was always paramount and the rooms around the courtyard were used as a type of poche. These buildings provided the model not only for countless other palaces throughout the world, but also many banks and office buildings, particularly in Britain and the United States.

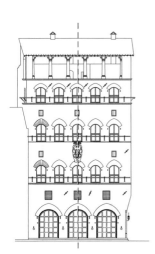

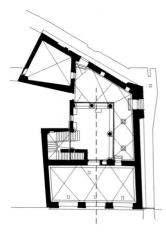

Palazzo Davanzati
Florence, Italy late 14th century
[Medieval, prototype of courtyard palazzo, masonry]

The Palazzo Davanzati is a medieval palazzo is made up of three residences that were combined together. It contains a central courtyard that organizes the plan. All the major rooms in the palazzo open out to the courtyard. This similarly occurs on the upper floors through a balcony system that circulates around the perimeter of the courtyard. This model of a palazzo with a central courtyard was to be replicated throughout the city, and indeed throughout the Renaissance. Though the rooms lining the courtyard often were compromised due to site constraints, the courtyard remained a perfect rectangle.

TYPOLOGY
PRINCIPLE | **179**

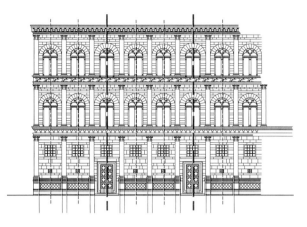

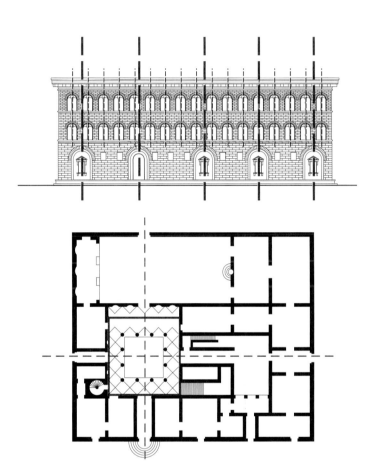

Palazzo Rucellai, Leon Battista Alberti
Florence, Italy 1451
[Renaissance, rectangular courtyard palazzo, masonry]

The Palazzo Rucellai is a quintessential example of an early Renaissance palazzo. The plan illustrates the importance of the courtyard typology. Due to the irregular nature of the site, the courtyard was configured as a rhombus, yet it still retained a sense of order and hierarchy over the surrounding rooms. This negotiation with the site became a large part of this building. The resulting courtyard appears off-center with the majority of the rooms placed on the front street side and along the right side of the courtyard. The true innovation of this palazzo occurs in the façade, where Alberti demonstrates some of the primary issues of the Renaissance. The overall organization of the proposed eight-bay façade was taken from the Coliseum in Rome. The stones and pilasters of the façade are purposefully very thin, resulting in a formal application that does not reflect load-bearing stone construction.

Palazzo Medici Riccardi, Michelozzo di Batolomeo
Florence, Italy 1460
[Renaissance, rectangular courtyard palazzo, masonry]

The Palazzo Medici Riccardi is considered to be one of the most important architectural monuments of the early Renaissance. It established much of the language and rules that were later considered critical. The courtyard, located at the heart of the palazzo, is a pure square surrounded by a perimeter of rooms. The courtyard façades are articulated through an open loggia below with the carved openings above. The exterior has three levels, each using different levels of rustication separated by stringcourses. The height of each level is proportioned appropriately as they ascend. The building is finished in a classical cornice. This and many of the other palazzi of the early Renaissance, are really a combination of medieval and Renaissance forms.

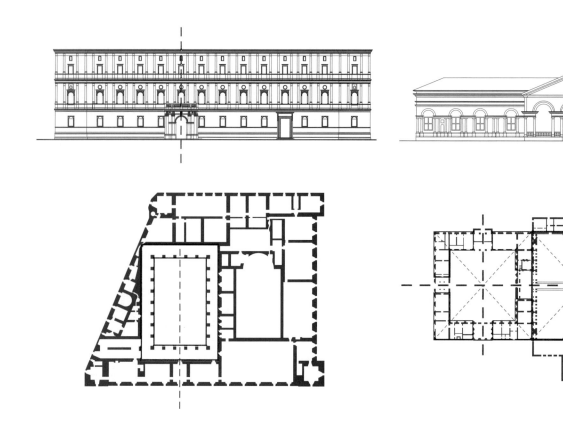

Palazzo Cancelleria, Donato Bramante
Rome, Italy 1513
[Renaissance, rectangular courtyard palazzo, masonry]

The Palazzo Cancelleria was the first palazzo in Rome that was constructed in the Renaissance style. The courtyard is much larger than those in Florence and was based on the geometric proportions of the golden section. The façade of the building is reminiscent of the Palazzo Rucellai in Florence and, along with the forty-four columns in the courtyard, was constructed from marble taken from the ruins of the theater of Pompeii. The façade is massive and asymmetrical due to the fact that it is encased within the palazzo. Directly next the courtyard is the fifth-century Church of San Lorenzo in Damaso. Bramante's ability not only to utilize the materials and columns from a former structure, but also to formally engulf one, continues to draw admiration.

Palazzo del Te, Giulio Romano
Mantua, Italy 1534
[Mannerist, rectangular courtyard palazzo, masonry]

The Palazzo del Te is a villa suburbana located just outside the center of Mantua, but readily adjacent to the city. It was built for Federigo Gonzaga by Giulio Romano. Its mannerist architecture and details were popular with academics and other architects. The plan, unlike the elevations, is very straightforward, forming a large square with a square courtyard in the center. There are two axes that cross each other in the courtyard, one extending from the entrance out to a larger court that was adjacent to the large stable building. The rooms are situated in an enfilade fashion around the courtyard and only joined to it visually. The elevations of the façades and the courtyards are a series of sophisticated experiments in the language of classical architecture. Romano took great pleasure in breaking the established rules of classicism in a perverse but playful fashion.

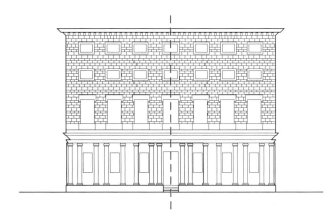
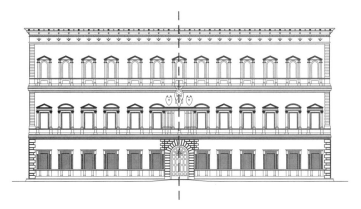
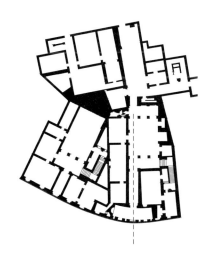
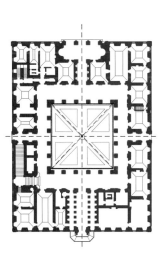

Palazzo Massimo, Baldassare Peruzzi
Rome, Italy 1536
[Renaissance, rectangular palazzo with curved façade, masonry]

Palazzo Massimo is unusual due to its convex façade, which is formed because the palazzo is constructed on the foundations of the Odeon of Domitian. The plan of this building is most ingenious as it regularizes a distorted site by using local symmetries to organize the rooms around the courtyards. Poche is used as a way to make up the irregularities as the different grids collide. The entrance is marked by a vestibule that worked as a way of relieving the pressure of the street immediately adjacent to the façade. There is an axis that joins the two regular courtyards to the entrance and thus the street. The overall plan language illustrates how an architect could carve out regular space from an irregular condition.

Palazzo Farnese, Antonio da Sangallo the Younger
Rome, Italy 1541
[Renaissance, rectangular courtyard palazzo, masonry]

The Palazzo Farnese exhibits the typical courtyard condition that was established in this typology during the Renaissance. The courtyard forms a square placed at the geometric center of the palazzo. The arcades forming the edge of the courtyard, which appear equal in elevation, are however smaller on the edge and consequently larger on the ends. This slight difference lends hierarchy to the main axis, but allows the void to remain a perfect square in plan. This reinforcement of a primary shape exhibits the importance of the courtyard as one of the most significant spaces of the palazzo.

House

The house is the most common building type and hence has a long and iterative history of typologies. Each culture has a number of typologies that have emerged from their rituals, climate, and place. Many typologies of the house cross time periods and cultural divides. One such typology is the courtyard house. This particular typology developed separately in many different spheres. In the Mediterranean, the Romans developed the Pompeii house with its courtyards, impluvium, and gardens; these courtyard houses actually served as the model for the Renaissance palazzo. In China and the Middle East, similar courtyard houses were developed to produce internalized exteriors of controlled private spaces. The courtyard house was also very common in South America and Mexico where the development of an indoor to outdoor living space was climatically and culturally driven. Typically, wherever it was warm, a courtyard house was used. The courtyard house also allowed light and ventilation into the heart of the house, thus making it popular across varied climates. Other typologies that have been utilized are the Shotgun House (Southeastern United States), the Dogtrot House (Southeastern United States), the Triple Decker (Boston), the Brownstone (New York), and the Ranch (the United States). Isolation of a culture results in different typologies. These typologies often occur as a simultaneous response to both climate and culture.

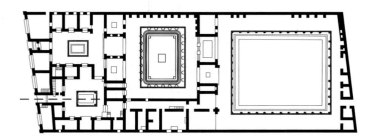

Pompeii House of the Faun
Pompeii, Italy second century BCE
[Roman, courtyard house, masonry]

As one of the largest and most impressive private residences of Pompeii, the House of the Faun is recognized as a paramount example of the courtyard typology through its organization around a series of atriums. Occupying an entire city block, or insula, the house has a storefront, or tabernae, that lines the street. Upon passing through this commercial zone, there is a series of four courtyards. Two are in the main body of the house and have typical impluvium and conpluvium openings in the ceiling with reciprocal indentations, or fountains, in the floor. One larger courtyard is surrounded by a colonnade. The final courtyard is a large garden space encircled by a colonnade that edges the site on three sides. Expanding in scale, and becoming ever more private, the courtyards act as subtractions from the main building, which infill the urban fabric, becoming the dominant spatial organizer within both the house and the city.

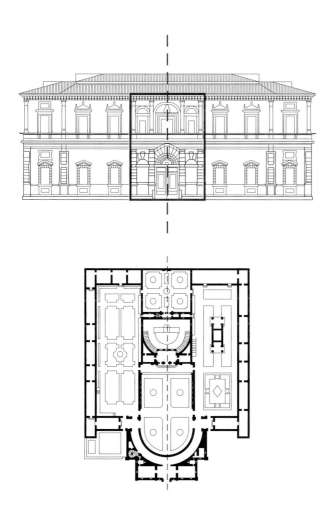
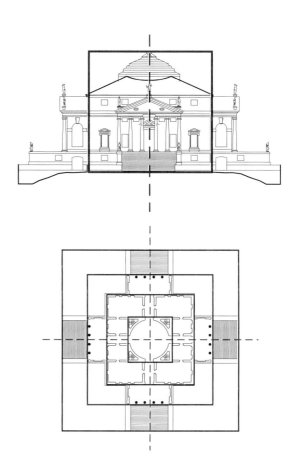

Villa Giulia, Giacomo Barozzi da Vignola
Rome, Italy 1553
[Renaissance, axial courtyard house, masonry]

Executed for Pope Julius III, Villa Giulia is a villa suburbana placed on the outskirts of Rome. It is an excellent example of a mannerist villa, and is thought to have been influenced by Bramante's courts at the Vatican. The villa was used for festive occasions and for the storage and display of Julius' art collection. The exterior courts or rooms dominate and control the form and composition. The rooms that might have been used in a conventional villa are all located in the urban casino, which is the first building encountered. They are arrayed around the semi-circular arcade and interestingly keep their geometry orthogonal despite their location. The nymphaeum (a monument to the nymphs typically containing water), which is located along the main axis and is recessed into the ground, also operated as a bath. In Villa Giulia, the project explores the fundamental relationship between architecture and the landscape.

Villa Rotunda, Andrea Palladio
Vicenza, Italy 1571
[Renaissance, multi-directional cross axis square, masonry]

Inspired by the Pantheon in Rome, Palladio had to resolve the reading of the object in a multivalent space. No longer bound by the confines of the urban fabric, the form needed to be fully resolved in the round to deal with the multiple vantages of approach and experience. To resolve this issue, Palladio relied on organizational geometries and symmetries. Derived from a circle inscribed within a square, the points of contact establish the cross axis. These points establish the center of an additional set of squares that half protrude from the body of the building and are demarcated by four identical temple fronts, equally reaching out along the cardinal axes. The center of the house is a domed cylindrical rotunda. The interior organization is mirrored to maintain order, balance, and a multivalent reading of the directionality and purity of the formal composition. The house is a masterpiece of order and geometric resolution.

| 184 | TYPOLOGY
PRINCIPLE | FUNCTION
ORGANIZATION | HOUSE
TYPE | PLAN
ELEVATION
READING | ARCHITECTURE
SCALE |

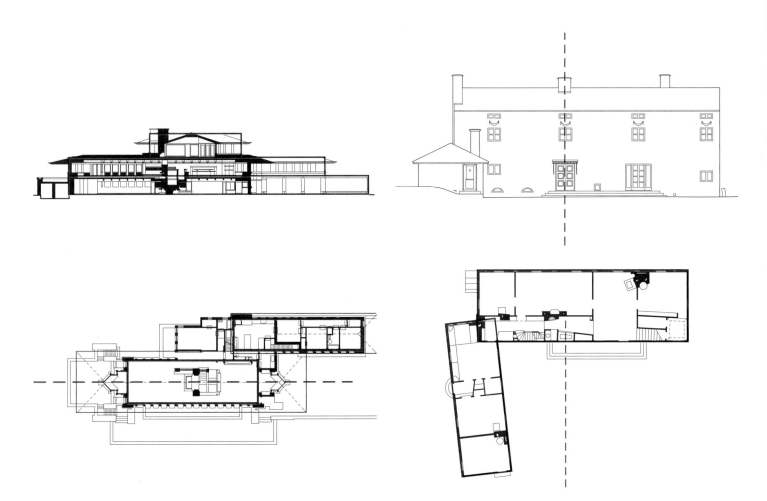

Robie House, Frank Lloyd Wright
Chicago, Illinois 1909
[Prairie Style, linear centrifugal house, masonry and steel]

The Robie House is a pinnacle example of Wright's prairie style house. Using sleek horizontal lines, pin wheeling spaces anchored by the hearth, deep overhangs that protect high glass windows, and built-in furniture that was positioned and designed to produce a holistic environment, the house presents a highly choreographed yet distinctly American modern architecture. Evoking the horizontality of the American West, the house established the dominance of the single family house typology that would prove influential in land-use patterns for the coming century.

Villa Snellman, Gunnar Asplund
Stockholm, Sweden 1918
[Neoclassical, single-loaded house, wood]

The Villa Snellman is a modest building only one-room wide with a small corridor. The ground floor of the villa contains the formal rooms and a hallway that joins them to the main living room. Situated at a slight angle from the main house is a smaller building, which houses the kitchen and servant quarters. In the plan this rotation makes little impact. However its shifting and twisting is instead read on the elevations where the window relationships between the upper and lower floors are articulated. It is as if the crank in the plan has forced the windows of the lower floor to move. Remarkably this house, despite its small size, kept many of the formal concepts of larger villas, such as a separate set of stairs for the servants, a smoking room, and the library on the second floor. This building was at once a precursor to both Postmodernism and Deconstructivism.

TYPOLOGY	
PRINCIPLE	185

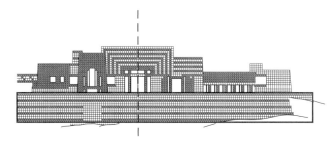

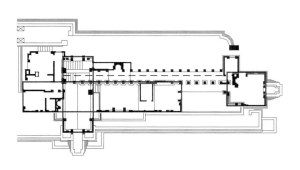

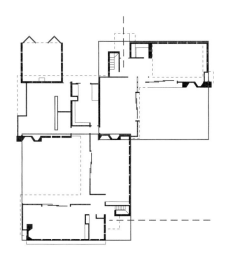

Kings Road House, Rudolf Schindler
Los Angeles, California 1922
[Modern, pinwheel plan house, wood and concrete]

Ennis House, Frank Lloyd Wright
Los Angeles, California 1923
[Modern, linear plan house, concrete masonry]

The Kings Road House emerged as a modernist experiment in single-family dwelling in the Southern California climate. Designed by Rudolf Schindler as a live-work house for two families, including his own, with a shared kitchen and a guest apartment. Employing a clever pin-wheeling plan, the configuration allowed for adjacency of the parts while providing them with individuation and privacy. The complex interlocking forms and honest use of material (through both process and tectonic) collaborate to blur the boundary between outside and in. Anchored with tilt-up cast concrete walls and balanced with delicate wood and canvas sliding-screen walls, the form of the architecture and landscape, as well as the blurred boundary of operable edges and transparent corners, further the enhanced interior and exterior living possible in the temperate Southern California weather.

The Ennis House is an exemplar of Wright's textile-block houses. Developed as an affordable site-built technology, the concrete masonry units served as structure, form, and ornament all within the same system. The module of the unit becomes a dominant figure throughout the building. Establishing an X, Y, and Z spatial module, the unit transitions from pattern to form. Buttressed walls, which serve as landscape stabilization on the steep site, anchor the form to the site, while the material forms aggregated masses and colonnades that evoke a primitive formalism to the architecture. The realities of the system proved both structurally very poor (especially in a seismically active region), and incredibly porous (causing severe complications from moisture).

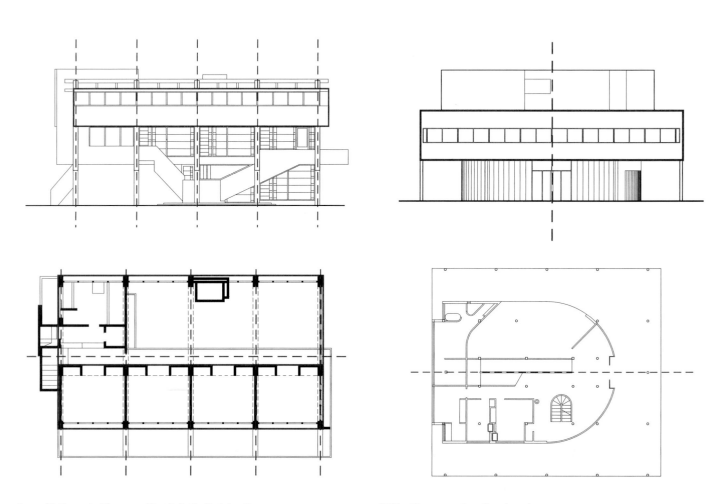

Lovell Beach House, Rudolph Schindler
Newport Beach, California 1926
[Modern, free plan house, concrete]

The Lovell Beach House is formulated as a series of parallel planes. Each of the parallel cast-in-place concrete wall planes develops as a repetitive marker. The system of support walls allows for suspended tubes of space to span from wall to wall. The structure, as expressed ribs, moves from inside to out, allowing a highly legible systemization of mass and space. The resulting form has a distinct figure and spatial reading from both inside and outside.

Villa Savoye, Le Corbusier
Poissy, France 1929
[Modern, free plan, concrete and masonry]

In Villa Savoye, the house is now conceived as "a machine for living." It removes all ornament and referential form. Relying upon the abstract simplicity of form, the house eliminates materiality for color, making the house primarily white with key contrasting colors. The organization and hierarchy of the historically quarantined and bounded formal spaces are blurred through the open plan. Spaces are allowed to flow into one another, even blurring the boundary of interior to exterior. Here the modern idea of house is now fully evolved.

TYPOLOGY
PRINCIPLE 187

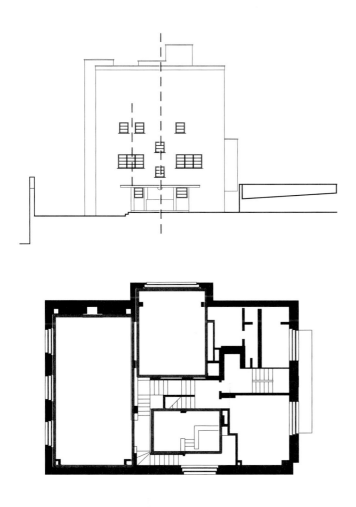

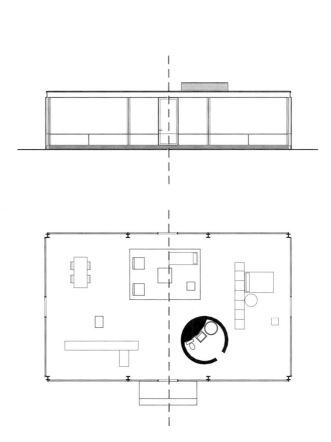

Villa Müller, Adolf Loos
Prague, Czechoslovakia 1930
[Modern, raumplan house, masonry]

Villa Müller uniquely develops a domestic organization based upon the raumplan. The house is three-dimensionally extended into discretely organized rooms that sectionally sequence, interpenetrate, and question the hierarchy of space. Embodied with ideals of economy and functionality, the intricacy of the interior sectional arrangement is masked on the exterior by a monolithic and primitive form, which is devoid of decoration or ornament. Materially this principle is reinforced with the interior having rich marbles, woods, and silk surfaces, whereas the exterior remains a neutral white.

Glass House, Philip Johnson
New Canaan, Connecticut 1949
[Modern, free plan house, glass and steel]

In The Glass House by Philip Johnson, the domestic application of the free plan is taken to its extreme. A large single room, divided only by one "core" containing the bathroom, allows for a continuity and consistency to the space. Functional subdivisions are made through furniture position, orientation relative to the site and the core, and the physical activity of social rituals. The transparent glass perimeter dissolves the boundary between inside and out and extends, orients, orchestrates, and composes the interior zones. The material allows for a merger between house and site, interlacing their dependence and experience.

| 188 | TYPOLOGY / PRINCIPLE | FUNCTION / ORGANIZATION | HOUSE / TYPE | PLAN ELEVATION / READING | ARCHITECTURE / SCALE |

 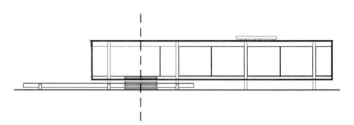

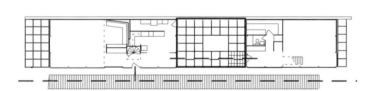 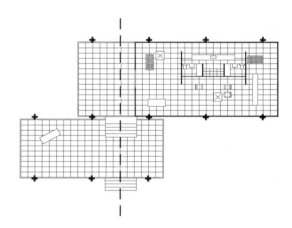

Eames House, Ray and Charles Eames
Pacific Palisades, California 1949
[Modern, semi-prefabricated open plan house, steel]

The Eames House, also known as Case Study House 8, was designed as their personal residence, containing living spaces and a home studio. It is a quintessential representative of the Case Study program, a series founded in rethinking the modern household, designing with new materials and construction processes, and implementing the project through an actualized built home. The Eames House developed out of the post-war production processes and the standardization of light gauge steel. Dissolving the solidity of wall in favor of the skeletal frame, the transparency of glass and the variable local composition within the standardized system provided the foundation of the composition. A simple bar in plan is broken to bracket a courtyard. The house engages the section of the site and the existing row of trees. The design in principle and execution embraces the new technologies of mass production and iterates a humanist yet machined form representative of modernity.

Farnsworth House, Ludwig Mies van der Rohe
Plano, Illinois 1951
[Modern, free plan house, steel and glass]

The "house" is engaged by Mies van der Rohe with an exquisitely reductive power in the Farnsworth House. Similar in premise to Philip Johnson's Glass House, though more refined in its detachment from the ground plane, the house dissolves the boundaries and presence of any conventional compartmentalization of function. Its transparent perimeter integrates the landscape fully as a compositional and ornamental foil. The primal nature of the form, rooted firmly in the rigor of its geometry, engages every aspect of living. The house represents a pinnacle of modernist reductivism.

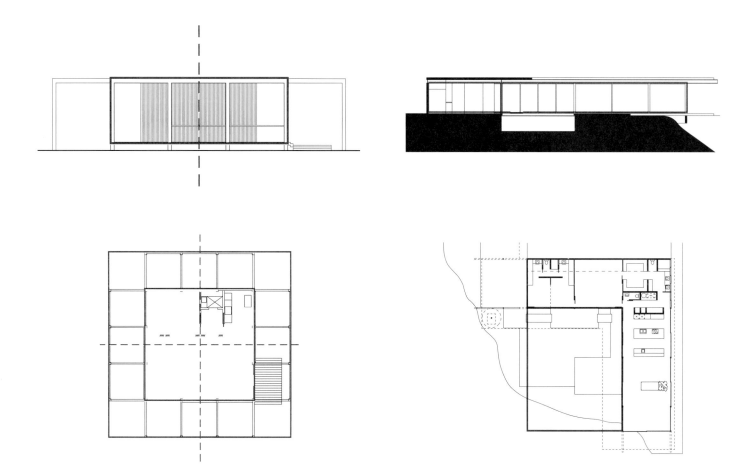

Walker Guest House, Paul Rudolph
Sanibel, Florida 1952
[Modern, free plan house, wood]

Built early in Rudolph's career, the Walker Guest House is a simple twenty-four-foot square. The interior organization is subdivided into a four square of bedroom, dining room, living room, and service (kitchen/bath). This four-part geometry plays off of the exterior three-bay geometry of the extended frame. Wooden panels with seventy-seven-pound steel counterweights shutter the entire building when closed. When open, the walls pivot up to reveal the transparency and operability of the infill panels beneath. Built lightly and detailed simply, the ingenuity comes out of the use of standard off-the-shelf pieces and the response of modernism to the climatic conditions of the site.

Case Study House No. 22, Pierre Koenig
Los Angeles, California 1960
[Modern, free plan house, steel]

In Case Study House Number 22, otherwise known as the Stahl House, Pierre Koenig uses powerful minimalist forms erected in steel construction to delicately define the house and seemingly float it through a cantilever on its dramatic site in the hills of Los Angeles. Weightless in appearance, the connection of interior and exterior spaces through the transparency and operability of the perimeter is further accentuated by the strong horizontality of the house through the extended eaves, connected ground planes, floating furnishings and expansive views. The house uses a simple "L-shaped" plan to bracket a swimming pool. The private bedrooms of the house are located to the back, with the public living spaces cantilevering over the edge of the ridge.

| 190 | TYPOLOGY / PRINCIPLE | FUNCTION / ORGANIZATION | HOUSE / TYPE | PLAN ELEVATION / READING | ARCHITECTURE / SCALE |

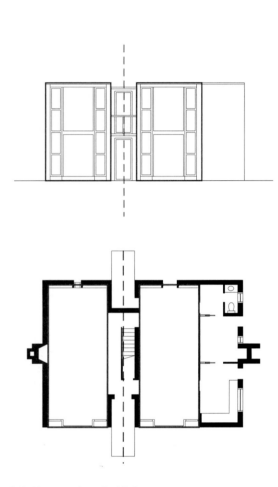

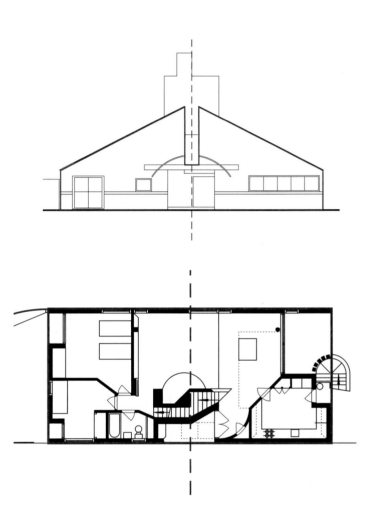

Esherick House, Louis Kahn
Chestnut Hill, Pennsylvania 1961
[Late Modern, planar open plan house, masonry]

Vanna Venturi House, Robert Venturi
Chestnut Hill, Pennsylvania 1964
[Postmodern, mannerist plan house, wood]

In the Esherick House, Kahn derives the forms from simple and monumental gestures of parallel, repetitive, rectangular figures in plan, and then reinforces this gesture with an intensely considered material execution. Using the weight and mass of masonry, Kahn thickens the walls to allow for the layered banding of programmatic spaces across the house. Two primary private spaces on the upper level are separated by a zone of vertical circulation and anchored on the end with the mass and density of the stacked service spaces. The entry occurs in the center space (a volume that is stacked with dining and entry on the lower level and bedrooms on the upper level). To the left, the stacked service spaces increase the density of the space. To the right, the living room is a dramatic double height that allows the space to expand. The clarity of plan and the sectional and material zoning of the house make it a sophisticated example of Kahn's late modern work.

As Robert Venturi's first important built house, the house for his mother was an expression of the sensibilities of postmodernism as laid out in his text Complexity and Contradiction *(1966). Deploying highly referential historical forms, but playing with their reference, context, and perception, the house uses a collagist compositional method of localized moves and collective interrelationships. Based on symbolic not spatial representation, the house as a whole references gable, pediment, and arch among its many architectural components. Deriving from Palladio's Nymphaeum at Villa Barbaro (back façade) and Michelangelo's Porta Pia in Rome (street façade), the house attains its legitimacy through its historical reference. Deploying a strict symmetry, the house achieves balance not through mirroring the composition, but by reiterating the modular components with hyper-equality.*

TYPOLOGY	
PRINCIPLE	191

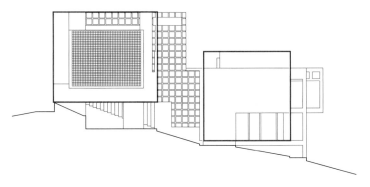
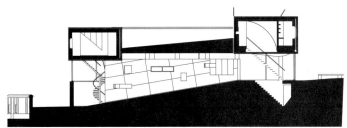
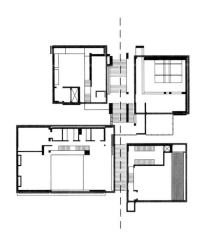
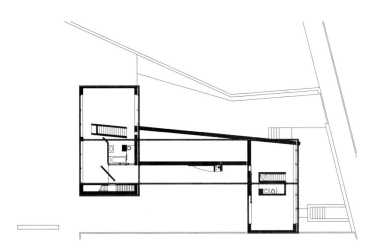

House X, Peter Eisenman
unbuilt 1982
[Deconstructivist, free plan house]

House X represents the pinnacle of postmodernism and marks a clear transition to post-structuralism and ultimately deconstuctivism. The unbuilt project is a three-dimensional composition that derives its form through a hybridization of geometric systems and philosophical intentions. Extending the principles of the earlier House I through House IV, House X is in many ways a more complicated, but also more complete, resolution of the thinking. Here, the principles of deconstructivism integrate typography, representing the architectural planes through type elements that create volumes and spaces made legible through the actual walls, windows, and voids of the structure. The highly conceptual and compositional premise of the unbuilt project fully extended the postmodernist agenda and commenced the deconstructivist movement that would emerge in the coming decade.

Villa Dall'Ava, Rem Koolhaas
Paris, France 1991
[Deconstructivist, free plan house, steel]

Villa Dall'Ava is a full extension of the compositional and material methods of postmodernism. Defined by the Corbusien five points (in reference to the many Le Corbusier projects in the surrounding neighborhood), the ribbon window, free-formed façade, pilotis, roof garden, and free plan are coyly played with. The façade is fractured and shifted in depth to bridge the site. The pilotis are elongated and wobbly to make a forest of columns that create a varied reference. The ground level is encircled in operable glass to enclose the space with transparency. The roof garden is fully extended into a roof-top swimming pool. The cleverness of the formal moves in conjunction with the collagist use of material make the house a highly postmodern expression of contemporary living.

Museum

The museum as a building type developed out of private collections. As a result, the first museums tended to be palaces or villas that remained active dwellings. With the advent of public galleries, many of these buildings were converted into museums proper. The Louvre in Paris and the Hermitage in Saint Petersburg are excellent examples of this evolution. The nineteenth century witnessed the emergence of the modern museum as an architectural project from the ground up. The museum became defined as a series of rooms situated in an enfilade organization, and typically lit from above. The museum as a building type has been challenged and altered more than any other typology. Its primal connection with the architectural promenade, a light- and space-based experience, and an innately intellectual user group allows for more architectural opportunity in the production of a cultural landmark.

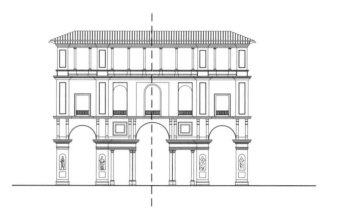

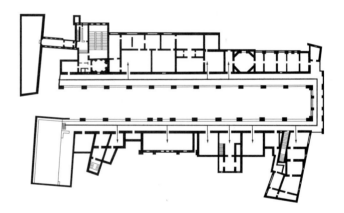

The Uffizi, Georgio Vasari
Florence, Italy 1560
[Renaissance, urban wrapper defining linear courtyard, masonry]

The Uffizi is a quintessential gallery that defines an urban edge. The single-loaded corridor establishes a lining perimeter to an urban room and connects the inner ring of discrete chambers housing the collection. Originally designed as offices for the Florentine magistrates, the building was converted to contain the paintings and sculpture collected by the Medici Family to become one of the first museums. The linear arrangement of the building allows for the galleries to be chronologically organized, synthesizing the curatorial organization with the architectural form.

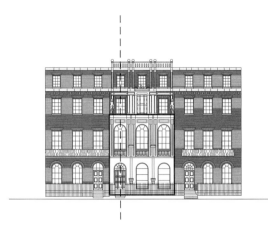

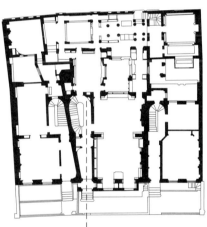

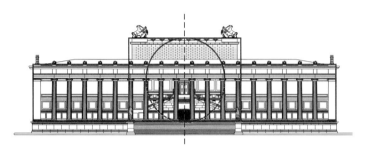

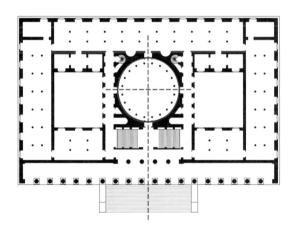

John Soane House, John Soane
London, England 1824
[Neoclassical, house as museum, masonry]

John Soane was appointed Professor of Architecture at the Royal Academy in 1806 and soon after began to arrange his extensive collection of books, casts, and models in his house in order to provide access to students before and after his lectures. The personification of his private collection transformed his house into a museum. The organization subscribes to the principles of the "wunderbox," which displays artifacts densely in a tightly packed space. Soane's constant tailoring and expansion of the space over the years deployed various methods to optimize the storage, display, and perception of the objects. Operable walls, interlocking spaces, and intricate installations make the experience of the architecture of the space equal to the experience of the architecture of the collection.

Altes Museum, Karl Friedrich Schinkel
Berlin, Germany 1830
[Neoclassical, central rotunda museum, masonry]

The central rotunda of the Altes Museum hearkens back to the form, figure, and effect of the Pantheon. Employing the rotunda in a classical manner, the museum establishes it as a primary organizational chamber for the museum as a whole. The room draws the viewer through the portico, past the stairwell, and into the majesty of its scale and experience. The axis continues to redistribute the visitor out the cross-axial circulation pathways into the subsequent hallways and galleries of the museum. The cellular, sequential nature of the architectural organization implies the curatorial method of discrete rooms with bounded themes and content.

| 194 | TYPOLOGY
PRINCIPLE | FUNCTION
ORGANIZATION | MUSEUM
TYPE | PLAN / SECTION ELEVATION
READING | ARCHITECTURE
SCALE |

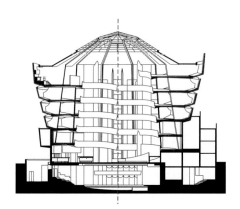

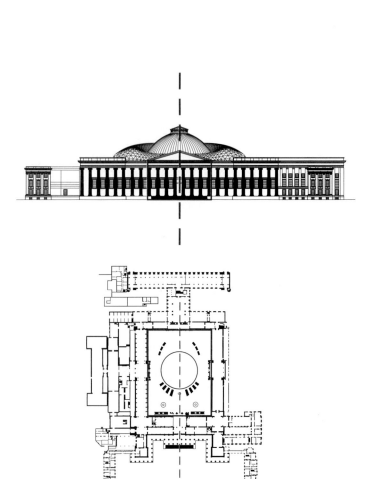

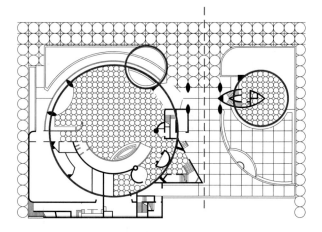

British Museum, Sir Robert Smirke
London, England 1850
[Neoclassical, central courtyard museum, masonry]

The Guggenheim Museum, Frank Lloyd Wright
New York City, New York 1959
[Modern, ramping spiral gallery around atrium, concrete]

The British Museum is considered one of the world's quintessential museums. Its monumental character and neoclassical language establish it as one of the major buildings in London. It houses objects from every corner of the globe. Architecturally it borrows from the concept of the palace, with a series of large rooms situated enfilade around a large courtyard. There are two main axes that cross in the center of the courtyard. Centrally placed on each side of the court are special rooms, or temples that re-emphasize the axes. On the main façade, two wings penetrate out, forming a plaza in front of the museum. The massive colonnade that dominates this elevation has set the visual typology for the museum ever since.

As Wright's last major work, the uniqueness of the formal composition is founded in the continuity of the promenade through the museum and its collection. The visitor is primed and intrigued by the form of the building and its juxtaposition with the rectilinear urban fabric of New York City and the anonymity of the seamless, white materiality. Upon entry, the visitor is compressed through the low ceiling of the lobby space and then released into a massive spatial explosion of the central atrium capped with a glazed roof. The visitor then takes the elevator to the top of the building and perambulates downward along the helix ramp. Artwork is displayed upon the continuous peripheral wall. Functionality somewhat compromised by the arcing wall surface, the sacrifice of function for the purity of the architectural concept established the precedent of the building being dominant to the collection and serving as an object in its own right.

	TYPOLOGY	
	PRINCIPLE	195

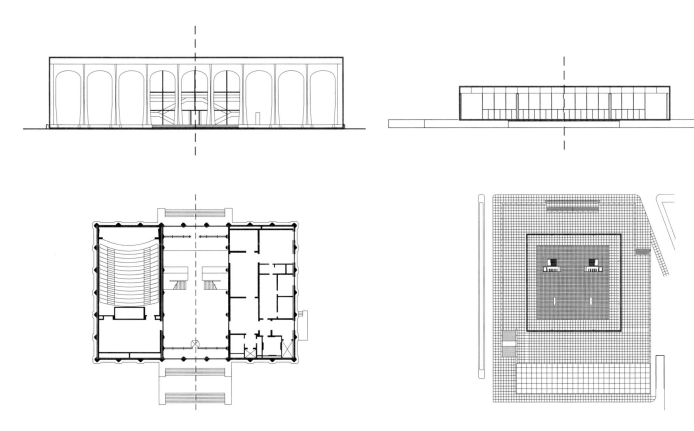

Sheldon Museum, Philip Johnson
Lincoln, Nebraska 1964
[Modern, central hall museum, concrete and stone]

The Sheldon Museum, located on the University of Nebraska campus, is a classically organized building. It is a symmetrical tripartite building, with the galleries situated on either side of a large entry hall. The exterior cladding and flooring materials are travertine marble, whereas most of the detailing is done in bronze. It stands as a building that synthesizes the power of classical architectural devices, such as symmetry, proportion, and materials, while deploying a language in the building that remains thoroughly modern in essence.

Neue Nationalgalerie Museum, Mies van der Rohe
Berlin, Germany 1968
[Modern, free plan - universal space, steel and glass]

As a definitively Miesian composition, the Neue Nationalgalerie deploys the glass and steel palette in conjunction with Mies's interest in universal space. Designed as a concrete plinth (which houses the functional requirements of the building and many of the galleries) with a glass gallery set atop, the building is defined by the massive steel roof supported on eight exterior perimeter columns. The perimeter is minimized and dematerialized through the large glass panels set back from the edge of the roof. Highly polished floors reflect the light and extend the spatial reading of the gallery into the surrounding context. The continuity of the space is amplified through the removal of interior partitions. The rigor and simplicity of the geometry, in conjunction with the scale and power of the material, make the gallery a tranquil, quiet, and timeless housing for the collection.

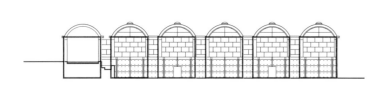
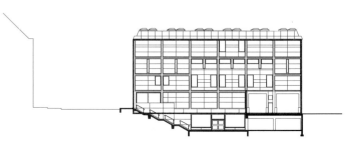
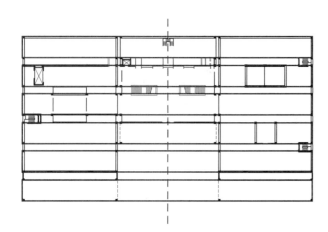
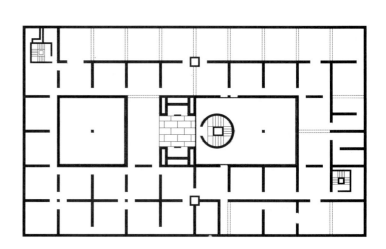

The Kimbell Art Museum, Louis Kahn
Fort Worth, Texas 1972
[Modern, parallel galleries, concrete]

Defined by the formal repetition of a series of parallel barrel vaults, the museum locally resolves the functional and effectual qualities through the individual bay and then collectively aggregates the bays to define the mat building. Opened at the top to allow light to bounce along the surface of the barrel-vaulted ceiling, the roof defines the galleries. Supported by concrete frames that establish three one-hundred-foot bays along the west face, the vaults aggregate into a mat field of layered repetition. Within this field a series of three internal courtyards are subtracted. Using simple referential Roman forms and a rich material palette of concrete, travertine, and white oak, the museum is the first to identify light as the primary moderator in a museum's performance. Developed as a ceiling, the museum becomes an exquisite facilitator for viewing the collection.

Yale Center for British Art, Louis Kahn
New Haven, Connecticut 1974
[Modern, atrium museum, concrete]

As Kahn's final project, the Yale Center for British Art is a largely internalized project. As an orthogonal box defined by the urban constraints of the site, the museum is spread across multiple levels. A three-dimensional structural concrete frame is visible on both the interior and exterior. Slate cladding panels are used as exterior infill, whereas the interior uses white-oak paneling. Intimate galleries washed with a diffused light create an interior facilitated by the cleverness of the sectional configuration and the roof articulation. The monumentality of the ordering system and skillful material usage create solemn and timeless spaces for viewing the collection.

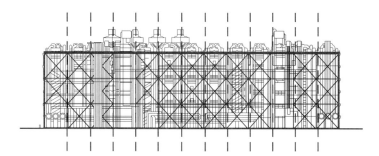
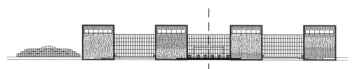
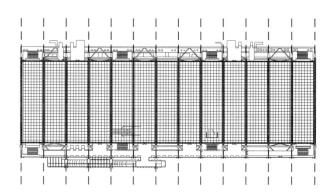

Centre Pompidou, Renzo Piano and Richard Rogers
Paris, France 1974
[Hi-Tech Modern, free plan museum, steel, glass, concrete]

Envisioned as a museum for the people, its design uniquely and distinctly left half the site open for a public plaza for congregation. To accommodate the square footage, the museum's height was doubled, allowing the height of the building to break the highly regimented datum of the Paris skyline and announce the presence of this new cultural institution. The building itself is turned inside out, with each system carefully color-coded to explicate the building and present the museum as an industrially performing object. Using blue for air, green for water, grey for secondary structure, and red for circulation, the building is a legible sign system. The externalization of the services allows for a fully free plan interior that provides infinite flexibility to the layout of the galleries. The vertical circulation is dramatically suspended off the surface of the building in a glass tube that terraces up the plaza elevation. This provides dramatic views of the city beyond while animating the elevation of the building with the activity and occupation of the museum itself.

National Air and Space Museum, HOK
Washington, D.C. 1976
[Modern, open atrium layered galleries, concrete and steel]

Housing the largest collection of historic aircraft and spacecraft in the world, the National Air and Space Museum is located on the National Mall in Washington D.C. alongside the other branches of the Smithsonian Institution. Designed as four marble-clad cubes (Tennessee pink marble to match the National Gallery) evenly spaced with tinted glass infill between them, the opaque volumes house the services, theaters, and inward-viewing galleries. The interstitial zones are massive chambers that provide the fly space for suspended artifacts. The west glass wall of the building acts as a massive door, allowing for the movement and installation of the giant artifacts. The overall dimension of the building matches the length of the National Gallery of Art located directly across the National Mall. Bridging the scale and monumentality of its context with the massive size of the collection itself, the National Air and Space Museum serves as an enormous, volumetric warehouse box, successfully defaulting to its collection and context for identity.

| 198 | TYPOLOGY
PRINCIPLE | **FUNCTION**
ORGANIZATION | MUSEUM
TYPE | **PLAN
ELEVATION**
READING | ARCHITECTURE
SCALE |

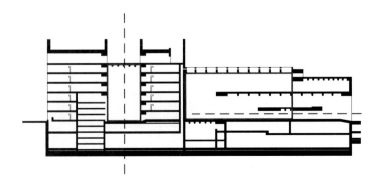

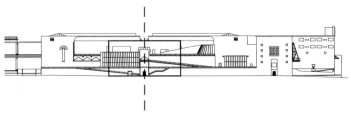

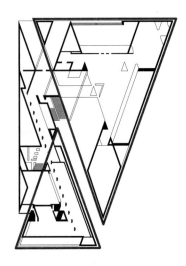

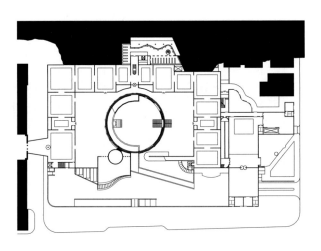

East Wing of the National Gallery, I.M. Pei
Washington, D.C. 1978
[Late Modern, central triangular atrium, concrete and stone]

A massive, multi-story central atrium establishes the heart of the museum. Peripheral enclosed galleries allow for more intimately scaled and traditionally partitioned gallery spaces. The scale and vocabulary of the architecture creates a highly performative abstract space for viewing the modern collection. The central multilevel open gathering and mixing space becomes a fluid atrium, connecting the artwork that is on the floor and also suspended from the ceiling. The multileveled and multi-positional circulation pathways allow for a complex and highly animated experience as one ebbs and flows in and out of the peripherally located galleries.

Neue Staatsgalerie, James Stirling
Stuttgart, Germany 1983
[Postmodern, linear enfilade, stone cladding]

In the Neue Staatsgalerie, Stirling commingles classical forms with a modernist free plan. The result is a spatial experience through discrete forms (defined by the postmodern methodology of historical collage) within a broader field of galleries and exhibition spaces. The mastery of the composition occurs through the careful development of the circulation. Producing an architectural promenade, the sequence through the building synthetically stitches it together, moving vertically through the site and curatorily through the exhibition.

TYPOLOGY PRINCIPLE | 199

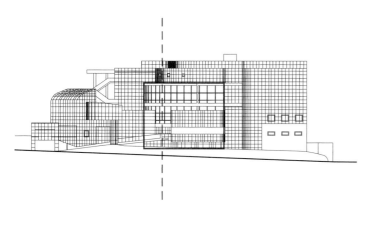

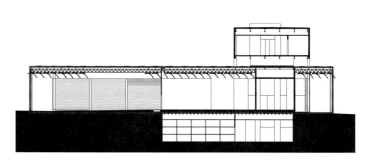

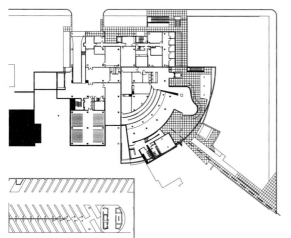

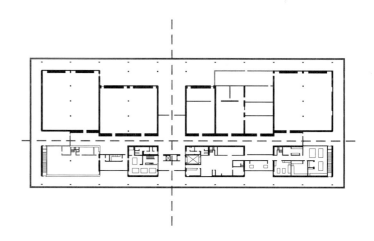

High Museum of Art, Richard Meier
Atlanta, Georgia 1983
[Late Modern, L-shaped galleries with circular atrium, metal]

The High Museum of Art uses a central atrium bracketed by an "L" shape of galleries. The circulation, much like Frank Lloyd Wright's Guggenheim Museum in New York City, develops a promenade that brings the visitor through the atrium, ascends to the top of the building through the elevator, and then slowly descends via a ramp that connects the various levels and galleries. The descending circulation is a switchback ramp sandwiched between the atrium and the front façade, allowing views both in and out of the building and allowing the visitor to animate the space. The landings for the ramp are the gallery floors. The distinctly composed relationship of wall and structure allows for the continuity of space while subdividing the collection into smaller gallery-sized spaces. The museum is one of Meier's most powerful works, operating as an urban icon as well as a very successful museum space.

The de Menil Collection, Renzo Piano
Houston, Texas 1986
[Hi-Tech Modern, double-loaded corridor, diverse]

The de Menil Collection is the first museum by Renzo Piano, establishing a model for which he has become synonymous. Like Louis Kahn, Piano dedicated the museum to the thoughtful production of naturally lit spaces. Designed with a highly articulated roof that blankets the largely one-story building, the galleries form as discrete volumes huddled below (connected with a central double-loaded corridor entered off a cross-axial lobby), yet always held within the boundary of the super-structural roof. The precast concrete light fins of the roof are designed to bounce the light, providing an even distribution of illumination while preventing any direct and damaging ultraviolet light from reaching the artwork. The delicate scale of the building, in the context of the broader urban plan that engages and homogenizes the surrounding residentially scaled buildings through unifying color, allows the de Menil Collection to sit elegantly as an institution within its hybridized Houston site.

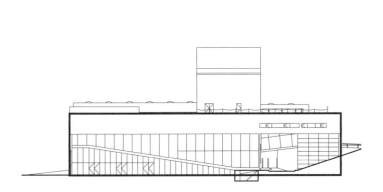
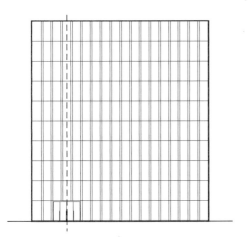
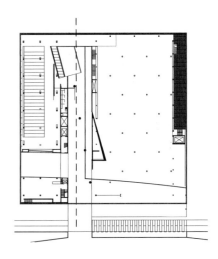
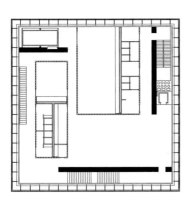

Kunsthal, Rem Koolhaas
Rotterdam, Netherlands 1992
[Postmodern, double-loaded sectional corridor, diverse]

The Kunsthal by Rem Koolhaas uses a ramp to connect the various galleries. But unlike the Guggenheim Museum in New York by Frank Lloyd Wright or the High Museum in Atlanta by Richard Meier, the Kunsthal centralizes the ramp, using it as the primary organizational tool with the galleries branching off of it. Designed with a split-level sectional relationship, the two sides of the ramp access galleries at each landing, allowing for a slow staggering of sectional ascension and preventing the length of the ramp from becoming insurmountable. The collective integration of sequence with the flexibility of the conventional gallery makes the museum a highly choreographed and integrated experience that maintains the potency of the parts yet intertwines the efficiency of the gallery and promenade alike.

Kunsthaus Bregenz, Peter Zumthor
Bregenz, Vorarlberg, Austria 1997
[Modern, multi-story free plan museum, glass and concrete]

The Kunsthaus Bregenz deploys a subtlety of spatial and material intention. Cubic in form and largely anonymous in function, the material cladding in a shingled, uniformly bayed white glass produces the enigmatic and solemn quietness. The inner plan, subdivided only by three carefully positioned concrete walls, maintains the continuity of the square figure of the perimeter. The practical function occurs outside these wall, whereas the exhibition occurs within them. As a museum, the space becomes a quiet and anonymous box for housing, displaying, and privileging the contents. The architecture defines itself through the subtleties of the tectonic resolution and the solemn continuity of the whole.

	TYPOLOGY
	PRINCIPLE **201**

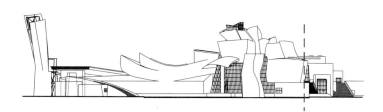
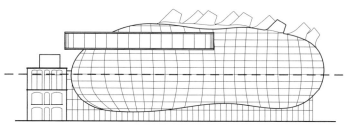

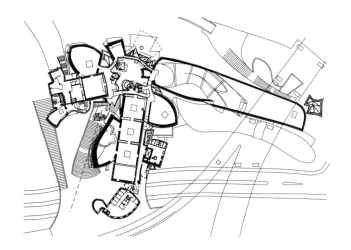
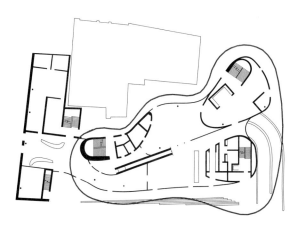

Guggenheim Bilbao, Frank Gehry
Bilbao, Spain 1997
[Expressionist Modern, free plan museum, metal and glass]

The Guggenheim Bilbao is the most overt example of the museum itself being deployed as a work of art. Visited as much for the container of the artwork as the collection itself, the Guggenheim Bilbao fully embraces the formal power of architecture to create a globally recognizable, architectural landmark and icon. Accelerating the formal and material expressionism of Frank Gehry's work to its zenith, the project is a destination building where the form of the museum itself is fully articulated as a visual object in its own right. Like many of Gehry's schemes, the program is subdivided into discrete elements based on function and gallery type. Each is rotated and relatively positioned then draped with interstitial cladding to create an even further divergent formal expression. From the orthogonal to the highly curvilinear, the forms and functions deploy varied organizational and hierarchical tactics, resulting in collagist forms and experiences governed by visually expressionist methods.

Kunsthaus Graz, Spacelab - Cook and Fournier
Graz, Austria 2003
[Expressionist Modern, free plan museum, plastic]

At the Kunsthaus Graz, the iconic quality of the geometry is deployed to produce an identifying and differentiating form. This established variation within the urban fabric produces a figure that demands attention through heightened curiosity. Subscribing to an organic and curvilinear formalism, the museum is clad in a dramatic, reflective blue plastic skin, which presents a stark contrast to the traditional form, height, and masonry material of the surrounding context. The museum extends the lineage of the museum typology through the employment of an iconic exterior form and furthers the agenda by developing the surface as a variably programmable and internally illuminated surface. Presenting the skin of the building as a composable experience, the impact and curation of the museum extends literally to the exterior of the building.

Library

The library, like the museum, developed primarily as a private space nested within a residence, church, or temple. The first libraries can be traced back to the Egyptians and were also common in Asia, Islam, Greece, and Rome. They were originally archives that held records for the city or state. During the Middle Ages libraries were typically part of and controlled by the church. Libraries like the Laurentian Library in Florence were some of the first libraries open to the public. The program most associated with the modern library emerged in the fifteenth century corresponding with the invention of the printing press and the ensuing mass production of texts. Libraries later became essential parts of universities and government complexes as collective resources of accumulated knowledge. The nineteenth century distinctively marked the beginning of the library as a singular building type dedicated to public use. This new type generated a series of functional and organizational components including stacks, reading rooms, and various administrative and support functions. Architects throughout the ages have enjoyed a tremendous amount of freedom as to how these particular functions and spaces were to be distributed, resulting in a number of different and powerful configurations within the typology.

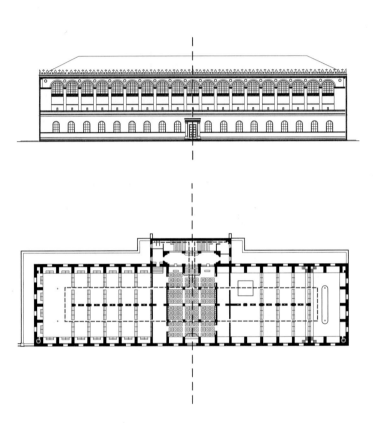

Bibliotheque St. Genvieve, Henri Labrouste
Paris, France 1851
[Neoclassical, vaulted reading room, masonry and iron]

The Bibliotheque St. Genvieve is dominated by its large reading room, which is based on the plan of the Basilica at Paestum. Both buildings have a series of columns that occupy a center axis, simultaneously expressing structure and a lack of an occupiable axis. Labrouste used this feature as a way of aligning the library, which he believed should be for the people, with the Basilica, which was also a building type intended for the citizens and not for the gods. The large reading room, which occupies the entire plan of the building, sits on the piano nobile and is surrounded by a double-walled system that contains all the book stacks. Labrouste executed this configuration as way of creating open access to the books while expressing them as the objects that literally support the walls of the library.

TYPOLOGY	
PRINCIPLE	**203**

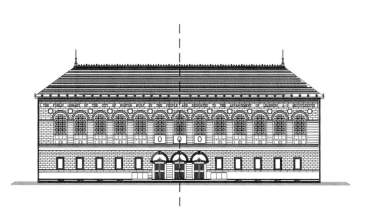

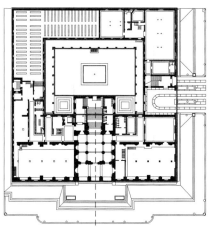

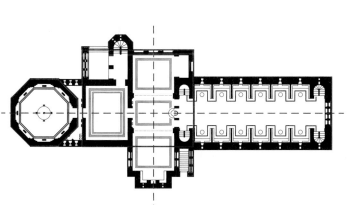

Winn Memorial Library, H.H. Richardson
Woburn, Massachusetts 1879
[Romanesque Revival, Latin Cross library, masonry]

The Winn Memorial Library is one of a series designed by H.H. Richardson and is representative of the broader collection. The plan in overall shape resembles a Latin Cross church. The stacks take the form of a long linear room (nave), and the reading room is located on an axis that is rotated at 90 degrees (transept). Beyond is located the picture gallery which also is the entry hall (altar) and finally the museum, a circular room adjacent to the picture gallery (choir). Unlike a church, these different rooms remain discrete and separate from each other. It is a representative study of how a typology of a different functional classification can be altered to solve the spatial issues of another functional type.

Boston Public Library, McKim, Mead and White
Boston, Massachusetts 1895
[Neoclassical, central courtyard library, steel, masonry]

Organizationally, the Boston Public Library is a palace with a second hierarchical piece added along its entrance elevation, projecting into the central courtyard. This central building houses the entry vestibule and the grand staircase. The structure of the entry façade, which faces Copley Plaza, is reminiscent of the elevation of the Bibliotheque St. Genvieve in Paris, whereas the actual arches owe a clear debt to Alberti's Tempio Malatestiano. The language of the courtyard is taken from Bramante's Cancelleria in Rome. McKim, Mead and White also used new forms of vaulting as a way of spanning the great distances within the building.

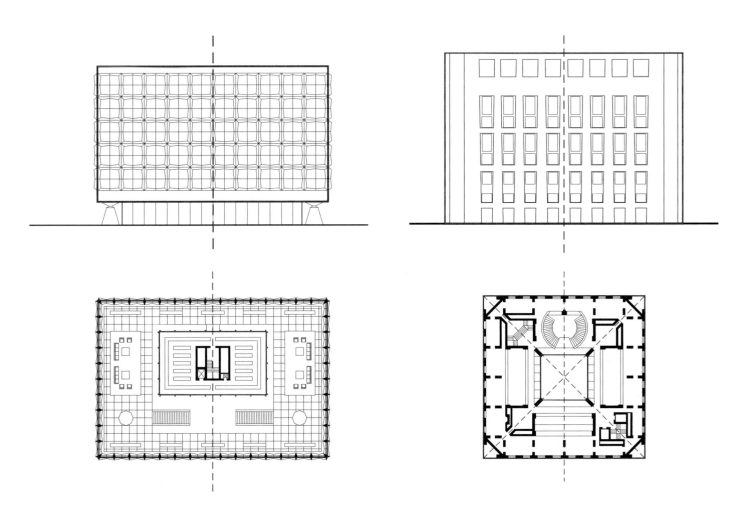

Beinecke Rare Book Library, Gordon Bunshaft, SOM
New Haven, Connecticut, 1963
[Modern, centralized glass stack, concrete and masonry]

The Beinecke Rare Book Library places a glass building within a translucent marble building. The six-story glass tower in the center of the library is devoted to stacks containing the rare book collection. It acts like a large display case. Surrounding this tower is the hovering translucent marble building, which acts as an enigmatic protective container. The space between the two contains the lobby and reading areas. The gap between the glass tower and the floor of the lobby emphasizes the separation between the two. Below the main floor there is a series of offices that enjoy a sunken open-air courtyard in the traditional arrangement of a cloister scriptorium.

Exeter Library, Louis Kahn
Exeter, New Hampshire 1972
[Late Modern, centralized square atrium, masonry and concrete]

In Exeter Library, two nested cubes set one inside the other, defining the functional organization. The perimeter masonry wall of repetitive windows establishes a ring of individual reading carols. Between this outer edge and the inner concrete square figure that defines the central atrium lie the stacks. The clarity of the diagrammatic organization and its relationship to light and material establish a simplicity of form with a complexity and power of detail and articulation.

TYPOLOGY	
PRINCIPLE	**205**

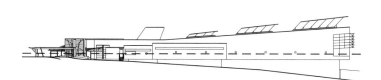

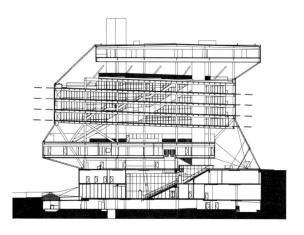

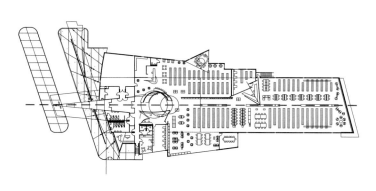

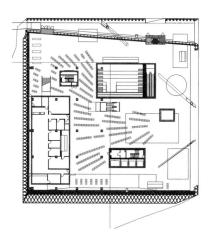

Buckhead Library, Scogin Elam and Bray
Atlanta, Georgia 1989
[Deconstructivist, free plan with central atrium, diverse]

As one of a series of public libraries designed across Atlanta by Scogin, Elam and Bray, the Buckhead library is a dynamic and uniquely formed building constructed on a limited budget. Firmly rooted in deconstructivist sensibilities of form and spatial sequencing, the building begins as an evolution of spaces and forms working on divergent scales and with varied agendas. Iterating from the exoskeletal front **porte-cochere**, through the lobby, crescendoing over the main desk with a dynamically decomposed shaft through the roof, it culminates in the cantilevered reading room looking over the skyline of downtown Atlanta. The forms reflect their functions but gain their specifics from the materials they are made of and the abstract composition of the formal effect. The rituals of the library are independently articulated as compositional disruptions expressed through dynamic highlighting forms.

Seattle Public Library, Rem Koolhaas OMA
Seattle, Washington 2004
[Postmodern, continuous free plan, concrete and glass]

The Seattle Public Library is unique in its typological innovations in both form and program as well as their interrelationship. Developing a series of reframes or entirely unique programs to the library, such as the transformation of periodicals into a "living room", these innovative functional ideas manifest in unique environments that collect in a functionally driven aggregated form. The sectional arrangement of the varied functional components is defined by the linear arrangement along a distinct experiential promenade. Layering the sequence to ease and orchestrate an interpenetration of spaces, the innovation of the whole extends into the subtlety of each piece. Exemplified by the reconsideration of the book stacks that slowly ramp and fold upward along a continuous floor plate, this arrangement allows for a singular reading of the space despite its sectional layering. The collective and individual organizational methodologies collaborate in their diversity to produce a form that is driven by the diagram of the program.

School

This building type has existed in one form or another since the dawn of civilization. One of the first schools was the Athenaeum building in ancient Greece where scholars, philosophers, and poets gathered to discuss their work. The Romans continued this tradition and expanded it to include schools that dealt with additional issues such as gladiator training. Like libraries, most schools in the Middle Ages were under the auspices of the Church. Later, universities began and the teachings of various subjects other than religion were instituted. The term "school" however has come to mean simply a building where people are taught. This can be and often was a single room, but can also be a large complex including a series of buildings that are joined together.

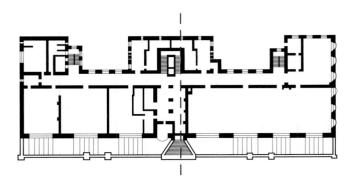

Glasgow School of Art, Charles Rennie Mackintosh
Glasgow, Scotland 1909
[Art Nouveau, single-loaded corridor, masonry]

The Glasgow School of Art ushered in the concept of the secular institution of learning. The large stone building uses local symmetries and the new technology of large glass surfaces to develop a new type. The organization is a single-loaded corridor with all the studios facing north to allow for each studio to be flooded with natural light from large windows. The important rooms such as the library and lecture hall acts as anchors at the end of the long hallway. This building is unique in that while it followed a simple organizational typology, it did not mimic the language of the palace or church as so many buildings had.

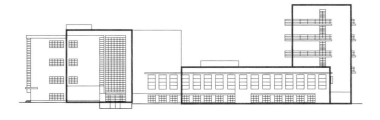
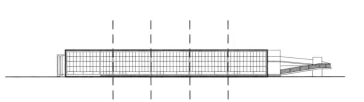
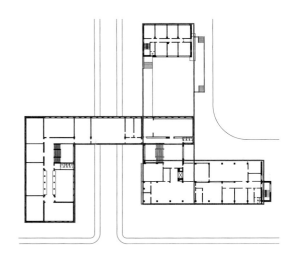
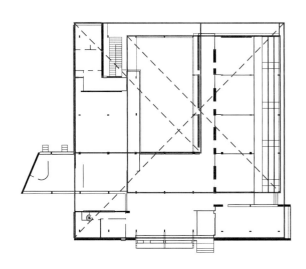

Bauhaus, Walter Gropius
Dessau, Germany 1932
[Modern, pin-wheel plan, glass, steel and concrete]

The Bauhaus complex is a pin-wheel form with three distinct projections. Each wing housed a different function: student housing, the technical school, and the workshop building. These programs were joined by a series of linear bridge elements that contained the offices and the auditorium. Beyond this organizational innovation, the buildings expressed new ideas in construction and formal language. The complex was seen as an embodiment of the teachings of the art and architecture school. Ideas of composition, transparency, and structure were all critical to the school and hence the building.

Asilo Sant'Elia, Giuseppe Terragni
Como, Italy 1937
[Modern, courtyard, concrete]

The Asilo Sant'Elia represents one of Terragni's most mature works. Based on a courtyard organization, the composition is complicated through the use of shifting rectangles that reoccur throughout much of his work. The plan of the Asilo, with its shifts, projections, and recesses, is reminiscent of the work of Radice the painter, who was a friend and colleague of Terragni. The outline of the building is a square that is then rotated on the site to allow the largest coverage possible. This shift permits the building to stand out from its immediate surrounding context and relate instead to the Bordello tower in the distance. There are four classrooms on the east side that have movable walls that can be adjusted to form one long space. On the opposite side of the courtyard is the refectory, whereas the indoor playroom dominates the front. The north side is open to the exterior garden beyond. It is a building that expresses both Terragni's architectural ideas and the educational pedagogy of the political times.

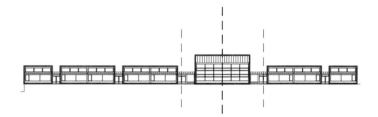
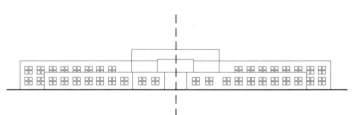

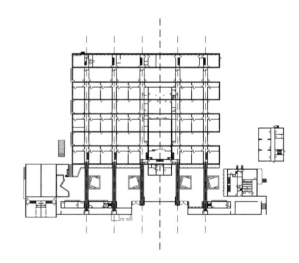
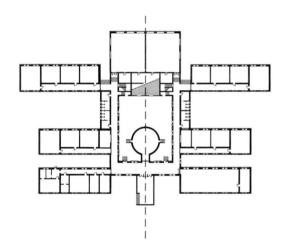

Munkegards School, Arne Jacobsen
Copenhagen, Denmark 1958
[Modern, multiple courtyard mat, masonry]

The Munkegards School utilizes seventeen small courtyards to give the classrooms individual and shared exterior spaces. There are five double-loaded hallways that run the entire depth of the school, linking together the various classrooms. Nestled between these classrooms are the enclosed courtyards. The hallways that penetrate the courtyards are glass, dematerialized to allow for spatial continuity through the building. The relentless connection of interior and exterior spaces dominates the composition.

Elementary School in Fagnano Olona, Aldo Rossi
Fagnano, Olona, Italy 1976
[Neo-Rationalist, symmetrical centralized courtyard, masonry]

This primary school has six wings that project out from a central building. The two front wings house the administration and the canteen; the other four wings contain the classrooms. There is a large gymnasium that completes the courtyard to the rear. Situated in the center of the courtyard is the cylindrical library. A large set of stairs descends into this courtyard and acts as a impromptu theater for the children. This school uses lessons from urbanism and its use of public space as a metaphor. The use of axes and particularly symmetry are felt at all times as one moves through the highly ordered complex.

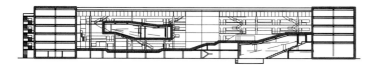
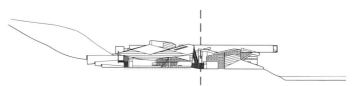
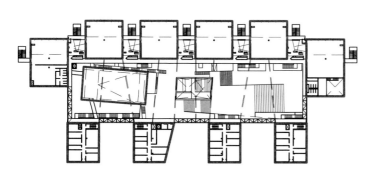
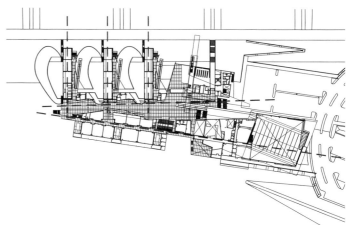

Marne School of Architecture, Bernard Tschumi
Marne les Vallée, France 1999
[Postmodern, dispersed objects within a building, metal]

This architecture school takes the form of a large room within which is a series of smaller dispersed objects. Projecting out of the large volume are the offices and seminar rooms on one side and six large studios on the opposing side. There is a continuous balcony corridor that connects the various pavilions. The floor of this large unprogrammed space is a series of undulating planes of stairs and platforms under which occur another auditorium and various service-oriented programs. The building is imagined as a voyeuristic collage of events that allow for a complete breakdown of traditional values and separations.

Diamond Ranch High School, Morphosis
Diamond Bar, California 2000
[Expressionist Modern, double-loaded exterior corridor, metal]

Diamond Ranch High School does not use a single ordering system but instead relies on expressionism and collage as a means of formalizing its concept. The architects use angles and shapes in an attempt to express the activity and exploration that takes place within the building itself. Despite this, the school operates through a series of linear elements that contain classrooms and a large headpiece that holds the gymnasium and administration. The section of the site allows for a complex embedding of programs in, on, and over the cascading landscape.

| 210 | TYPOLOGY
PRINCIPLE | FUNCTION
ORGANIZATION | PRISON
TYPE | PLAN
ELEVATION
READING | ARCHITECTURE
SCALE |

Prison

Since the beginning of civilization people have seen the need to incarcerate others. Early prisons were typified by the castle dungeon or simply a rudimentary cage. They were often large rooms where the inmates were left to fend for themselves. Often these dungeons were not even separated by age or gender. Not until the nineteenth century did prisons, as we know them today, become commonplace. During the Victorian era, prison reform brought about important changes, such as one man to each cell and countless other human rights reforms. Prisons as a typology are invariably a series of small cells that can be arranged in a number of ways. Other larger spaces, such as eating halls, gymnasiums, hospitals, religious facilities, and schools are now part of this ever-changing typology.

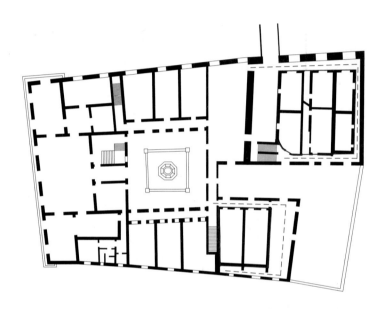

Palazzo delle Prigioni, Antonio Contino
Venice, Italy 1614
[Renaissance, single-loaded perimeter corridor, masonry]

Located directly adjacent to the Doges Palace in Venice, this prison is considered one of the earliest. The basic organization is a series of eighteen vaulted cells that are positioned along a single-loaded corridor. The corridor has windows into a central courtyard. The cells have no windows, but do receive indirect light from the corridor. The cells themselves are constructed of Istrian stone, whereas the walls and floors are covered with thick planks of wood. Each cell was given a special name, such as Goleotta, Vulcana, and Leona.

	TYPOLOGY	
	PRINCIPLE	**211**

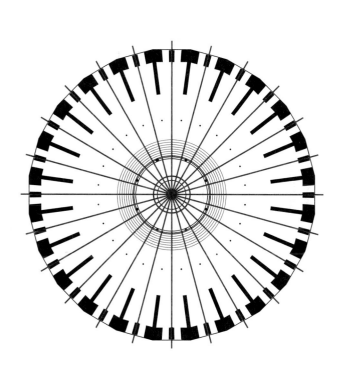

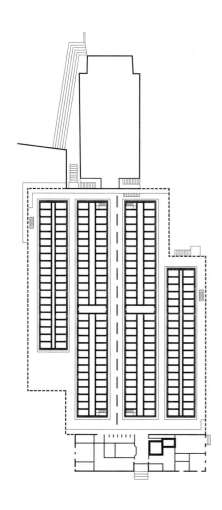

Panopticon, Jeremy Bentham
unbuilt 1785
[Neoclassical, radial plan for visual surveillance, masonry]

This prison type developed out of the desire to establish visual control over the prisoners. Typically there was a tower placed at the center of the complex, within which the guards would be able to visually control the space. The individual prisoners were placed in series of perimeter cells that constituted the outer wall of the complex. This typology allowed a minimum of guards to maintain control of many more prisoners. This hierarchy required constant good behavior of the inmates as one never knew when one was being watched. Its striking circular form also recognizes this prison type.

Alcatraz Prison, Major Reuben Turner
San Francisco, California 1912
[Neoclassical, dispersed objects within a building, concrete]

The main cell house at Alcatraz was essentially a very large room containing a number of other buildings that housed the actual cells. These smaller buildings were designed so that two rows of cells backed up to each other with a party wall serving as a mechanical shaft. The cells either looked across a corridor to another series of cells or they had a view across to a series of windows that looked on to San Francisco Bay.

Theater

Theater began in Egypt around 2500 BCE as a relationship with religion, whereas Western theater originated in Greece around 500 BCE. Architecturally the Greek play was performed in amphitheaters, a typology that continues even today. Formally, the amphitheater was embedded into the topography as a semi-circular group of stone seats rising from a central stage. This radial organization was based on an optimization of sight lines, establishing a performative efficiency that remains pertinent even to contemporary theaters. This aspect of the theater typology's functionality is powerful in its determination of form. The early Greek Theaters continue to resonate in the form and functionality of every movie theater or auditorium used today.

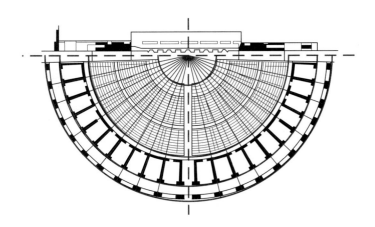

Theater at Ostia Antica
Ostia, Italy 12 BCE
[Classical, semi-circular outdoor theater, masonry]

The Theater at Ostia Antica was a typical Roman theater. It was located near the market square at the culmination of the Decumanus. It was semi-circular in shape allowing for excellent viewing lines and an intimate atmosphere. The circular shape ensured that the patrons were never too far away from the performance, ensuring optimization of sight line and acoustics. It was constructed of stone and could accommodate up to four thousand people. This type of theater was common during the Roman Empire and set the standard for theaters for centuries.

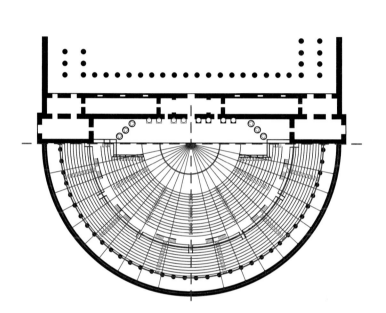
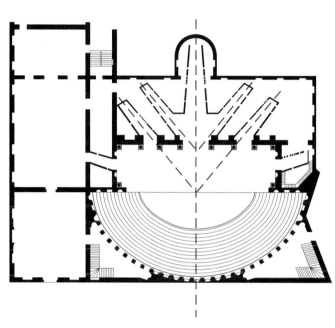

Odeon of Herodes Atticus, Herodes Atticus
Athens, Greece 161
[Classical, semi-circular outdoor theater, masonry]

The Odeon of Herodes Atticus, situated on the slopes of the Acropolis, is similar to other Roman theaters in its shape and function. It was enclosed by a large high wall acting as a backdrop for the musical and theatrical performances. Originally this three-storied wall supported a wooden roof, which sheltered the orchestra or stage.

Teatro Olimpico, Andrea Palladio
Vicenza, Italy 1585
[Renaissance, semi-circular indoor theater, wood and masonry]

The Teatro Olimpico, built for the Academia Olimpica in Vicenza, reconstructs the semi-circular form of the ancient Roman theater. Palladio placed the wooden theater within an existing building, extending the idea of the stage well beyond a backdrop and into a three-dimensional city. The late Renaissance fascination with perspective is articulated through Palladio's orchestration of multiple false perspectives down five different street scenes. The proscenium takes the form of a large, articulated wall that is richly decorated with architectural details and sculptures of the members of the Academy. It is the only surviving Renaissance theater and is often credited with the preservation of classical plays as it remains one of the last sites to appropriately conduct them.

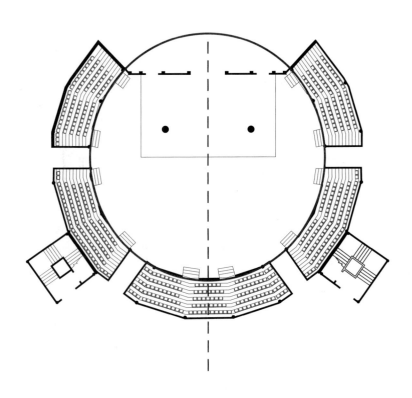

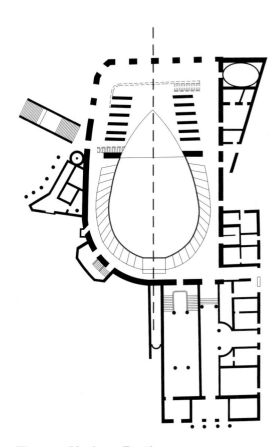

The Globe Theater, James Burbage
London, England 1599
[Tudor, outdoor circular theater in the round, wood]

The Globe Theater was a twenty-sided, open-air, three-storied building that created a theater in the round. There was a large stage that occupied the middle around which the poorer citizens gathered. There were three tiers of stadium seating defining the edge and providing a vantage for wealthier customers. The stage was raised approximately 5 feet to allow the actors to enter from beneath. There was a suspended roof over the stage that similarly allowed actors to descend from above. The theater could house up to three thousand spectators at a single performance. This theater configuration was closely associated with Shakespeare as it was designed and constructed by his playing company "Lord Chamberlain's Men."

Fenice Theater, Meduna Brothers
Venice, Italy 1837
[Neoclassical, horseshoe plan, wood and masonry]

The Fenice Theater in Venice is a quintessential nineteenth century opera house. Using a horseshoe plan, seating was placed on the bottom level with another five levels of box seating surrounding the perimeter. The interior is resplendent with gold leaf decoration and ornament. This fantasy of architectural form and color worked well with the fantasy of theater and became synonymous with opera house interiors throughout the world. The theater was completely destroyed by a fire in 1996 and reconstructed by the postmodern architect Aldo Rossi to the original splendor of the 1837 design. This was a highly controversial action that brought a great deal of attention to the project in terms of ideas of renovation and reconstruction.

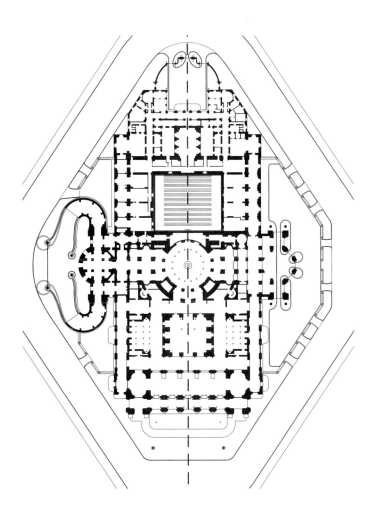

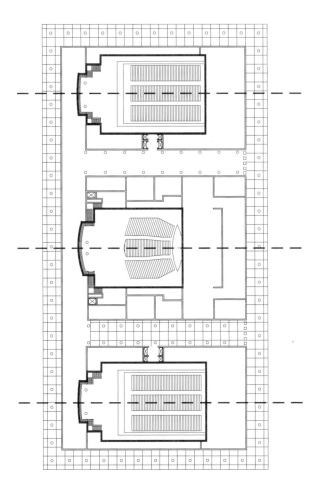

Paris Opera House, Charles Garnier
Paris, France 1874
[Neo-Baroque, horseshoe plan indoor theater, masonry]

The Paris Opera House is considered one of the most recognizable of all theaters. It is the epitome of grandeur, monumentality, and style. The entry sequence set up by Garnier is still one of the most exhilarating architectural moments of Paris. The building entirely occupies a diamond-shaped city block. The entry begins through a colonnade that subsequently leads to a larger entry hall resembling the narthex of a church. The next room, and possibly the highlight of the building, is the stair hall, considered by many to be the zenith of the neo-baroque movement. At the culmination of this sequence is the horseshoe-shaped theater itself. Cutting directly through the building at the proscenium is a series of walls that bisect the entire complex. Due to the complexity, innovation, and flexibility of configuration, the actual stage and its support spaces are larger than the seating area.

The Kennedy Center, Edward Durrell Stone
Washington, D.C. 1971
[Late Modern, multi-theater performance hall, metal and stone]

This complex, which is dramatically sited on the Potomac River, is a series of three main performance spaces and five other minor performance spaces. The building is designed as a large box that contains three smaller boxes nested within. This concept was developed to help mediate the sound created by the airplanes landing at nearby Ronald Reagan Airport. The three main spaces — the Concert Hall, the Opera House, and the Eisenhower Theater — are separated by the Hall of Nations and the Hall of States. The main façade of the Center, with its extensive colonnade, contains the large grand foyer facing the Potomac River. This space, with its monumental proportions, red carpet, and hand-blown chandeliers, creates a ceremony fitting its performances.

216	TYPOLOGY	FUNCTION	THEATER	PLAN	ARCHITECTURE
	PRINCIPLE	ORGANIZATION	TYPE	READING	SCALE

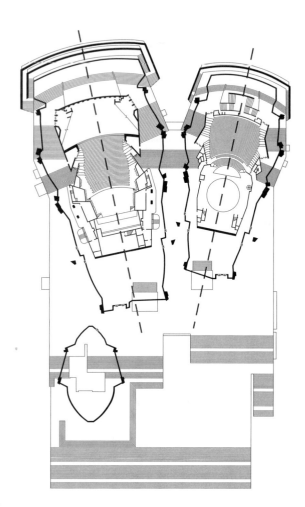

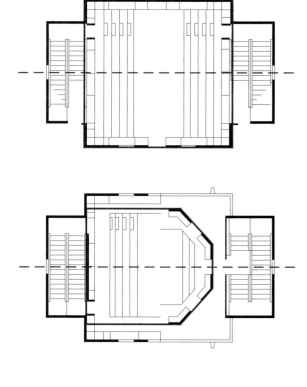

Sidney Opera House, Jorn Utzon
Sidney, Australia 1973
[Structural Modern, multi-shelled theaters, concrete]

The Sidney Opera House is instantly recognizable with its large concrete shells that dominate the exterior form of the building. These shells produce a series of interlocking layers set on a monumental podium. The two figures define the separate performance spaces. The concert hall and the opera theater are located under the western and eastern groups of shells, respectively. The form of the complex was heavily criticized as the shells remained purely formal, having no connection to the acoustics or function of the internal performance spaces. The project was renown for being completed ten years late and costing fourteen times its original budget.

Il Teatro del Mondo, Aldo Rossi
Venice, Italy 1979
[Postmodern, floating theater, steel and wood]

The Teatro del Mondo was a floating theater (set on a barge) designed for the Venice Biennale of 1979. Rossi used a traditional geometric form rendered in bright colors to contrast against the stone and brick buildings of Venice. The square plan had bleacher seating on both sides of a centrally located stage. Above the cubic form, an octagonal balcony looks down on the stage. The interior and exterior of the small building was clad in wood attached to a steel frame. The Teatro del Mondo recalled the early floating theaters of Venice.

	TYPOLOGY	
	PRINCIPLE	**217**

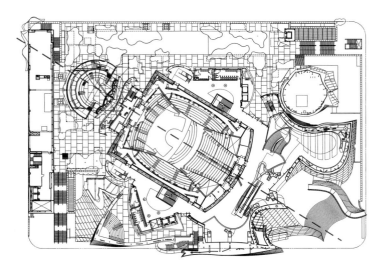

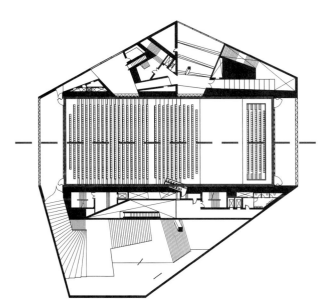

Disney Concert Hall, Frank Gehry
Los Angeles, California 2003
[Postmodern, concert hell, steel]

The Disney Concert Hall is a pinnacle example of formal expressionism. The external form is dynamic and complex, whereas the concert hall itself is quite traditional in shape and form. Gehry understood that the problems of acoustics and reverberation had been considered over the years and that the physics of these issues had been solved. In response, the design is a beautiful wooden box shrouded in a series of expressionistic metal layers. The congregation and circulation spaces happen between these two layers.

Casa da Musica, Rem Koolhaas, OMA
Porto, Portugal 2005
[Postmodern, concert hall, concrete with stone panels]

The Casa da Musica is a multi-faceted box set in a large plaza. The exterior form is not developed from either acoustical or contextual concerns, but rather from a purely formal agenda. The main concert hall is buried within the building and is completely disguised from the exterior. The entire building is structured by two massive parallel walls. The large concert hall spans between these two walls, whereas much of the rest of the program is cantilevered from them. Select windows open into the public lobbies, washing the spaces with light and governing the exterior articulation. A roof top garden, open only to the sky, is carved out of the distorted box. This project is diagrammatically unique with a strictly orthogonal concert hall housed within a sculptural box.

Office / High-Rise / Commercial

The typology of the commercial building began when men began trading some 150,000 years ago. Originally there was no building necessary for this action; instead it occurred wherever it was deemed most appropriate or convenient. One of the first known markets was actually the beach, as vendors from different places would naturally meet there to exchange goods. The center space of any urban area typically took on aspects of a market, resulting in the "market square." From an architectural standpoint there were many types of structures that were used for commerce. These were often as simple as carts or tents that housed the vendor's goods. Later they became more complex and took on more architectural significance, such as the bazaar or the souk found throughout the Middle East, or the markets of ancient Rome. Along with these markets came the office, or a place where business could be coordinated and accomplished. Like the markets and bazaars, this action could essentially take place anywhere. However, as civilization advanced, buildings were constructed solely for this purpose. There were temples and basilicas in Rome, and churches, halls, and palaces during the Renaissance. Other building and functional types such as stock exchanges emerged. One of the first buildings developed precisely for this function was the Uffizi in Florence. This building originally served as the office building for the Medici family. As commerce supplanted religion as the most important aspect of society, so the office building replaced the church in architectural significance.

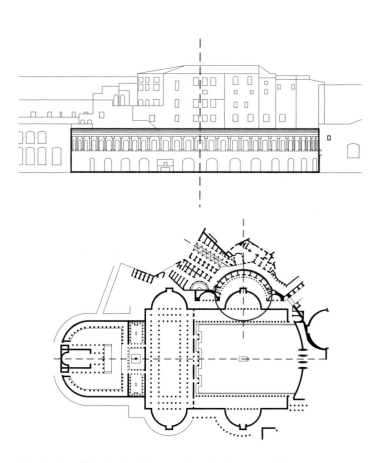

Markets of Trajan's Forum, Apollodorus of Damascus
Rome, Italy 112
[Renaissance, semi-circular indoor theater, wood and masonry]

The Markets of Trajan form a complex and significant part of Trajan's Forum in Rome. The semi-circular market form faces outward toward the forum, acting as an apse and backdrop. Above and behind the major façade of the semi-circular portion is the market proper. It is composed of a series of stalls that are on either side of a linear central hall. This building is critical in that it is really one of the first structures constructed specifically for commerce. Its multilevel architecture was a precursor to the modern mall. The Market's of Trajan are one of the few Roman public buildings meant to be seen as an exposed brick building and not one faced in marble.

TYPOLOGY PRINCIPLE **219**

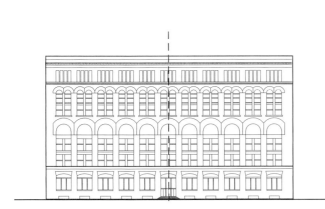
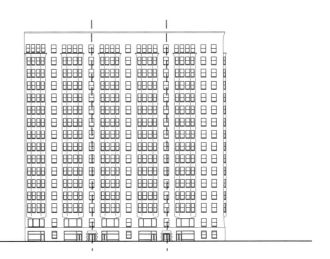
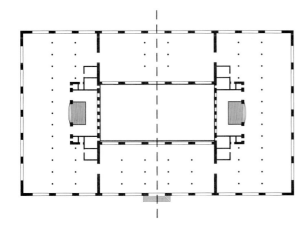
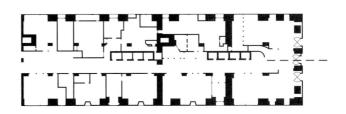

Marshall Fields Building, H.H. Richardson
Chicago, Illinois 1887
[Richardsonian, commercial warehouse, cast iron and masonry]

The Marshall Fields Building illustrated how a large commercial structure could be designed without referring to historical imagery. The building appears to be a large, solid block of stone, which at the time, was considered extremely modern. Richardson allowed the exterior stone walls to express their structure and materiality. This articulation became the building's ornament. The plan is a variation of a palazzo plan, employing symmetry and a large central space. The structural system was a grid of cast-iron columns combined with structural masonry walls.

Monadnock Building, Burnham and Root
Chicago, Illinois 1891
[Richardsonian, double-loaded linear office building, masonry]

The Monadnock Building is often considered the first skyscraper. It was sixteen-stories high but was constructed using traditional load-bearing masonry construction. The combination of the building's height and the compressive loads dictated six-foot-thick brick walls at the base. To accommodate this thickened base, the construction method resulted in the graceful, curved vertical lines that become the visual identify this building. The mass of the wall forced the lobby (traditionally one of the largest and grandest spaces in this building type) to be a cramped but still beautiful marble hallway.

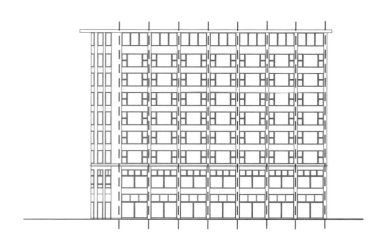
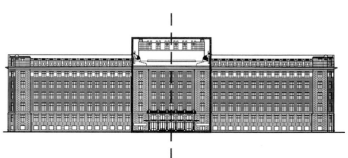
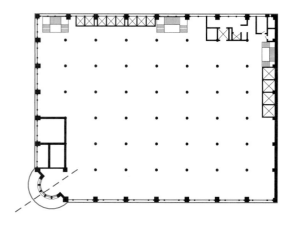
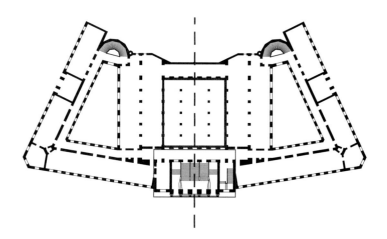

Carson Pirie Scott Building, Louis Sullivan
Chicago, Illinois 1899
[Early Modern, department store, steel and masonry]

The Carson Pirie Scott Building utilized the rational grid of the frame as both a structural and formal compositional figure. The two lower floors were used as a department store, requiring large spaces, interrupted only by the structural column grid. The exterior was an expression of its frame structure with a traditional tripartite elevation treatment. The base was designed to maximize its lightness and delicacy as an enlarged ornamental element that worked well with its consumer function. The middle portion is simply expressed as a structural frame faced in white ceramic tiles.

Postal Office Savings Bank, Otto Wagner
Vienna, Austria 1912
[Early Modern, reductive form with technical expression, diverse]

The Postal Office Savings Bank used ornament derived from new industrial materials to define its articulation. The plan has a straightforward classical organization. It covers an entire city block with three courtyards that provide natural light to the many offices and rooms embedded within this massive building. The ground floor of the central courtyard is the glass-covered indoor banking hall. This hall, through its detailing, uniquely expresses the ideas of the building. The columns are exposed steel and the lighting fixtures and heating vents are exposed and articulated as ornamental features. The floor and ceiling are constructed of glass, which produces a vertical spatial explosion through the building. The exterior of the building was completely covered in stone with aluminum fasteners exposed. This articulated detailing became Wagner's signature tectonic expression.

TYPOLOGY PRINCIPLE **221**

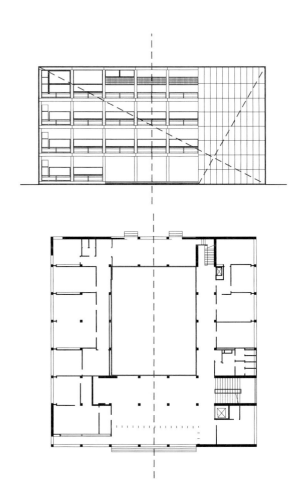

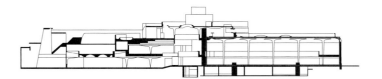

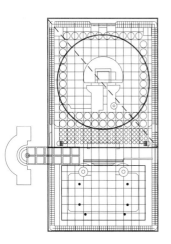

Casa del Fascio, Giuseppe Terragni
Como, Italy 1936
[Italian Rationalist, urban object, masonry]

The Casa del Fascio represents the zenith of the Italian Rationalists' contribution to architecture. It forms a stark white geometric figure that naturally forms a contrast against the mostly medieval urban form of the city. As the headquarters of the Fascist Party, its siting, directly adjacent to the Duomo, set up a dialogue between the church and state. The plan of the building is based on the palaces of the Renaissance, with the various offices replacing the domestic rooms of the palace. Terragni used a grid structure that could be manipulated to allow for larger conference rooms. The courtyard was enclosed, which allowed for even larger ceremonies. The building is faced in white marble, allowing it to stand out amongst the other earth- tone buildings. This building functioned not only as an office, but also as an elegant architectural essay of propaganda.

Johnson Wax Building, Frank Lloyd Wright
Racine, Wisconsin 1939
[Modern, open hall with streamlined form, concrete and masonry]

The Johnson Wax Building has a highly streamlined exterior that was defined by its rounded corners and raked joints elongating the masonry coursing and the movement of the façade. Even the form and function of the furnishings, designed by Wright and manufactured by Steelcase, reinforce this streamlined quality. The functional organization is a large open floor of secretarial positions encased at the perimeter with privatized offices. Meeting rooms for management wrap the surrounding mezzanines. The open hall of the "Great Workroom" was made possible by massive, concrete **dendriform** *columns that taper as they go upward. The misalignment of their circular ceiling forms were infilled with glass tubes that provided natural sky lighting.*

| 222 | TYPOLOGY
PRINCIPLE | FUNCTION
ORGANIZATION | OFFICE
TYPE | PLAN ELEVATION
READING | ARCHITECTURE
SCALE |

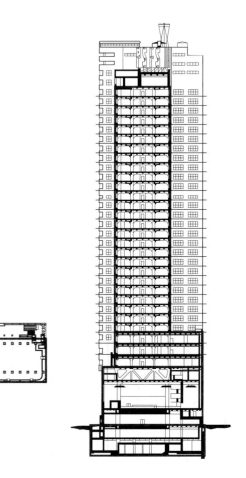

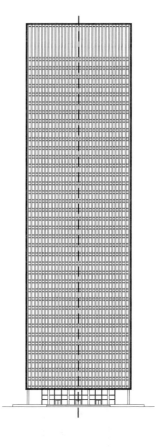

PSFS Building, Howe and Lescaze
Philadelphia, Pennsylvania 1932
[Modern, office tower, steel frame, stone cladding]

The PSFS Building essentially established the model for the modern skyscraper. The various parts are separated and individually expressed. The core of the building, containing the elevators, stairs, and bathrooms, was rendered as a glazed black brick tower. Extending out of this service tower is the office block which is expressed by the vertical columns that dominate the elevation. The **spandrels** *are faced in a gray brick to subordinate their presence. The ground floor is cantilevered out from the structural columns to allow an almost all-glass elevation. By rendering each function differently, the architects were able to bring a new functionally driven language to the building type.*

Seagram Building, Mies van der Rohe
New York, New York 1958
[Modern, free plan with central service core, steel and glass]

The Seagram Building developed the modernist high-rise that came to typify the corporate office building. Its form and detail have been mimicked and iterated in nearly every city on the globe. Developing the minimalism of the field and celebrating the industrialized forms innate in the materiality, the open plan and centralized office core make the spaces highly subdividable. The exterior articulation of the structure and the elimination of the variation between base, middle, and top made the elevation a uniform field that perceptually extends infinitely. Mies desired a full expression of the structure to honestly represent the systems, but fire codes prevented the steel from being revealed requiring its encasement in concrete. To reinstate the visual legibility, the I-beams were artificially reapplied to the surface of the building. The resulting type created a functionally flexible and highly formal personification of the modernist office.

TYPOLOGY PRINCIPLE **223**

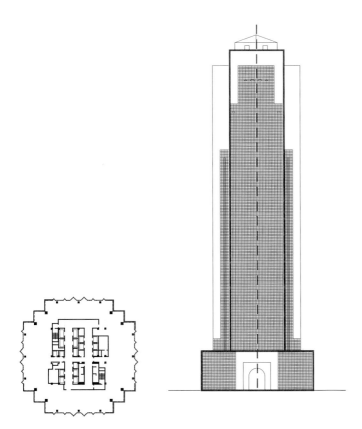

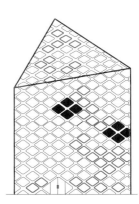

Transco Tower, Philip Johnson and John Burgee
Houston, Texas 1983
[Postmodern, square plan with crenellated corners, glass]

At the time of its construction, the Transco Tower was built as the world's tallest building outside of a city center. It is a postmodernist hybrid composition that mimics a hybrid of another civic medium-rise building in Texas as well as Johnson's own earlier IDS Center in Minneapolis. Transposing the material, scale, and site strategy, the tower is a unique combination of two separate towers superimposed upon one another. The floor plan adopts the conventional centralized service core and a perimeter shape with exaggerated **crenellation** *at the corners to optimize the number of "corner offices." The functionalist practicalities in combination with the referential form and the location outside of a city center make it a postmodern pinnacle of office design.*

Prada Headquarters, Herzog and deMeuron
Tokyo, Japan 2002
[Postmodern, polygonal plan with diagonal grid skin, glass]

In this definitively object-based building, the form is rendered to engage five sides. Blanketed in a diagonal grid, the infill panels and their varied curvature (flat, concave and convex) optically affect the experiential reading of the surface. As a hyper-extension of the display window, the optical dynamism of the surface (on both the interior and exterior) is animated as the viewer moves through space causing light and perception to bend and respond based on position. The interior is open through staggered floor plates to create sectional continuity. This space is further exaggerated through the cores and diamond-shaped tubes that subdivide and structurally engage the collective shape. The heightened reliance on experience, surface, and form advance the office/store type as a hybridized yet highly composed building.

Parking Garage

The parking garage typology does not typically command the attention garnered by other building typologies. Across its lineage, the actual size of the automobile has not changed significantly, hence the function of this typology has remained relatively constant. As a result, the first parking garages and the newest ones are not architecturally or spatially that far apart. The interesting aspect of these parking garages is the lengths to which architects have gone in their desire to raise the type to something beyond mere building and to articulate this functionally driven architecture. Most parking garages continue to fall into the category of a functional box. Some architects, however, have expanded the role of the parking garage into a symbol of technology and beauty.

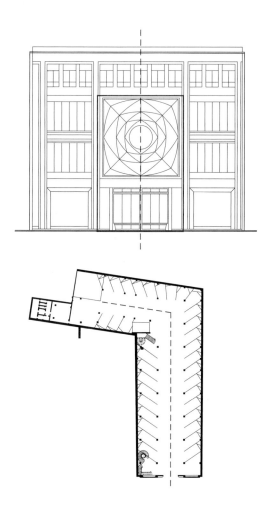

Garage de la Societe, Auguste Perret
Paris, France 1905
[Early Industrial, L-shaped parking structure, concrete]

The Garage de la Societe was the first parking garage, setting the standard for all ensuing parking garages. Perret resolved the functional requirements in a very straightforward fashion. He realized that angling the cars would achieve the necessary spatial requirements of the automobile and also fit into the tight confines of the urban infill site. The garage had flat floors, requiring a massive elevator to move the cars vertically. The second and third floors were accessed by two movable bridges that connected to the elevator. This allowed the central portion to be three floors high and completely lit by natural light from the skylights. The elevation expressed this spatial concept through a large, all-glass centralized window and used the concrete frame to express the structure within.

TYPOLOGY	
PRINCIPLE	**225**

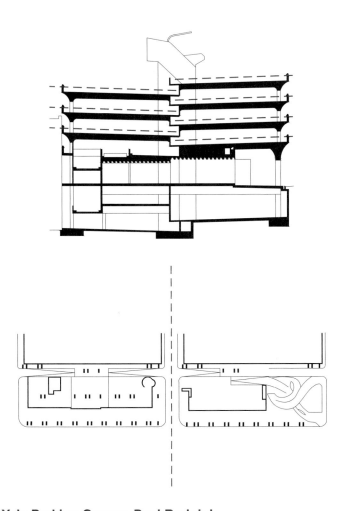

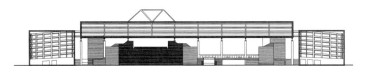

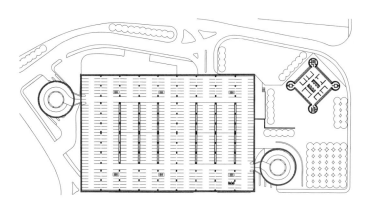

Yale Parking Garage, Paul Rudolph
New Haven, Connecticut 1963
[Expressionist Modern, Multilevel parking structure, concrete]

Rudolph stated that he wanted people to question the function of this building. The visual language has since become synonymous with the parking garage. The structure, which is over eight-hundred-feet long, truly developed the parking garage as a piece of architecture rather than merely an engineered building. The structure is completely cast-in-place concrete, which is overtly referenced in its construction. Even the futuristic light fixtures are articulated in concrete. The scale of the building is massive and undeniable, creating a dominating presence in the city. The parking garage is lifted off the ground and accessed by a series of ramps, allowing the lower floor to house a collection of shops and retail spaces. Ultimately, Rudolph used the referential forms and shapes of highway and interstate infrastructures to create and define this unusual building.

New Haven Parking Garage, Roche Dinkeloo
New Haven, Connecticut 1972
[Modern, parking garage over theater, steel]

The New Haven Parking Garage was placed on top of the New Haven Coliseum, as the site would not allow for subterranean parking. This problem required an ingenious formal solution. As a result, the parking garage dominates the project, requiring large, circular ramps spiraling to the roof, which evolved as iconic symbols of modernism in New Haven. The plan was an extremely simple grid structure that extended to the ground except for where the garage hovered over the Coliseum. Here, a large structural beam, from which the bays were suspended, was installed above the garage. The corten-steel material of the building was a signature of the building's industrial technology.

Campus

The word "campus" is traditionally used to describe a park-like setting that contains a series of buildings that are all part of an institution of higher learning. A campus will typically have academic, administrative, and recreational buildings as well as housing, all positioned and composed as an interrelated collection. The word campus was initially used to describe the land surrounding Princeton University in New Jersey. Later, this typically American term was used in Europe to describe more urban conditions. In the United States, the campus has proven to be one of the more inventive cultural productions. As a result, some of the most important buildings and associated spaces in the U.S. are located on college campuses.

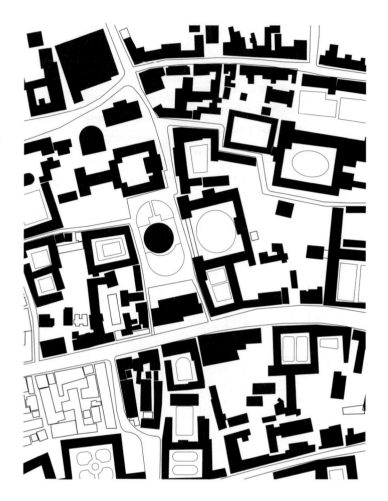

Oxford
Oxford, England 1585
[Gothic, campus integrated into city fabric, masonry]

Oxford, one of the oldest universities, is not set on a typical campus. Instead the university is completely integrated into the fabric of the city of Oxford. It is impossible to draw a line dictating where the university and the city are separated. The university is comprised of a series of colleges, each with its own independent facilities. The colleges are always structured around a central courtyard, or green, and are almost without exception designed in the Gothic style. These characteristics set the tone for many other ensuing universities around the world.

TYPOLOGY	
PRINCIPLE	227

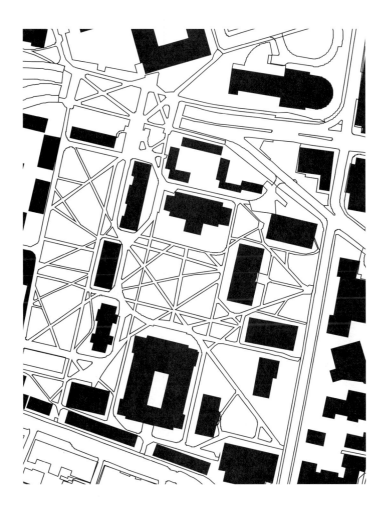

Harvard University
Cambridge, Massachusetts 1636
[Diverse, campus integrated into urban fabric, masonry]

The campus of Harvard University is centered on Harvard Yard, but diffuses into the surrounding urban context of Cambridge. Establishing some congruity through the use of organizing quadrangles, the density of the urban condition and the long chronology of Harvard's development and evolution result in a unique and diverse collection of historic buildings that represent three hundred years of architectural styles. Spanning from origins in early Georgian architecture through the dominance of Romanesque buildings by H.H. Richardson (Sever Hall), the campus equally boasts modern buildings by Le Corbusier (The Carpenter Center—the only building by Le Corbusier in the United States), Jose Luis Sert, James Stirling, Charles Gwathmey, and Renzo Piano.

Princeton University
Princeton, New Jersey 1756
[Neo-Gothic, nodal college courtyards, masonry]

Princeton University is similar to Oxford in that the colleges are typically Gothic in style and are configured around a central courtyard. Like Oxford, a series of colleges form a critical part of the university from both an architectural and social standpoint. Students are assigned to a respective college upon arrival and reside there throughout their stay at Princeton. Each college has its own dining hall, library, and study rooms. Unlike Oxford, Princeton is set apart from the town by the division of Nassau Street. This division later became known as "Town and Gown" and described a condition whereby a college was segregated from a city or town by a road or street.

| 228 | TYPOLOGY
PRINCIPLE | FUNCTION
ORGANIZATION | CAMPUS
TYPE | PLAN
READING | URBAN
SCALE |

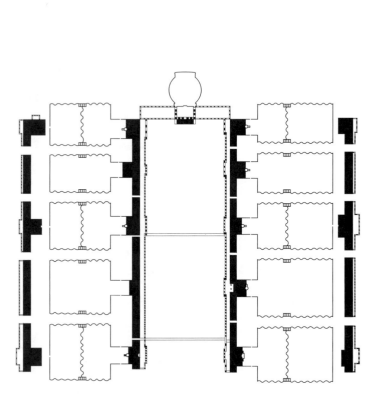
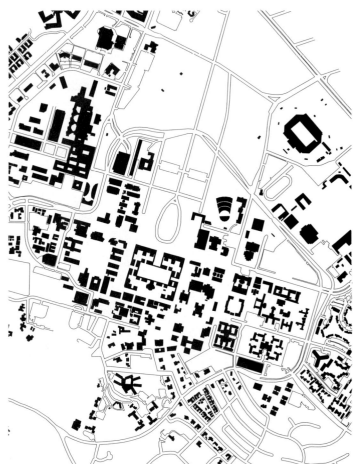

University of Virginia, Thomas Jefferson
Charlottesville, Virginia 1826
[Neoclassical, axial central lawn, wood and masonry]

The University of Virginia was conceived of as a self-sustained academic village. Jefferson designed the University to have one hundred students and ten professors. The students would live in fifty rooms that occupied either side of the Lawn. The professors would live in ten different pavilions that separated the rooms. The Lawn is a large, multilevel grass area surrounded by a single-loaded colonnade that connects the rooms and pavilions. Jefferson used the buildings as architectural lessons, deploying the classical language to articulate various compositions. Some are highly original, while some are more derivative in their form and language.

Stanford University, Frederick Law Olmstead
Palo Alto, California 1888
[Spanish Colonial, courtyards and quadrangles, masonry]

Stanford University was begun in 1888 with a design by Fredrick Law Olmstead and the leadership and vision of the founder of the University, Leland Stanford. They saw the plan of the university as a model of simplicity and clarity. The primary north-south axis connected Palm Drive with the Memorial Court, the Inner Quad, and the Memorial Church. The second, east-west, axis extended through the Inner Quad and was to connect a series of additional quadrangles, as the University grew. Originally there were to be three main smaller courts, with four larger quadrangles on either side along the east-west axis. Unfortunately the clear plan of Olmstead was gradually lost over the years and today only the Oval at the entrance and the main court are still intact.

	TYPOLOGY	
	PRINCIPLE	**229**

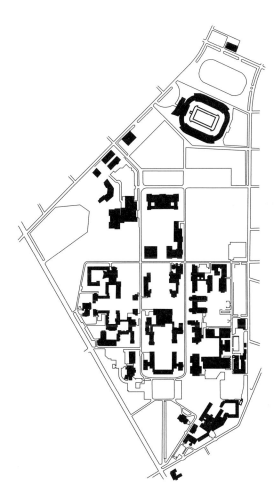

Rice University, Ralph Adams Cram
Houston, Texas 1911
[Neo-Byzantine, central quadrangle, masonry]

Founded on a the central building of Lovett Hall (named after the first University president), Rice University's impressive sallyport begins an axis that runs from the city grid of Main Street through the primary quad and terminating at the library. Four flanking buildings created the formally planned main quadrangle that is repeated several times along the same axis. These quadrangles are encircled by a central-campus loop road. A secondary ring of buildings includes the specialized laboratories and dormitories that fill out the remainder of the campus. The hierarchy of the plan and the consistency of the material palette (St. Joe brick with an overly thickened mortar) make for a unified and formal university campus.

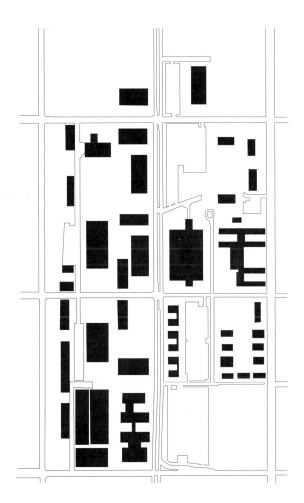

Illinois Institute of Technology, Mies van der Rohe
Chicago, Illinois 1943-1957
[Modern, free plan campus, glass, steel and masonry]

The rise of Nazi power in Germany forced the closure of the Bauhaus in 1933 and encouraged its director, Mies van der Rohe, to depart. He relocated to Chicago to be the head of the architecture school at Chicago's Armor Institute of Technology (later renamed IIT). Upon his arrival, he was immediately asked to aid in the planning and design of the campus and its initial buildings. Fully deploying the fluidity of universal space at the scale of the campus, the buildings are sited to allow space to flow around them and through open quadrangles. The architectural vocabulary of a tight steel frame with coplanar glass and masonry infill skin established the compositional rule set for the architecture and highlighted the flexibility of the modern agenda with the intellectual agenda of the University. The twenty campus buildings designed by Mies van der Rohe (notably including Alumni Hall, the Chapel and Crown Hall— the school of architecture) are individual extensions of the spatial and material intentions of the campus and his broader work.

230 | FORM
PRINCIPLE

FORM	**231**
PRINCIPLE	

Form

04 - Form

Form is perhaps the most dominant and immediate aspect of architecture. Its derivation and visual associations present it as the identity of a project and the culmination of all of the collective localized decisions made during the design process. Form is the governing factor for categorization; the origination, justification, and clarity of its evolution is essential.

The following categories of form are named for their primary driving factor:

Platonic Formalism Platonic forms are essential building blocks derived by the pure geometric expression of cube, sphere, and pyramid. As primal geometric figures, they intrinsically hold a high level of abstraction derived from artificial and external constructs relating to their geometric formulation and cultural reference.

Functional Formalism Form as a result of function is a core principle of modernism. Allowing for a direct translation of programmatic necessities to govern the formal expressive articulation of the architecture creates a legibility of form to use. This embrace of practical needs into formal articulation derives from an intrinsic association with an inevitable essentialism rooted in function.

Contextual Formalism Contextual formalism relies on the immediacy of site and the details of place to impact form. This requires engaging vernacular traditions founded in indigenous materials, traditions of construction and responses to local climate. Relying on local precedent, the reiterated forms are culturally ingrained and chronologically iterated, modified from generation to generation and evolved with layered improvements and adaptations. **Typological Formalism**, a subset of contextual formalism, deploys the referential lineage of the functional type to focus upon the historical context of a function and the evolution of programmatically associated precedents. Defined by the historical traditions of organizational and functional formalisms, the established figures of typological precedent-based formalism allow for a legibility of a building's use by the lineage it references.

Performative or Technological Formalism Performative formalism responds to a functional agenda governed by such things as structure, material, environment (light, wind, orientation, and position), and energy. The translation of these functional forces into specific formal responses is the essence of performative formalism. The complexity and customization resulting from these requirements allows for a formal variability and intricacy, which permits a hyperarticulation of responsive functions in the form.

Organizational Formalism Organizational formalism derives its primary form from the morphology of the organizational system deployed. Primarily a plan-based mechanism, the form becomes a derivation of the extruded horizontal organization. Thus a centralized scheme tends to have a radial form, a single-loaded corridor tends to have a linear form, and so on. This association of organizational systems with form permits a legibility of the building's use in the final composition. The repetitive lineage of these associations develops the legibility of organizational typological formalism based on the repetitive and iterative evolution of precedent.

Geometric Formalism Geometric formalism refers to a form-making process that relies on geometry, and the subscription to its rules, principles, and orders of organization, proportion, and repetition to derive form. Grids, golden sections, even Le Corbusier's unique Modulor, all rely on a belief in truths found in ancient numerology, ideals that proclaim modules, proportions, and interrelations as rules. **Symmetrical Formalism** is a subcategory of geometric formalism. Using a mirror line (or a series of mirror lines), the reflective nature of the plan establishes an axis and an immediate hierarchy. The reflective mapping of one side creates an intense ordering that allows for a mental map of the entire composition, providing a constant awareness of one's individual position in relation to the building's compositional whole. **Hierarchical Formalism** is similarly derived from geometric formalism. Dependent on the ordering, proportional systems, and modules, the forms result in distinct hierarchies of the interrelationships of the parts. Provided through scale, position, length, ornament, or other methods, the base geometries establish hierarchical formalism.

Material Formalism Material formalism derives from a design-based relationship with the physical qualities of the matter engaged. The final legibility of material is governed by diverse manufacturing processes and their resulting material modules, as well as fabrication technologies that manipulate raw materials into building components. Tectonic systems govern the geometry, expression of assemblies, construction requirements, and practicalities that administer the installation of material. The localized detail and its collaboration with the collective assembly determine material formalism.

Experiential Formalism Experiential formalism is derived from a generative application of the perceptive manner in which objects are visually and physically engaged. Primarily derived through perspective or model-based representations, these processes engage the perceptive experience of a choreographed, chronological sequence as the body moves through space. As the experiential form-finding process is rooted in perception, the results of its compositional methods are often a seemingly irrational whole, as the collective composition is determined through locally decided relationships as opposed to overarching systems. **Sequential Formalism,** as a subset of experiential formalism, uses circulation as a basis for the perceptual sequence. Guided by the architectural promenade through the building, the resulting sequence both defines the perception of the architecture and assumes responsibility for its formal generation. **Axial Formalism** is a subset of sequential formalism and often a resultant of symmetrical and hierarchical formalisms. Focused on a linear circulation sequence that establishes a dominant path through the form's organization, the line typically becomes an anchor for other geometric moves, thus producing a hybridization of geometric, symmetrical, hierarchical, experiential and sequential formalisms. An example of such a project is Beijing, China's Forbidden City, which maintains a dominant central axis that establishes the sequence and hierarchy through the spatially and sequentially layered complex. Symmetrical about this axis, the experience is anchored in the regimented geometry of the axial line.

In contemporary architecture, there is a divorce of function and form. In looking towards more complex geometries, often justified through contextual or performative criteria, there has evolved an hyper-variable geometry whose complexity is possible through advanced digital processing and parametric design processes. There are often references to biomimicry in these forms, producing scripted systems that allow for responsive iterations and variations. The mass customization of components (structural, tectonic, or otherwise) presents unique localized actions within larger fields of systematized field effects.

Regardless of categorization or classification, form is essential to architecture. It holds a spiritual and physical quality that bridges the systematized rationale of the architect and the visceral figurative response of form.

Platonic Formalism

Plato's theory of objects defines forms by their essence. Not visually determined, but conceptually defined by the closeness to the purity of their mental depiction, Platonic Forms describe the primal three-dimensional geometric forms: sphere, cube, and pyramid. Their idealism is significant due not only to the purity of the form, but also to the clarity of the system(s) that determines their form. As a result, the shape of the object governs not just the visual purity but also the conceptual purity of the forms origination.

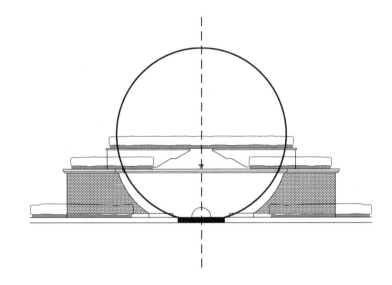

Newton's Cenotaph, Étienne-Louis Boullée
Unbuilt 1784
[Neoclassical, heroic architecture parlant, masonry]

SPHERE: In Newton's Cenotaph, Boullée deploys the sphere as an idyllic form representative of Newton's concept of geometric and mathematical purity. It uses the conceptual geometry of the sphere to provide for an effectual design through the representation and signification of the celestial constellations. The sphere as an equally centrifugal form simultaneously defines a center and an edge. Its visual experience as both an exterior form and an interior space is rooted in the purity of the geometry and the equal simplicity and complexity of its perception.

FORM
PRINCIPLE **235**

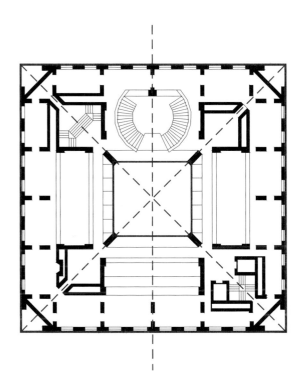
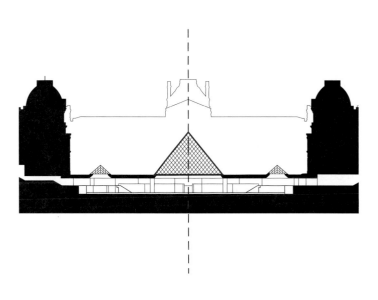

Exeter Library, Louis Kahn
Exeter, New Hampshire 1972
[Late Modern, centralized square, masonry and concrete]

CUBE: *In the design of the Exeter Library, Louis Kahn utilizes the Platonic figure of the square, ultimately culminating in the cube, to produce and isolated and inscrutable figure. This cubic shape is used as a way to organize the library both in a functional aspect as well as a formal one. Kahn organized the building by using two separate cubes, one nested within the other, to produce areas that correspond to the functional attributes of a library such as stacks, reading areas, and more honorific atrium. The exterior cube is constructed of brick and reflects the repetitive construction techniques common to that system. The inner concrete cube, which defines the atrium, is constructed from concrete and is penetrated by large circles (another juxtaposing Platonic form) radiating from the center point of each elevation and dematerializing the surface to translate the reading of the planar enclosure to become frame-like. Overall the cube is deployed in a multivalent manner that capitalizes on the many attributes of this most powerful figure.*

Louvre, I.M. Pei
Paris, France 1989
[Postmodern, monumental pyramidal form, steel and glass]

PYRAMID: *When confronted with the daunting task of adding onto the Louvre, a historical and museological icon on both the architectural and urban scales, Pei employed the pyramid to add to the dominant axis and self-resolved symmetrical form of the existing space. By creating a new entry located in the center of the courtyard, Pei was able to create a single point that reorganized the circulation and hierarchy of the entire complex. The pyramidal form is deployed as an internally self-resolved and pure geometry that juxtaposes with its context through the foreign shape. This contrast is further heightened by the dematerialization of the form through an intricately engineered and detailed glass skin. This translation of materiality contrasts the technology of the glass against its stone masonry surroundings and the traditional typology of the pyramid as a solid mass.*

Functional Formalism

In functional formalism, the function of the building is expressed in the form of the architecture. This equation is most overtly recognizable in industrial buildings such as factories or power plants. Here the building is more "engineered" than architecturalized, so there is an economy and essential quality to every move. The translation of "use" into an associatively legible, three-dimensional architecture requires a sophisticated understanding of program and type. Functional formalism is rooted in a foundational principle of modernism and led to the popularized idea of "form following function." In cases where functional requirements are strict, this equation proves useful and successful. However, when the functional requirements of the building are more pedestrian, it tends to produce rather simplistic and mundane structures. Despite these shortcomings, functional form, when used carefully and with proper intent, remains a viable type of form making.

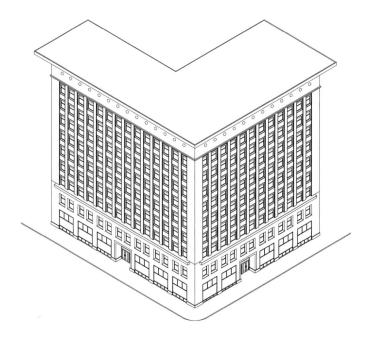

Wainwright Building, Adler and Sullivan
St. Louis, Missouri 1891
[Early Modern, open plan with core, steel fame and masonry]

The Wainwright Building was Adler and Sullivan's first skyscraper. Understanding it as a new building typology, the designers emphasized the verticality in the elevations to stress the significance of its form. The building is articulated using the classical tripartite divisions of base, middle, and capital. The building satisfies the concept of functional form by developing a new language based purely upon the skyscraper as a type and the office building as a function. The building was not disguised as anything else, but rather derived its articulation from its function.

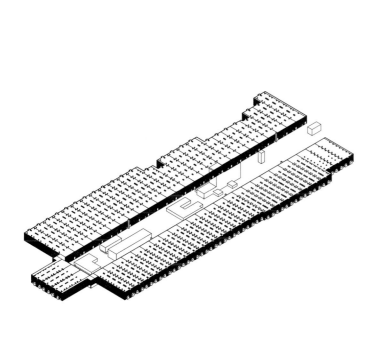 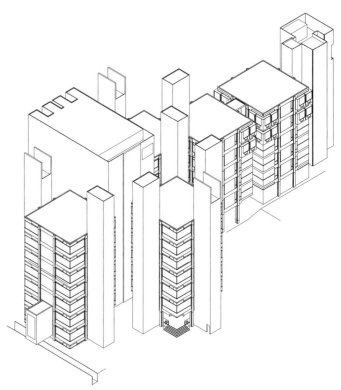

K-25 Plant, SOM
Oak Ridge, Tennessee 1945
[Modern, U-shaped uranium enrichment plant, concrete]

When this massive plant was completed in 1945, it was the world's largest building, measuring one-half mile in length by one thousand feet in width. It is a remarkable example of functional form because its dimensions were explicitly dictated by the programmatic processes housed inside. The uranium enrichment at K-25 was done by gaseous diffusion that functionally required extremely long, straight lines. This process also required tremendous amounts of electricity, hence the plant's location near the Tennessee electrical power plants. The building is defined as a manifestation of functional necessity and siting relative to form.

Richards Medical Center, Louis Kahn
Philadelphia, Pennsylvania 1961
[Late Modern, centralized laboratories, concrete and masonry]

In the Richards Medical Center, Kahn separated the various functions of the building's program and then articulated the functions in a way that differentiated them by material or shape. The resulting complex form becomes a direct manifestation of the functions of the building. Due to its programmatic function as a medical research center, this design had very specific requirements. Kahn established the organization and the architectural nuances to address and celebrate these requirements. The flexible laboratory organization, compartmentalization of programs, and integration of building systems became hallmarks of the project and Kahn's design methodology as a whole.

| 238 | FORM
PRINCIPLE | **CONTEXTUAL**
ORGANIZATION | | AXONOMETRIC
READING | ARCHITECTURE
SCALE |

Contextual Formalism

Contextual formalism refers to forms that are derived from the surrounding circumstance(s). Contextual elements such as shape, height, texture, ornament, and urban strategy can serve as guiding forces that aid in the derivation of form. Before modernism, this method was quite common, even to the extent that it was often practiced unconsciously. Later, during postmodernism, it was very fashionable and was utilized uncritically, resulting in some unfortunate, dogmatic, and predictable buildings. When rigorously applied, contextual form remains one of the most powerful and useful tools for any architect. To a certain degree, it makes common sense to derive form from the surrounding environment. A referential form will not be identical, as it inevitably undergoes a series of transformations, allowing for localized changes. However, the remaining references (such as height or material) will still offer legible connections to the context and original model.

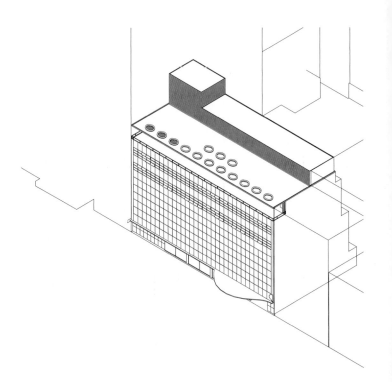

The Museum of Modern Art, Edward Durrell Stone
New York, New York 1939
[Modern, International Style museum, glass and steel]

The Museum of Modern Art in New York City is a modernist deployment of contextual form. Despite its minimalist and modern styling, the building obeys many of the contextual tenets set up by the older historical buildings surrounding the site. Mimicking numerous features of the contextual precedent, the building is sympathetic to building height, relationship to the street, use of honorific materials, and the outdoor courtyard. Because of these traits, the building intricately relates to adjacent buildings and the larger urban context.

FORM PRINCIPLE **239**

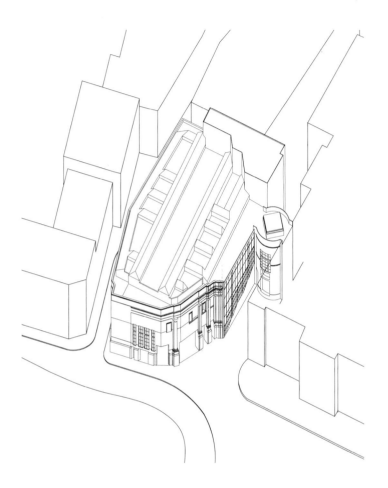

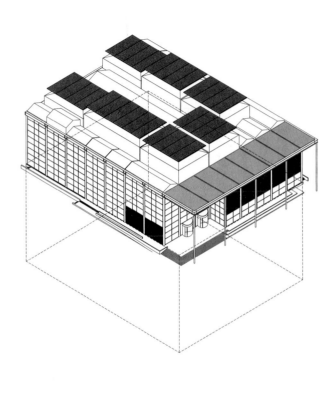

Sainsbury Wing, National Gallery, Venturi Scott Brown
London, England 1991
[Postmodern, iterative contextual façade, stone and glass]

Venturi Scott Brown's Sainsbury Wing of the National Gallery in London is a critical example of contextual form. In this addition, the contextual form is relegated to the major façade that looks onto Trafalgar Square. In this element, Venturi has attempted to coalesce a number of the surrounding buildings into one façade. The result is a fragmented exercise in classical forms and language.

Carré d'Art, Norman Foster
Nimes, France 1993
[High-Tech, modern Maison Carrée, glass and steel]

When Foster began this building, the intent was to refurbish the square occupied by the new museum and the original Maison Carrée. The Carré d'Art is nine stories tall with five of them below grade. This contextual decision allowed the new museum to match the building height of the surrounding buildings and integrate with the scale of its environment. Though the building uses a radically divergent material palette of glass and steel, and houses a unique function, it maintains a contextualism through the scale, geometry, and proportion of form.

Performative Formalism

Performative forms are derived from systems that have functional responsibilities. Forces such as sun (heat and light), wind, water, and active mechanical systems, each have a series of requirements, responsibilities, and technologies that can influence and determine form. The forms can adopt either passive or active systems. An ever increasing need for efficiency and ecological responsibility in buildings, coupled with the expanding capabilities and complexities of design and building technologies, performative formalism provides expanding opportunities for responsive systems and forms.

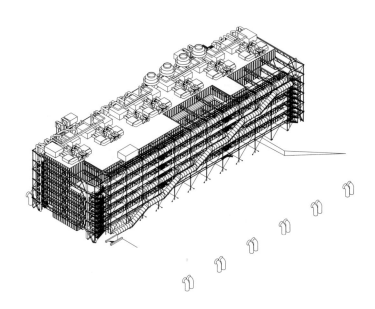

Centre Pompidou, Renzo Piano and Richard Rogers
Paris, France 1974
[High-Tech, free plan museum, steel, glass, concrete]

Envisioned as a museum for the people, Piano and Rogers uniquely and distinctly turned the design inside out to reveal typically concealed functions and systems. Each layered system is color-coded to explicate the building and present the museum as an industrially performing object. Using blue for air, green for water, grey for secondary structure, and red for circulation, the building is a legible sign system. The externalization of the services allows for a fully free plan interior that provides infinite flexibility to the layout of the galleries. The vertical circulation is dramatically suspended in a glass tube that terraces up the surface of the building. This provides views to the plaza below and the city beyond while animating the façade of the building with the activity and occupation of the museum itself. The functional performance of the building is exploited to define the composition of its form.

FORM PRINCIPLE 241

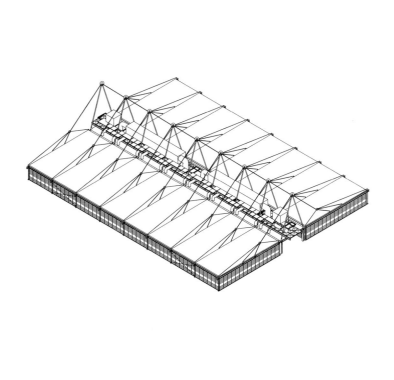

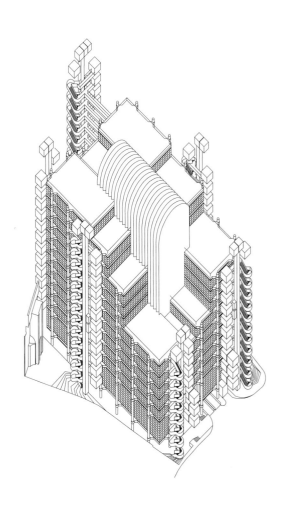

PA Technology Center, Richard Rogers
Princeton, New Jersey 1982
[High-Tech, exposed suspended roof, metal and glass]

At the PA Technology Center, Rogers continues an expressive externalization of the building's systems to define the building's form. Formulated with a desire for an unobstructed floor plan to allow for maximum continuity and flexibility, the structure and ventilation systems are suspended above the roof. Heroic masts extend dramatically above the roof with elongated suspension cables supporting the deck. Axially feeding the spaces below, the mechanical systems incrementally flair and change position. Color is deployed to highlight and codify the discrete systems. The perimeter is simple and anonymously dematerialized through a thinly mullioned glass grid further, emphasizing the dynamism of the performatively expressive roof.

Lloyd's of London, Richard Rogers
London, England 1986
[High-Tech, free plan with encircling services, stainless steel]

Lloyd's of London derives its performative form through its organization. The building's functions are dispersed to the edges, allowing for an open floor plate that provides maximum flexibility. The peripheral location of bathrooms and stairs are inverted from the traditional centralized core, eliminating the traditional dependence upon skin and form to produce the identity of the building and instead presenting the functional workings of each individual piece as the formal determinant. Uniformly clad in stainless steel, each function is personified through highly practical yet descriptive forms. The resulting reading of the building is one that expresses the **anthropomorphic scale** of the individual functions while aggregating their repetition and collection to create a systematized composition derived from the practicalities of need.

Organizational Formalism

Organizational formalism refers to forms that are produced by their organizational system. Often the system is simple and recognizable. This allows the form and the organization to reinforce one another. Organizational formalism differs from functional formalism in that the latter is driven by programmatic use while the former is governed by the system's planning. Organizational form can exist regardless of function. The two can be intrinsically associated, or can be uncoupled and independent. For example, there can be buildings that share an organizational type (such as a linear organization), but be completely different in their function (such as an office or hotel). Organizational formalism is general in the figures that it derives, typically requiring another layer of information and influence to understand and govern the architecture.

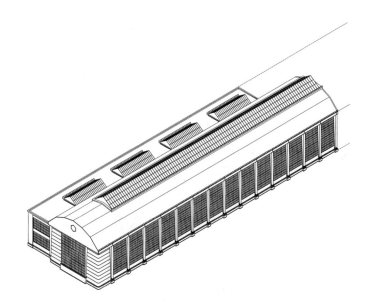

AEG Turbine Factory, Peter Behrens
Berlin, Germany 1910
[Modern, three-hinged steel arch with curtain wall, diverse]

The form of the AEG Turbine Factory was based on the dimensions of the turbines. As a result the building is scaled to respond to the spatial requirements for building and manufacturing the turbines. For instance, the height of the roof was dictated by the dimensions required to operate a crane within the space. Functions that did not require that scale and type of space were housed in the lower wing that was placed to the side. The entire complex was set up on a grid structure that allowed maximum flexibility in plan, with the size and form of the buildings being the deciding factor in their formal articulation.

	FORM
	243
	PRINCIPLE

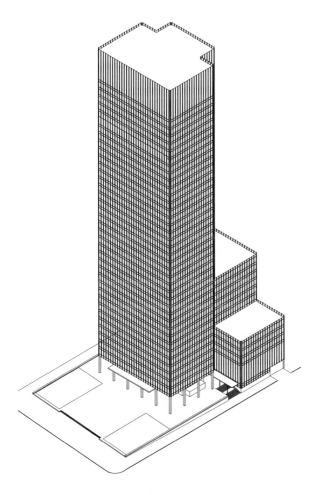
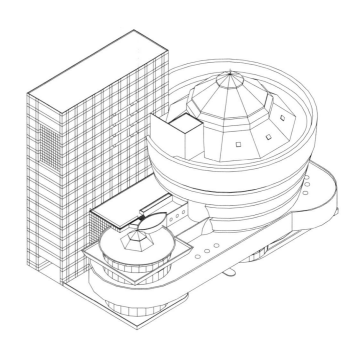

Seagram Building, Mies van der Rohe
New York, New York 1958
[Modern, free plan with central core, steel and glass]

The Seagram Building is a fundamental example of organizational form. As one of the first skyscrapers that did not rely on historical references, it derived its simple block form from the sectionally layered and repetitive floor plates. It is a combination of a form derived by the buildable envelope of the site, the flexibility of the free plan, and a regularized structural grid. The clarity of the architectural language used is so essential and clear that is has been widely mimicked becoming intertwined with the definition of the modern skyscraper.

The Guggenheim Museum, Frank Lloyd Wright
New York City, New York, 1959
[Modern, ramping spiral gallery around atrium, concrete]

The Guggenheim Museum in New York is an extremely clear example of organizational form. The main gallery is formed by the spiral ramp that is clearly expressed in the exterior form. Wright deployed this formal organizational aspect as the form generator for the building as a whole. The building contrasts its surroundings with the dynamic curves of the spiral ramp to heighten its presence and elevate the formal power of its geometry.

Geometric Formalism

Geometric formalism emerges from an intensive determination of geometry to regulate and systematize modules, dimensions, structure, or figuration to produce form. Relying on the purity of geometric principles, the formal systems are highly dependent on their regulatory controls. With increased capabilities through computational modeling, the expanded field of emergent nonlinear geometries allows for an increasing complexity of potential forms. The resulting collection of forms are so geometrically determinate that their description and perception are conceptually and visually intertwined.

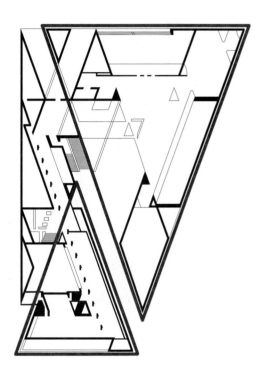

East Wing of the National Gallery, I.M. Pei
Washington, D.C. 1978
[Late Modern, central triangular atrium, concrete and stone]

L'Enfant's neo-baroque plan of Washington, D.C., dictates the shape and form of many buildings in the city. The most notable being the East Wing of the National Gallery by I.M. Pei, where the resultant triangles of the urban plan are transformed into the very essence of the architecture. Pei used the collision between L'Enfant's radial streets and the rectangular mall as a vehicle to construct a geometric system that was adopted and accelerated. The site is further divided into a series of separate triangles, resulting in a conglomeration of solids and voids that are constantly challenging each other. The triangle as a geometric form is manifest even at the smallest detail, resulting in a triangular waffle slab and pyramidal skylights in the central atrium. The geometry fully pervades the conception and perception of the building at all scales.

	FORM	**245**
	PRINCIPLE	

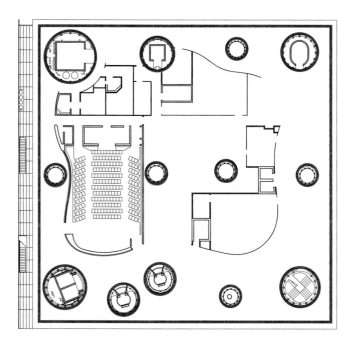 

Sendai Mediatheque, Toyo Ito
Sendai, Japan 2000
[Postmodern, free plan with nodal cores, glass and steel]

The Sendai Mediatheque, by Toyo Ito, utilizes the powerful geometry of the square as a way of establishing the base shape of the building through the use of a transparent glass perimeter boundary. Posed against this purity are the dynamic, variably deformed tubular cores that run vertically throughout the building. The uniform geometric purity and simplicity of the square enclosure allow for the striking contrast of the organicism and structural filigree of the internal towers. The combination of these two separate systems permits the reading of the structure as object against the backdrop of the pure form of the envelope. This geometric contrast is continually referenced as the rectilinear box is experienced within the context of the curvilinear interior towers. The geometric formalism establishes a datum through the square envelope and then plays off that datum with the localized variation in the curvilinear cores.

21st Century Museum of Contemporary Art, SANAA
Kanazawa, Japan 2004
[Postmodern, circular gallery with rectilinear rooms, glass]

The 21st Century Museum of Contemporary Art is designed through plan. The simple geometric organizational system is variably orchestrated through size, position, and material to determine their cumulative effect. It is compositionally determined by the primary geometric shapes, nested within an even purer figure of the circular perimeter. The inner chambers are extruded figures of varying sizes, heights, and positions, driven by their internal geometries scales and needs. Aggregated as a cloud, collected and bound by its evaporative circular edge, the boundary form is dematerialized through the careful use of glass. The building produces layered views of the discrete geometries of the galleries, mixing the purity of the geometry with the effect of their minimal readings.

Material Formalism

Material formalism refers to projects whose form is dictated by the matter/material used. Embracing practical concerns such as the physical, structural, and tectonic capabilities of a construction system, the form becomes a derivative collaborator with the material decisions. Often emerging out of cultural traditions and vernacular sensibilities, and at the same time responding to local availabilities of material, traditions of construction, function, and climate, the specific method and attitude of a material's use and application to determine a form requires a design sensibility that synthesizes space, form, and matter.

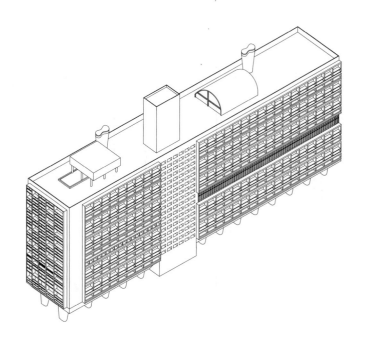

Unite d'Habitation, Le Corbusier
Marseille, France 1952
[Modern, double loaded corridor, concrete]

The Unite d'Habitation is built out of the sensibilities available to its concrete construction. Built in Marseille, given limited existing building technologies, concrete was adopted for its abundant supply, durability, ease of construction, and multiplicity of potential applications. The flexibility of the cast-in-place concrete allowed a single material to be simultaneously deployed as structure, finish, wall, column, floor, brise-soleil, and furniture. The fluid formalism, dependent only on the shape of the vessel, allowed diverse formal types, both singular and repetitive in occurrence. With panelized and board-form casts, the surface/finish of the material determined the building's identity and reveals its construction sequence and method. The color, shape, texture, and function of the composition emerge from Le Corbusier's embracing the capability of the material.

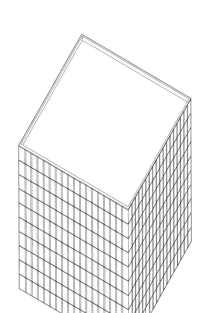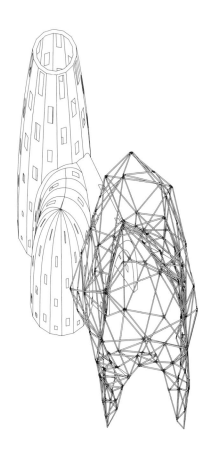

Kunsthaus Bregenz, Peter Zumthor
Bregenz, Vorarlberg, Austria 1997
[Modern, multi-story free plan museum, glass and concrete]

Peter Zumthor's Kunsthaus Bregenz displays a singular presence shaped by the meticulous and uncompromising use of its material. The building's enigmatic quality is created by the use of white glass that is deployed in a shingled fashion over the entire exterior of the figure. This repetitive material module forms a relentless pattern that dominates the language of the architecture. The quiet, anonymous box serves its function well as a museum, allowing the contents to be privileged and displayed appropriately. The architecture of the Kunsthaus is clarified through the careful and considered use of a singular exterior materiality and the subtleties of the tectonic resolution.

Light Frames, Gail Peter Borden
Silver Lake, California 2010
[Material Ultra-Modern, two material pavilions, PVC and metal]

Light Frames is a materially driven installation for Materials & Applications gallery. Consisting of two pavilions, each is formally determined by a collaboration of material with physics. One-inch electrical conduit was used in a variably triangulated configuration to produce a shadow frame tower. Vinyl fabric was used in a double-walled, geometrically formed inflatable chapel that projects light. The two layers are then repetitively linked with four-sided truncated pyramids that create funneling apertures. The dialogue between form and material produces a light effect and spatial experience. The collective composition is a poetic and truly "material" architecture.

Experiential Formalism

Experiential formalism is one of the most powerful yet elusive types of form generation as it requires an emotional component for its derivation. This type of form is typically only understood through the inhabitation of the building. The use and, in particular, the chronological movement or passage through the space, dictate the form. Space, materiality, sequence, and narrative can all collaborate to heighten the experience. There are no clear factors that mark a building as an architecture that represents experiential form, rather it depends on the orchestration of experience to govern the system.

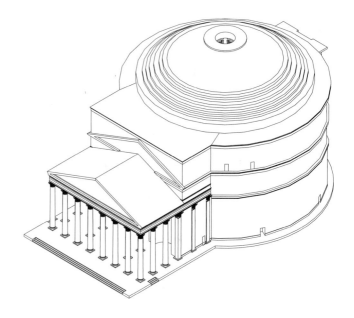

Pantheon
Rome, Italy 126
[Classical, temple, masonry]

The Pantheon is a seminal example of experiential form. It relies on the perfection of the circle and the sphere as a way to communicate. The circular drum of the exterior is conspicuous within its context, hinting at an interior of unrivaled experiential qualities. The sequence begins in the porch, or portico, of the building, where the observer is confronted with the ancient columns and heavy wooden beams above. Passing through the thick bronze door, the inhabitant is faced with a geometrically perfect space: a massive sphere with an oculus that connects the building to the sky. The materials, scale, shape, and form all coalesce into a magnificent combination of experiential perception.

FORM PRINCIPLE **249**

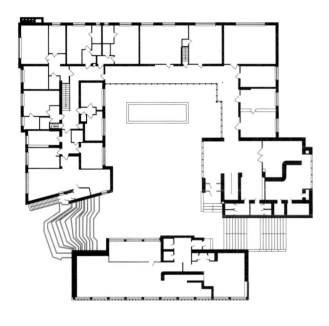

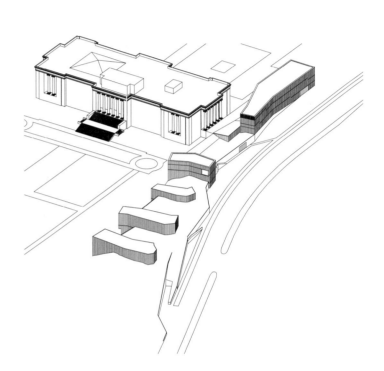

Saynatsalo Town Hall, Alvar Aalto
Saynatsalo, Finland 1952
[Modern, courtyard with experiential sequencing, masonry]

The Saynatsalo Town Hall is one of Aalto's most iconic buildings. The experiential aspect of the project begins with the isolation of the building itself. The building sits alone in a forest setting that immediately highlights the contrast of built and natural form. A large set of stairs deliver the user into the courtyard, which is the heart of the building. Though closed off from the forest, the inhabitant still has views out to the wilderness. The materials of the project—brick, wood, and copper—play an overt role in articulating the experiential nature of the building. The site, form, configuration, materiality, and sequence all collaborate to produce an experiential building that emerges from a sensitivity to the **genius loci** *of the site.*

Nelson-Atkins Museum of Art, Steven Holl
Kansas City, Missouri 2007
[Postmodern, underground museum with light boxes, glass]

Holl's addition to the Nelson-Atkins Museum of Art is an example of experiential form achieved through the careful use of site planning and materiality. The original, classically inspired museum is a concise rectangle against which Holl contrasts the use of five variably positioned and scaled forms. These large glass forms are collected along the west side of the museum site and cascade down a series of terraces away from the original museum. Faced in translucent glass, the forms act as a series of over-sized light boxes. Their anonymity of detail and internal illumination at night provide an inverted and arresting experience. Minimal landscaping creates the effect of glowing sculpture resting on an open lawn.

Sequential Formalism

Sequential formalism refers to a series of events that derive a form that is experienced through a particular chronology. This methodology emerges from a desire to control the user's experience in a manner that strengthens the perception of the building. The resulting form can resonate on the exterior, but often this type of formalism is an internal phenomenon. Sequential formalism is closely related to experiential formalism in that they both attempt to direct the user experience by a way of series of frames, or rooms, and impart a narrative that conveys a feeling or atmosphere.

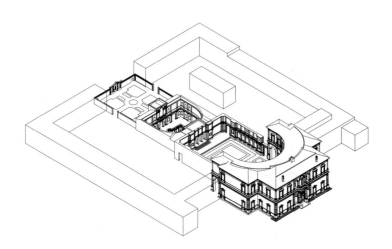

Villa Giulia, Giacomo Barozzi da Vignola
Rome, Italy 1555
[Late Renaissance, layered courtyard house, masonry]

Villa Giulia, through the use of a series of elevations and courtyards, uses a specific sequence as an architectural ideal. The courtyards are employed as a series of functions, such as a garden and a nymphaeum, whereas the elevations are a series of historical references that explain the history of architecture (see Lineages in Chapter 02). At times, the inhabitant follows a central axis, while at others is forced off the axis and placed in a position of voyeur, rather than participant in the formal symmetry of the composition. Each courtyard and gate is completely different and allows for the memory of each to figure prominently in the overall sequence.

	FORM	**251**
	PRINCIPLE	

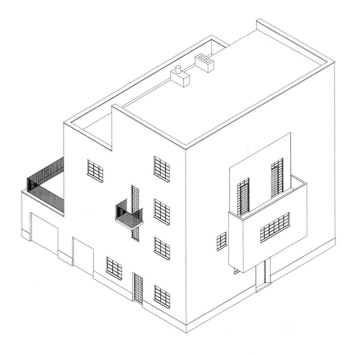 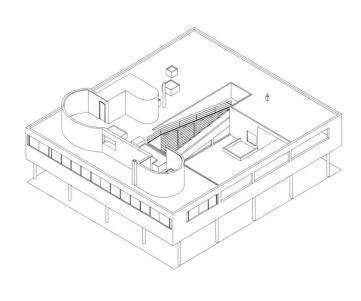

Villa Moller, Adolf Loos
Vienna, Austria 1927
[Modern, Raumplan, sectional arrangement of spaces, masonry]

The Villa Moller, by Adolf Loos, exploits the Raumplan as a planning methodology to achieve a sequential formalism. This formalism is experienced only from the inside and is actually completely denied on the exterior. Loos concentrated on this concept of the Raumplan as a spatial and formal generator with his domestic structures. It allowed the user to move in directions that were organized through section rather than axially as was more common. The center room, or foyer, on the piano nobile of the Villa Moller acts as a starting point from which the user can access any of the major public rooms of the villa. The sequential forms generate a complex and rich interior, which is however completely denied by the stark, surreal, and symmetrical exterior.

Villa Savoye, Le Corbusier
Poissy, France 1929
[Modern, free plan, concrete and masonry]

As a villa in the periphery of Paris, the sequence of Villa Savoye begins with the approach of the automobile. The automobile is drawn around the building (dictating the shape and scale of the ground floor wall arc), depositing the user at the main entrance before continuing around and parking in the garage. Upon entering the glass lobby, the user is confronted with a ramp, which is really the device around which the sequence takes place. The ramp slowly ascends to a foyer on the piano nobile that continues directly to the living room. The private rooms are nestled behind this foyer and remain private and separate from the sequence. The living room connects to the terrace through a massive sliding glass wall. From the terrace, an exterior ramp ascends to a roof garden, which includes a solarium with a picture window that acts as a **belvedere**. This sequence is carefully organized and simultaneously revolutionary and traditional.

252 FIGURE/GROUND
PRINCIPLE

Figure / Ground

05 - Figure / Ground

The rendering of figure to ground is a representational technique primary to the reading of space. Addressing the dialectical relationships between solid and void, positive and negative, open and closed, occupiable and un-occupiable, the juxtaposition of these contrasting conditions provides an alternative way of diagramming, analyzing, and visualizing architecture.

Applicable to space, material, structure, or nearly any system, the figurative inversion of the juxtaposing elements set in contrast with one another allows for a graphic clarity of the condition. The dramatic inversion of the two conditions increases their contrast and allows for legibility of a potentially otherwise subverted formal condition. As a graphic technique, the legibility of the system is dependent upon the scale of the representation (from the scale of the wall in an architectural composition to the scale of a building in an urban composition) and what is depicted as positive and negative (the wall or the space in either black or white).

Nolli map of Rome Giambattista Nolli was an architect and surveyor who dedicated his professional life to the documentation and cartographic representation of Rome. The map of Rome, drawn in 1748, was the first iconographic map to represent the city through the divide of public and private space. Showing the public realm as a continuous open figure while representing the buildings as poche, the resulting drawing illustrated the levels of ownership and control of space. It provided a reading of the city of Rome based on the privatization of ownership and the continuity of the public realm. The representation interlocked the city streets, piazzas, and detailed plans of civic and religious facilities as a continuous figure while representing privatized spaces as opaque bodies. The result was a graphic map that allowed a re-presentation of the city and its hierarchies.

Poche The term poche comes from the French word for "pocket." Referring to a method of graphic representation where the thickness of a wall or building is filled in with black to emphasize the continuity of the in-between and directly emphasize the edge of the figure, this method establishes a clarity to the balance between the geometry and the space. Prior to the nineteenth century, there existed an intuitive thickness to architecture. Lacking the quantitative engineering, structural knowledge, and material science so integrated with contemporary architectural design, architects were unable to predict the structural performance of elements, which therefore suffered material inefficiencies. Designers struggled with a need for massiveness to provide fortification for defensive purposes and a desire for permanence, and yet limited ability to address weathering. There was an endless supply of inexpensive labor due to social inequities and monarchical political systems. The associated social class system that provided an intense serviced-to-served relationship required dual systems of circulation for separate service requirements. These factors and more provide a genealogy to the categorization of poche and the relationship of scale, function, and use of wall thickness.

Positional Poche(s)

The extension of the graphic method of representation across scales and in various picture planes allows for the principles of the graphic condition to be applied across a selection of drawing types. Including both plan and section, each can be implemented at the scales of both architecture and the urban fabric.

Plan Poche The use of poche in plan at the scale of architecture allows for an engagement with the tectonic thickness of the wall. The implication of material, usage, hierarchy, and structure become evident through this representation.

Sectional Poche The sectional use of poche translates the principle of plan into the section. The result permits the graphic representation of the sectional engagement of topographic ceilings and inclined panels. The complexity of resulting cross-sectional opportunities provides for increased spatial complexities and varied vertical transitions and sequences.

Urban Poche Urban poche refers to readings of the urban fabric as specific as the Nolli plan, or more generically between the open and closed spaces of the urban field. In certain scenarios, the complexities of topographic or environmental features, in conjunction with the constructed landscape, allow for varied readings of urban density and methods of development and growth relative to geographic context (either horizontal or vertical).

Functional Poche(s)

Derived from their application, functional poches refer to the rationales for the architectural thickness that provides their figuration. Within or along this thickness, the resulting use of the mass and surface defines the architectural and spatial implications of the functional types.

Defensive Poche Defensive poche refers to the thickness determined by the need to provide defensive fortifications. The increased depth of the poche and the massiveness of the construct allows for the scale of the typically un-occupiable to expand to a scale that becomes large enough to physically inhabit. The result is the ability to create space in a subtractive manner: one carved from the mass of the poche.

Monumental Poche Monumental poche emerged from the desire for permanence at a heroic and thus intensely hierarchical scale. Monuments, such as the Pyramids at Giza, create a solidity that allows for the interior spaces to be made through a removal of material.

Structural Poche Structural poche graphically articulates the **cross-section** of the structure. As a varied articulation, the poche allows the expression of the hierarchy of the structural system within the mass of the wall as a whole. Thus the poche articulates the legibility of the tectonic and structural organization within the continuity of the enclosure.

Service Poche Service poche illustrates a thickened zone of functional spaces, governed by a class structure of a service-and-served social hierarchy, or simply denoted by the functional elements related to need and function (core) as opposed to ceremony. The density of these functions, carved out of the solidity of the graphic representation of the boundary, allows for individuated moves to be aggregated in larger super-structural figures.

Material Poche Material poche refers not only to the final material presence in an architectural project, but also the fabrication and construction sequencing engaged in the project's derivation. The Brother Claus Field Chapel by Peter Zumthor is a project derived from the solidity of its process of fabrication. Logs were propped to create an inner figure. An outer conventional **formwork** then encased the stacked log figure. The space between the two forms was then filled with cast-in-place reinforced concrete. The inner logs were then ignited, allowing them to burn within the fire-resistant concrete. The resulting fire consumed the inner logs, leaving a spatial void of the chapel space. This remainder is a material process poche.

Positive / Negative

The balance between positive and negative space is the fundamental relationship between object and space. The legibility of any form is made possible through the interrelationship of the physical formal "positive" and the resulting spatial "negative." The two together create a completed whole, defining a total that is full. In architecture this interrelationship is even more significant as the scale of building makes the formal the visual and physical artifact, but it is in fact the residual space that we can occupy. This balance between the making of object and form and the resulting spatial production of space results in an ongoing dialogue between the actual and the implied. Figure and field are visual terms derived from a graphic method of representation. Collectively they provide a method of presentation of both object and space in dialogue with one another. The method introduces the variable dialogue of mass. This can occur either at the thickness of the wall relative to the opacity of architecture or at the scale of the building when engaging the urban fabric. The solidity of the graphic denotation of fill is termed poche; the mass of the denoted "between" is the solidity that bridges the lines and blurs the articulation of the interior, representing only the impenetrable solidity and the figured edges of inner and outer. The result is a balance of what is "in," what is "out," or what is "between." The balance of these figures is primary to the interrelation of form and space.

Nolli Plan of Rome

In 1736, Giambattista Nolli began to construct the most accurate and up-to-date map of Rome. This plan was executed in response to a commission from Pope Benedict XIV. Twelve years later, Nolli had finished one of the great plans of the millennium. The "Nolli Plan," as it is typically known, is based largely on an earlier plan by Bufalini, but with some notable differences. First, Nolli changed the orientation of the map from east to magnetic north, reflecting an acceptance of north as a universal orientation for most maps. Second, the Nolli Map was much more accurate. It was so accurate, in fact, that it was the Nolli plan that was used until the 1970s by the government planning commission. Last and most important, the Nolli Map separated not only built form from open space, but also public space from private space. The Bufalini plan used the figure ground representational technique as a way of illustrating the distinction between what was built and what was open. Buildings were rendered as black solids and the open space was left white. Nolli took this separation a step further and included any public space as a white space as well. There was no distinction between the exterior space and the interior space in the map, resulting in a diagram of the city that explored new ideas of space and ownership. It represented a completely new way of conceptualizing a city. Suddenly the city was no longer merely represented by buildings, streets, and piazzas. Now the city could be seen as an interconnection of space including both interiors and exteriors. Hidden spaces, such as courtyards and alleys, suddenly began to contribute to the knowledge of the city. Nolli was also instrumental in allowing typography to show in the plan, thereby acknowledging the famous hills of Rome and showing how they contributed to its planning and organization.

Plan Poche

Plan poche occurs at the scale of the wall within a building or at the scale of the building within the city. Its articulation denotes the difference between the void of the publicly accessible and the privatized interiority. Plan poche creates boundary, prevents access, and creates opacity. The depiction of the graphic boundary emphasizes the form and figure of the resulting spaces created. The balance between the two is the mediated representation of space—one that is three-dimensionally conceived but two-dimensionally depicted.

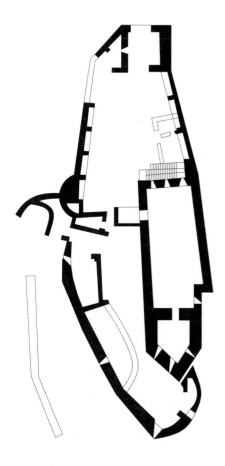

Medieval Castle - Castle of Lastours
Lastours, France 13th century
[Medieval, defensive towers, masonry]

The towers and fortifications at Lastours represent a typical plan poche situation. The castle was constructed exclusively of stone, reflecting the technology of the time. As a result, the walls were extremely thick to accommodate the height, stability, and defensive capability. The tower keep was the most defended area of the castle and as such was the portion of the castle with the most massive poche. The poche of the walls represents the innately defensive and protective nature of these structures. To achieve the necessary height, the walls had to be a certain thickness, hence the large mass that allowed the spatial consideration of architectural poche.

FIGURE/GROUND PRINCIPLE 261

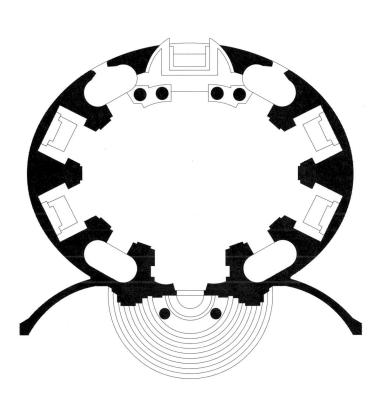

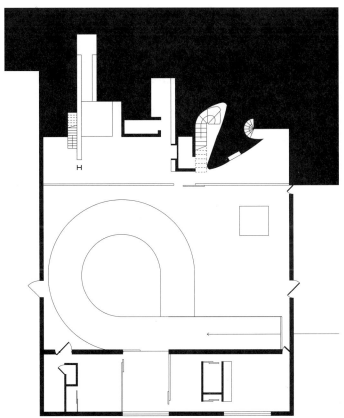

Sant'Andrea al Qirinale, Gian Lorenzo Bernini
Rome, Italy 1678
[Baroque, short-axis elliptical plan, masonry]

Sant'Andrea al Qirinale uses plan poche as a combination of space and structure. Surrounding an oval sanctuary is a series of chapels embedded in the poche of the thickened wall supporting the dome and the church itself. The removal of the chapels from the sanctuary had previously occurred; however the method of sinking them into the wall was unique, resulting in an ingenious and dynamic solution. The chapels themselves are focused not on the center, but instead on the two foci of the oval, allowing for a multi-centered space that has a direct relationship with the poche.

Bordeaux House, Rem Koolhaas
Bordeaux, France 1998
[Postmodern, additive and subtractive spaces, diverse materials]

The Bordeaux House in its organizational structure subscribes to the same principles as the Maison de Verre, juxtaposing the heroic scale of the large, open, primary living space with the subtractive and independently formed service spaces. The relationship of the free plan to the carved organicism of the poche establishes contrasting spatial scales and formal environments. The balance of the plan poche, hyper-figured through individuated forms, is juxtaposed with the open field of the free plan, providing a spatial and formal contrast.

Sectional Poche

Sectional poche extends the sculpted figuration of wall, illustrated in plan poche, to include the relationships of ground (floor) and ceiling. The result is a sculpting of space in the Y-axis, permitting the dynamic expression of vertical space. Here the divergence of the wall planes described in plan now have the opportunity to engage all planes allowing for transitional ramping levels, topographic ceilings, and walls that engage the body and space to a thickness that can allow occupation. The significance of poche is the ability for the two surfaces of a single solid to be divergent, allowing independent governance to their form and rationale.

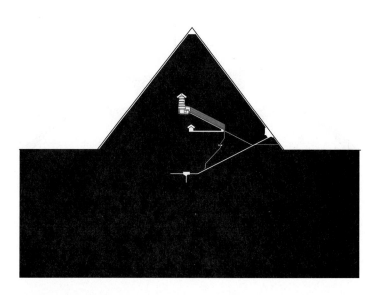

Pyramid at Cheops
Cairo, Egypt 2560 BCE
[Egyptian, pure Platonic pyramidal form, masonry]

The Pyramid at Cheops is a combined product of its material and function. As a tomb for the pharaohs, the monument to their lives was intertwined with their religious beliefs about the transference of the soul to the afterlife. The elaborately constructed pyramids use massive, dry-stacked masonry units. The result of the construction is a form with a largely solid middle. A series of chambers located within the larger pyramids houses the mummified remains of the pharaoh himself with additional spaces for his servants and a collection of riches and artifacts. A series of tunnels offers air and circulation access for the priests and workers overseeing the entombment. The section is a dramatic massive poche with very intimate voids nested deep within.

| | FIGURE / GROUND |
| | PRINCIPLE **263** |

Notre Dame du Haut, Le Corbusier
Ronchamp, France 1955
[Expressionist Modern, free plan church, masonry and concrete]

Ronchamp is a ground-breaking synthesis of modernist principles, expressionist forms, and programmatic functions traditional to the church. Using a combination of construction techniques including massive, stucco-clad, masonry walls built of the rubble of a church previously on site and cast-in-place concrete to form the figurative roof, resulted in a structural poche of the walls that allowed and required the choreography of mass. Cleverly deploying the thickness of the side wall, Le Corbusier developed an aperture system that varied the dimension of the opening on the interior and exterior wall, resulting in truncated pyramidal projections between the two. This funnel through the mass of the wall, in combination with colored stained-glass inserts, creates an effectual wall that both projects an ecclesiastical, referential quality to the light while dissolving the massiveness of the wall itself.

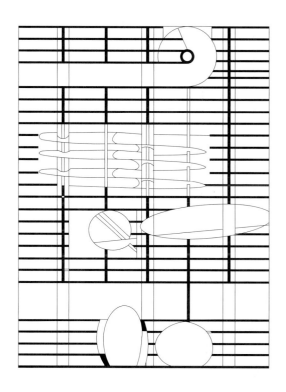

Très Grande Bibliotheque, Rem Koolhaas, OMA
Paris, France 1989
[Postmodern, cubic book field with subtracted spaces, unbuilt]

The Très Grande Bibliotheque deploys the conceptual and graphic sensibility of poche to address the programmatic necessities of the building. Translating the mass traditionally assigned to the constructed thickness of the building itself, Koolhaas extends the visual technique to include the book stacks. Using the solidity of the book and the density of the tightly stored shelving system, the building is conceived as a uniform solid of storage. From this "mass" the primary spaces are subtracted. Figuratively shaped, each of the ceremonial functions are translated into discrete forms that then carve themselves from both the solidity of the poche and the boundary of the figured stacks themselves. The result is the subtractive operation of making space. By first filling the building and then choreographing the careful removal to define the publicly accessible spaces, the building celebrates the practicalities and actualities of its function while engaging a uniquely "negative" space-making methodology.

Elevational Poche

Elevational poche refers to a unique but powerful method of subtractive determination of the façade. Often evident through an elevation's material, depth, and shadow or relative position to its contextual fabric, this type of poche can occur on many scales and levels. The thickening of the elevational poche allows for a layering of depths and dimensional recessions, which in turn allows for a contrast that can be engaged as part of a specific conceptual idea or performative concern, such as climate or materiality. The depth of shadows that are created makes the building more three-dimensional, increase its reading and monumentality, and produce a space within the surface of enclosure.

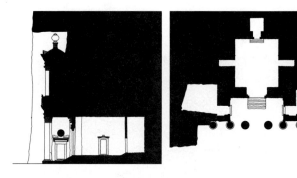

Treasury at Petra
Petra, Jordan 100 BCE
[Eastern Hellenistic, temple, stone]

The Treasury at Petra is one of the iconic examples of elevational poche. The façade of the building is literally cut out of the mountain that surrounds it. This elevation is completely monolithic, heightening the reading of the continuity, depth, and subtractive figuration all innate to elevational poche. The contrast between the rough stone of the natural rock and the delicate carvings of the classical ornament intensifies this extraordinary moment of architecture. The building at Petra, through the use of elevational poche, questions the typical assumptions and formal legibility of both construction and structure.

FIGURE / GROUND
PRINCIPLE
265

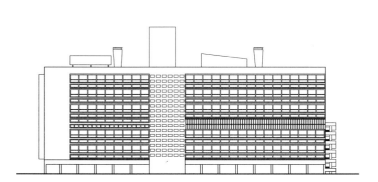
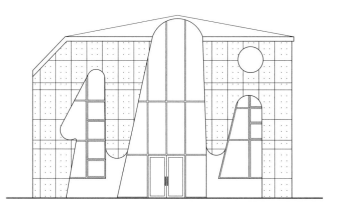
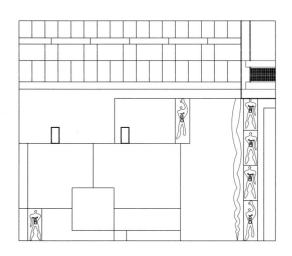
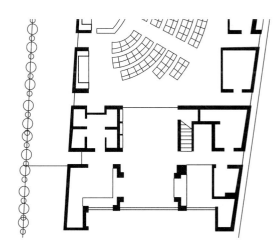

Unite d'Habitation, Le Corbusier
Marseille, France 1952
[Modern, double-loaded corridor, concrete]

The Unite d'Habitation building uses elevational poche at a variety of scales and depths. The largest and most striking gesture is the vast repetitive field of the brise-soleil. This external frame is constructed of cast-in-place concrete that intrinsically engages a positive-negative relationship of formwork and cast figure as doppelgangers of one another. The continuity of the resulting monolithic reading creates an illusion that the collective cellular frame is carved from a single giant block of concrete. The resulting shadows and deep occupiable space that are created provide protection from the sun, but they also contribute the overall complexity of the building, ultimately being responsible for the image of the building. At a smaller scale, Le Corbusier embedded the negative of the modular man and other imprints in the surface of the lower façade as ornamental and proportional elements.

Harajuku Church, Ciel Rouge Creation
Tokyo, Japan 2005
[Postmodern, open plan church, steel and concrete]

The Harajuku Church carries the section of the sanctuary through the glass wall of enclosure to directly engage the space of the sidewalk. This extrusion results in a clear and effective elevational poche that makes apparent the most important articulated component of the church: the particular section that results in a two-second delay of sounds within the church. By carrying this element out of the façade through an elevational poche, the architect has managed to highlight the critical aspect of the building as a formal, effectual, and spatial generator.

Urban Poche

Extending the representational and conceptual technique of poche to the urban condition, the combination of the subtractive method of space-making with the scale of the urban operation allows for complex collective opportunities for space-making. Levels of construction technology, amounts of subtractive and additive processes, and rationales for the use and resulting spatial methods (for instance, climatic, geometric, geologic, and defensive) vary with deployment. The resulting urban spaces with passages, collective spaces, and diverse spatial modules (all formed subtractively out of the urban fabric), present a unique method of space-making and, as a result, an equally unique spatial reading and experience. The Nolli Plan, as previously discussed, is a representational reading of the mass of the city where the impenetrable nature of delineated public and private space is read as open or closed. The subtractive spatial tectonic at the urban scale is selectively represented through the following examples.

Goreme, Cappadocia, Turkey
7th century BCE
[Anatolian, cave town, dug out city, subtractive earth]

Although referred to as an underground city, the communities of Cappadocia likely served as temporary shelters rather than as permanent hidden cities. The initial pre-metal tool caves were hewn early and expanded by the Christians escaping invaders. Defined by discrete entryways that give way to elaborate air and waste shafts, along with wells, chimneys, and connecting passageways, the collection defines a massive, urban, subterranean complex. The upper levels were used for living quarters; the lower levels were used for storage, wine-making, flour-grinding, and worship in simple chapels. Everywhere, walls have been blackened from the use of torches. The collective complex used an entirely subtractive method to create the space desired for occupation, allowing localized form-finding through dimensional necessity or functional practicality.

Gila Cliff Dwellings
Silver City, New Mexico 1300
[Mongollon, wall dwellings, masonry]

The Gila Cliff dwellings of southwestern New Mexico are a series of interlocking buildings that form a community built by the Mongollon peoples within the cliff alcoves. Built as a combination of subtractive spaces, found or carved within the rock surface, and additive spaces, built onto the natural condition with masonry units, the entire compound boasts over forty-six rooms housing ten to fifteen families during its peak. The architectural method of engaging the natural condition and building onto it allowed for both subtractive and additive methods to be deployed. The resulting composition is one of true urban poche, hybridizing the natural organic condition of the cave with the rationally planned orthogonal additions of the constructed.

Barrio Troglodyte, Guadix, Spain
16th century
[Troglodyte, dug-out city, subtractive earth]

The Guadix Area is typified by a large portion of the town living in underground cave houses. Known as Barrio Troglodyte, the subterranean dwelling developed as a combination of the climate (the natural insulation properties of the ground allow for an escape from the Granadin summer heat) and the geology (the soil allows for easy cutting while remaining stable). Massive chimneys were erected to bring fresh air and light into the subterranean spaces. Entryways and towers were typically painted white to demarcate their presence, provide a formality, reflect light, and establish a constructed hygiene to the natural spaces. The subtractive method allows for the creation or expansion of the dwelling through excavation—a conceptual and literal removal of material to define space. The collective urban composition creates a subterranean complex that is never fully visible, but is uniquely formed and spatially derived from its construction method.

Military / Defensive Poche

Military or defensive poche was commonly developed by necessity as a method of stopping projectiles during combat. Designed and orchestrated to halt attackers, it became the foundational formal methodology in the development of many early cities or military posts. It operated under the simple idea that a thicker, taller wall was more difficult to scale or penetrate. As a result, the mass of the wall became an architectural element. Fortifications began to take on their own architectural quality due to their size and materiality. The subsequent designs resulted in structures that tended to have long life spans as their sheer size allowed them generations of use and weathering and prevented their deterioration. The flak towers of Berlin (as representational artifacts of a whole generation of World War II military bunker structures built in Europe and the United States) stand even today as seemingly indestructible figures.

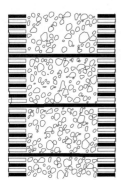

Roman Wall
Rome, Italy, 378 BCE
[Roman, defensive wall, masonry]

The walls of Rome were constructed during two stages. The Severian Wall was constructed by the Roman King Servius Tullius, while the Aurelian Wall was constructed under the Aurelian emperors. These walls were an average of approximately twelve feet in thickness and twenty-four feet in height. Encircling the city of Rome, their design and presence provided protection for hundreds of years. Their mass reflected their function in both a literal and abstract sense. From a functional standpoint, the walls were difficult to scale and were effective in their ability to withstand bombardment. In the abstract sense, they symbolically represented the power of Rome to any potential invaders. There were numerous gates that penetrated the walls at critical points of the city road system. In many places the wall was hollow, allowing it to also act as a circulation route from one guard tower to the next.

		FIGURE / GROUND	
		PRINCIPLE	**269**

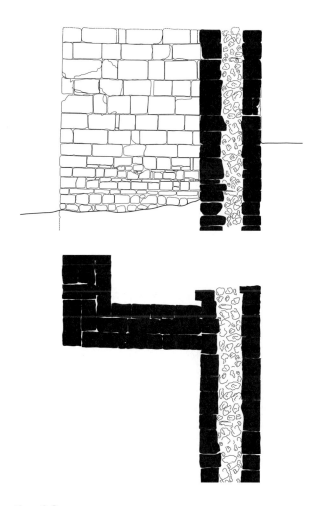

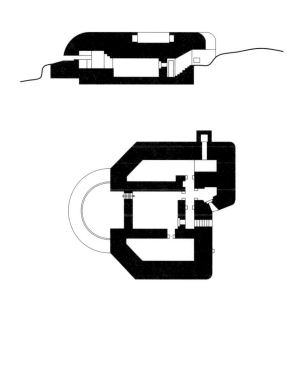

Castle of Canossa
Reggio Emilio Romagna, Italy 940
[Medieval, Castle, masonry]

There are many types of castles, but they were most prevalent during the medieval era when localized security was essential. The mass of the structure, developed for defensive purposes, creates a defensive poche typically focused on the exterior walls. These walls served as the main defense perimeter and were therefore thicker and taller for protection. These castles were typically designed around a courtyard surrounded by a large, thick wall connecting rounded towers in the corners. These towers and the accompanying walls were the elements that developed the use of poche. Subtractions were carved from their mass for defensive purposes to include moments for archers, lookouts, movement, or escape. The interior walls were typically much thinner and did not employ poche as a design element.

Bunker
Brittany, France 1940
[Military, semi-subterranean shelter, reinforced concrete]

During World War II, many bunkers were constructed throughout Europe. The German Army constructed the majority of these bunkers along the Normandy coast. These bunkers were constructed of massive amounts of cast-in-place, reinforced concrete. They were designed to withstand direct bombardment of projectiles and bombs. Typically they were semi-subterranean with openings for guns or cannons. The poche in these structures was slightly different than in early fortifications in that now steel was being used as a way of strengthening the walls. This type of poche, fueled by more technical engineering, increased material understanding; more sophisticated programmatic needs allowed for a more plastic approach to design. These bunkers were built to such specifications that even today many are still standing.

Monumental Poche

The use of the many of the aforementioned poche types can be reclassified as monumental poche when deployed for the production of a monumental effect. Referring to structures developed for their formal, commemorative, or cultural significance, they are often tied to the identity of a particular city, event, or political movement. In these projects, the poche is deployed as a scale-and-mass device to create a large and imposing element that immediately announces and demands attention. This poche provides a sense of grandeur and power that represents its significance. Often these buildings were designed for other purposes and over time have in fact come to represent other entities. Edinburgh Castle, for example, began with the use of defensive poche for functional purposes but now is considered an example of monumental poche as it has become an icon of the city.

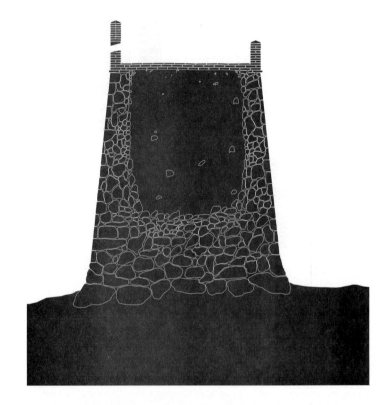

Great Wall of China, Qin Shi Huang
Northern China, 220 BCE
[Chinese, defensive wall, masonry]

The Great Wall of China extends over eight-thousand kilometers across northern China. It was originally built as a defensive wall to protect northern China from invading nomadic tribes. The wall is approximately twenty-feet thick and often reaches heights of thirty feet. As the largest structure ever built on earth, it is a singular piece of monumental poche and has become the representative symbol of a nation. Unlike many poche projects, which are often hollow on the interior, the Great Wall of China is completely solid.

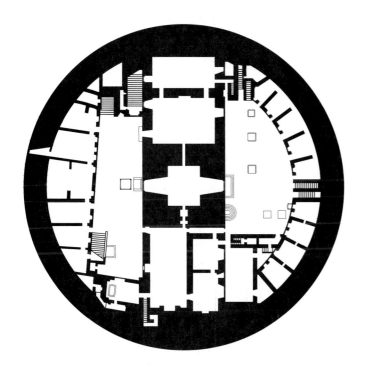
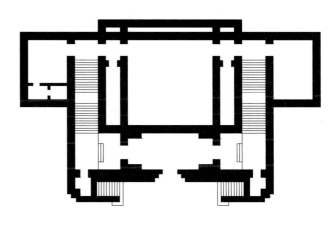

Castle Sant Angelo, Hadrian
Rome, Italy 139
[Classical, Mausoleum, masonry]

The Castle Sant Angelo, located near the Vatican in Rome, represents an example of monumental poche similar to Lenin's tomb in that the Castle was formerly the mausoleum of the Emperor Hadrian. In this building, the majority of the poche takes the form of the massive base that supports the drum. The original building was faced in white marble and its roof was planted with cypress trees, creating the typical "cypress hill" that is synonymous with burial. Over the centuries, the complex has changed to accommodate many varied functions, including a fort, a castle, and a residence for the pope. Within the castle are massive structural walls and dark rooms that inspired Piranesi to construct his famous "Prison" etching series.

Lenin's Mausoleum, A.V. Scusev
Moscow, Russia 1930
[Constructivist, tomb, marble]

Lenin's Mausoleum in Red Square is an example of monumental poche as it relies on the ziggurat (a typologically monumental model) as a symbol of its function as a burial mound. The monumentality of this building is readily recognized by the building's scale, materiality, and location. In reality, the building is not as massive as it seems. Instead, a tortuous circulation path leads the visitor to a below-grade chamber where the body lies in state. The hallways leading to the burial chamber actually comprise most of the poche. The monumentality is increased by the fact that the building was also used as a rostrum for the leaders of the old Soviet Union during state events and parades.

Structural Poche

Structural poche refers to the thickened wall or floor to provide the necessary material mass of the building to perform structurally. Determined through the engineering of the physical forces acting on the structure, the shape and scale is determined scientifically. The resulting scale and mass is then architecturally addressed through its poche. Considering the diverse components and segmental construction necessary to accomplish the architectural member, the poche allows for the unification of the assembly system to present a collective mass and continuous profile to the aggregation. The structural poche is determined by necessity but engaged and celebrated from a visual, compositional, and ultimately descriptive architectural agenda.

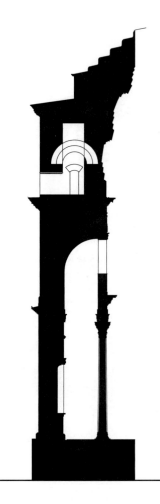

Pantheon
Rome, Italy 126
[Classical, temple, masonry]

*Built with a massive concrete dome concentrated on a ring of **voussoirs** that form the oculus, the dome is supported on eight massive barrel vaults that transfer to eight massive piers in the drum wall. The tapering thickness of the concrete dome, in conjunction with the coffers, removes the weight of the heroic span. These removals from the ceiling and interior walls are done to remove weight or simply to eliminate mass where not structurally needed. The resulting voids in the perimeter wall are translated into ornamental apses for decorative and commemorative statuary. The massive wall required for the self-supporting structure illustrates the carved poche of the structural mass, making a legible expression of the design and engineering principles at play.*

FIGURE / GROUND PRINCIPLE 273

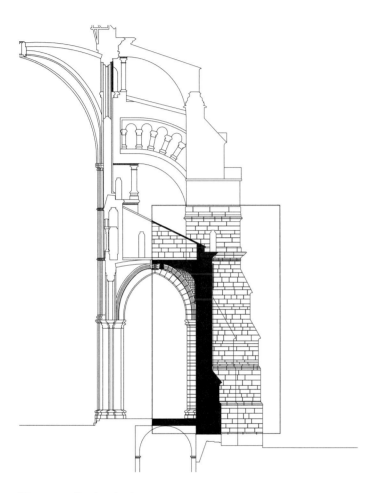

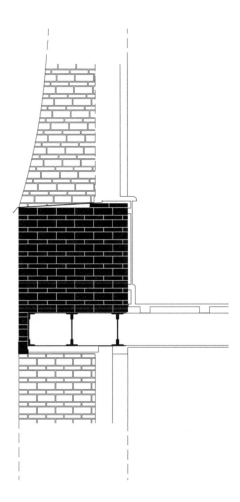

Chartres Cathedral
Chartres, France 1260
[Gothic, Latin Cross, stone masonry]

Chartres Cathedral represents the pinnacle of the High Gothic period. Using the Latin Cross in plan, the extension of height and the dissolve of the wall to introduce light translate the solidity and mass of the wall to a delicate tracery of skeletal structure. The increased height of the nave reached a structural limitation with the associated wind loads. As a result, the secondary external structure of the flying buttress (that was to become a hallmark of all Gothic cathedrals) was developed to provide the necessary lateral bracing. With the wall now articulated as a structural skeleton, the large glazing infill allowed for greater lighting effects to illuminate the spiritual spaces as well as assume a responsibility for telling biblical stories through the extensive decorative stained glass. The development of the rose window at the head of the transept allowed resolution of the roof vault while giving hierarchy to the tripartite façade.

Monadnock Building, Burnham and Root
Chicago, Illinois 1891
[Richardsonian, double-loaded linear office building, masonry]

The Monadnock Building is often considered the first skyscraper. It is sixteen-stories high but was built using traditional load-bearing masonry construction. The building's structural walls are constructed of brick. The combination of the building's height with the compressive loads dictated six-foot-thick walls at the base. The thickness of the wall was maximized to optimize a balance between the building's overall height and the physical crushing of the material. To accommodate this thickened base, the construction method resulted in the graceful, curved, vertical lines that become the visual identity of this building. Due to the mass of the wall, the lobby, traditionally one of the largest and grandest spaces in these buildings, is here a cramped but still beautiful marble hallway. Its limited size, light, visual openness, and physical access are all by-products of the necessary structural mass of the wall itself. This project illustrated the maximum limitation of load-bearing masonry as a high-rise construction technique.

Indeterminate Poche

Indeterminate poche is a combination of structural, material, and defensive poche, deriving from their collective performative responsibilities, but executed in an era of indeterminacy. Constructed in a time before physical analysis of structures was possible, these projects are dominated by rules-of-thumb, intuition, and tradition. Unable to physically detail to prevent weathering, precisely design and optimize for weaponized military assault, and precisely predict the material properties or performance, these indeterminate poches were simply made massive to ensure their performance. Their scale is dependent upon the construction technology and an overcompensation in size to allow for inaccuracies in the specific performance. The wall could lose a few inches due to erosion and spauling and the structure would not be impacted. The inconsistency of handmade or hewn materials could be overcome by simply building in redundancy. The projects form their mass through a predictive intention, but one ultimately cloaked in indeterminacy.

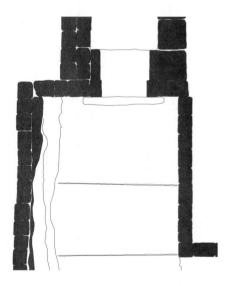

The Lion Gate of Mycenae
Greece 1250 BCE
[Ancient Greek, ornamental relief above lintel, masonry]

Built as the entry gate to the city, the Lion Gate is particularly significant for the bas-relief ornamentation carved in the triangular stone atop the **lintel**. *Representative of the general structural forces in overall shape, its presence is highlighted through the ornamentation of its surface. The Lion Gate uses a traditional massive post and lintel to create an opening in the protective stone wall. Created of massive dry-stacked stones, the walls are loosely shaped and rugged in appearance. The quantifiable strength and stability of the wall was not predicted through any mathematical calculations but rather through the traditions, availability of raw materials, and the scale of the construction emerge as determining factors. The overestimation in all of these aspects allowed for a buffer to deal with unforeseen weathering, and structural and defensive requirements.*

FIGURE / GROUND	
PRINCIPLE	**275**

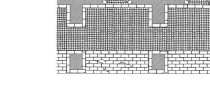

Aurelian Wall
Rome, Italy 275
[Roman, clay masonry military wall, masonry]

The Aurelian emperors' construction of the walls of Rome was an emergency measure responding to the Barbarian invasion of 270. Built as massive defensive figures, the walls were constructed using traditional Roman techniques. Instead of a solid wall of brick or stone like the Lion Gate of Mycenae, the Roman wall was a facing of elongated, clay, thin "Roman" brick filled with rubble, mortar, and packed clay. The hybridized construction of the wall allowed for a refined surface and an intrinsic flexibility to the width of the wall. The resulting thickness of poche was sized to over-accommodate the structure, defense, and weathering necessary to ensure stability and longevity. The determination of the thickness was not predicted with the use of physics and mathematics, but instead, proportion and estimation were deployed, making it an indeterminate poche.

Medieval Wall
Canterbury, England 1050
[Medieval, massive defensive fortification, masonry]

The Medieval wall, as typified by the Canterbury Wall, was built primarily for defensive purposes. Made massive to withstand the siege of invaders, the design is entirely protective. Large enough to be occupiable both within as well as on top, the thickness is accomplished through multiple **wythes** of parallel masonry with infill. Its design was for strength and permanence. Crenellated on top, to provide protection for and from archers, and anchored by corner towers for added hierarchy, range, and vantage, the position and configuration provides a barrier. The engineering was done through practical knowledge and tradition, not predictive mathematical analysis. As a result, the section, both in terms of material construction and building performance, is an indeterminate poche.

Hierarchical Poche (Service / Served)

Hierarchical poche emerged from the need for the organization of a building's service elements to remain hidden within the public spatial sequence of the plan. Service poche can have multiple configurations including two of its most common: a centralized solid or a linear-edge element. It is critical as a functional plan-making element, organizing both use and form through a conceptual graphic technique. The juxtaposition of this service "fill" against the hierarchy of the public rooms allows a heightened relationship of these interdependent parallel systems.

Villa Rotunda, Andrea Palladio
Vicenza, Italy 1566
[Renaissance, villa, masonry]

In the Villa Rotunda, Palladio hides the functional service components in the poche and mass of the building's base. In his comparison of buildings to the human body, Palladio suggested that the more pedantic parts of a building (service functions) should be hidden away in the same way the base parts of the body are also hidden. In the Villa Rotunda this is illustrated on the main floor where the four stairs, which are situated symmetrically around the rotunda, are all subsumed within the poche. This hierarchical poche was typical of Palladio's villas and palaces.

 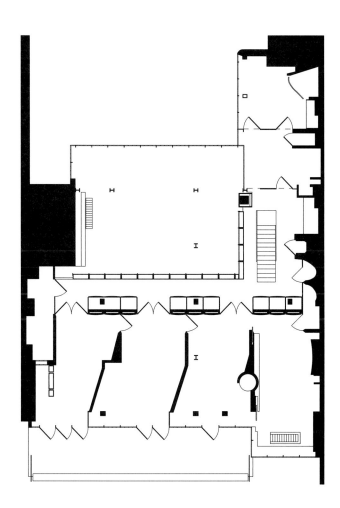

Homewood, Edwin Lutyens
Hertfordshire, England 1901
[Arts and Crafts, House, masonry and wood]

Like many of Lutyens' residential projects, Homewood utilizes the concept of poche in a way that allows the service elements to remain hidden within the interior. Functional necessities such as stairs, kitchens, and bathrooms are removed from view to allow the hierarchical or served rooms to be readily recognized and the service areas to recede into poche. This balancing of service and served relationships is used throughout Lutyens' career and can be seen in many of his residential projects. This expression of hierarchical poche is typified by the aggregated banding, or gathering, of service elements as an organizational and compositional effort to remove them from sight.

Maison de Verre, Pierre Charreau
Paris, France 1931
[Modern, additive and subtractive spaces, glass and metal]

In the Maison de Verre, Pierre Charreau used hierarchical poche on the side party wall of the infill house. This edge is thickened and then carved with removals that become curvilinearly defined, service-based support spaces. This spatial condition, though common in historical examples, was revolutionary in this modernist context. Like Renaissance architects, Charreau chose to place the service elements (bathrooms, elevators, and closets) within this poche. Despite this formal articulation, these elements were not base in the sense of their design as they were given equal attention and articulation. It was instead their function and cultural hierarchy that regulated them to be part of the poche.

Material Process Poche

Material process poche refers to the intrinsic thickness of the physical material and the process by which something is built. Defined either through the solid-to-mass relationship of the material itself or the void-to-surface of an assembled system, material process is the visual rendering of the segmented and tectonic as a continuous spatial boundary. Lumping the componential and individuated nature of constructed and aggregated pieces into a singular and unified whole, the graphic process is effective in revealing the macro-scaled intention of a design. It reveals the implication of a material process on the outcome of the final form and ultimately the final spatial legibility produced by the form.

Main Temple at Chichen Itza
Yucatan, Mexico 250
[Mayan, stepped pyramid with axial staircase, masonry]

The Main Temple at Chichen Itza uses a primitive technique of massive, stacked, cut stones with astounding levels of accuracy. Studded with intricate shallow ornamental carvings, the articulation is largely topical. The overarching form develops from terracing at various scales. It occurs in the massive formal steps of the stone coursing and the anthropomorphic scale of the centralized stair that repeats on each of the four faces. Achieving intense levels of precision and control over massive units, the tolerance of construction is incredibly small, accomplishing impressive accuracy even by contemporary standards. The resulting material poche that is created through the construction technique of aggregated and stacked terracing units creates a solidity and permanence to the composition. Emerging out of the technique by which the temple was made, the resulting formal associations of solidity, mass, and material create a poche of section. The solidity of the surface defines a synthesis of material and program.

FIGURE / GROUND PRINCIPLE 279

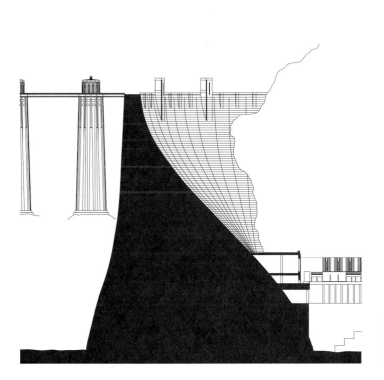

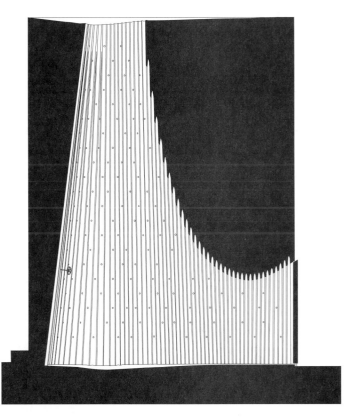

Hoover Dam, Frank Crowe
Black Canyon, Boulder City, Nevada 1936
[Modern, concrete dam for hydroelectric power, concrete]

Hoover Dam is defined by the massiveness of its scale. Materially determined by the engineered responsibility of structural continuity and impermeability, concrete as a "solid" material (one that joins tectonically on the microscopic scale through the chemistry of hydration) provided the necessary final material properties. The primitive nature of concrete as a material that is mixed from its raw ingredients and thus "made" on site, allowed for the necessarily massive scope and scale of the undertaking through sequential construction. The overall shape of the dam arose from the geometric functionality of physics in conjunction with the natural context of the rock walls and water flow. The heroic scale of the undertaking and the material selection required a careful understanding of construction—so much so that the exothermic reaction of hydration required a complicated system of material delivery and cooling. The resulting material poche defines the thickness of the cross section, allowing for the creation of functional access spaces to be nested within its mass.

Brother Claus Field House, Peter Zumthor
Wachendorf, Eifel, Germany 2007
[Postmodern, material process formed chapel, concrete and lead]

The Brother Claus Field House is a contemporary worship space that emerged out of the limited resources and specific construction skills of the remote community it was to serve. Founded in process, the chapel is an organic plan formed by the construction logic. Derived from a series of lean-to logs that define the organic perimeter form as a series of tangential points, the outside is formed conventionally with smooth faceted surfaces. The space between the two was then filled with layers of concrete, determining the material poche. The inner logs were then burned out of the figure to reveal the void of the sanctuary space and produce both a scalloped surface to the walls and a distinct charred patina. The blunt honesty and informal legibility of the material process of construction produce a simultaneously abstract yet primitive form and experience. The form of the poche is derivative of the process by which it was made and the material logic of its construction, typifying it as an example of material poche.

Programmatic Poche

Programmatic poche refers to the solidity formed by the thickened mass of functional support spaces. Present in every building, the functional service-based program requirements assume different organizational configurations. Depending upon organization, configuration, and aggregation, the use of program to create formal densities introduces programmatic poche. The use of the density of program allows for planametric or sectional programmatic poche, but each is dependent upon the practicalities and functionalities of the service programs.

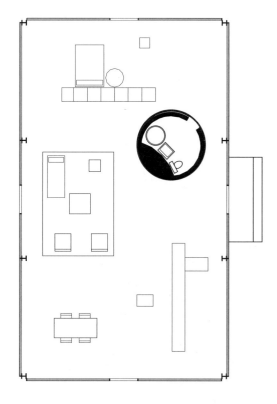

Glass House, Philip Johnson
New Canaan, Connecticut 1949
[Modern, free plan, steel and glass]

In The Glass House by Philip Johnson, the free plan is accomplished through the singularity of the space. A large single room, divided only with one "core" containing the bathroom and fireplace, allows for a continuity and consistency to the space. Subsequent functional subdivisions are made through furniture position, orientation relative to the site, and the core and the physical activity of social rituals. The effect of the free plan is further emphasized through the uniformly transparent façade. The glass dissolves the boundary between inside and out and extends, orients, orchestrates, and composes the interior zones. The material allows for a merger between house and site, commingling and interlacing their dependence and experience. The programmatic poche comes from the solidity of the core box containing the service elements, contrasting the dense "closed" with the "open" free plan of the remaining space.

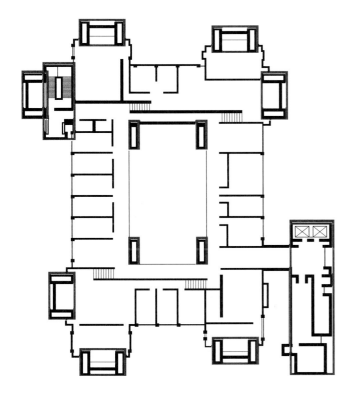 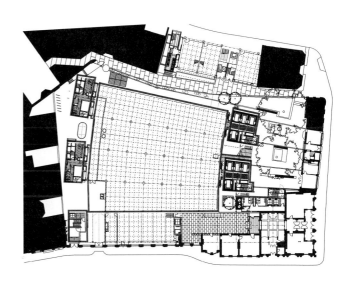

Yale School of Architecture, Paul Rudolph
New Haven, Connecticut 1963
[Brutalist, centralized architecture school, concrete]

The Yale School of Architecture is typified by its Brutalist form and material. Monolithic in appearance, the visual effect is furthered by the continuity of the concrete surfaces and the textured bush-hammered finish. The dense fortress quality of the composition is expressed through the programmatic aggregation of the service functions around the perimeter. The result is a largely opaque series of vertical tower-like figures surrounding the central open studio spaces. The programmatic poche produces a planametric, sectional, and elevational formal effect through the gathering of specific functional programs as an encircling perimeter that produce an opaque, articulated, functionally driven exterior.

Lloyd's of London, Richard Rogers
London, England 1986
[Hi-Tech Modern, free plan with encircling services, stainless steel]

Lloyd's of London derives its performative form through the organizational plan diagram. The building's functions are dispersed to the edges, allowing for an open floor plate that provides maximum flexibility. The peripheral functional location inverts the traditional centralized core and eliminates the dependence upon skin, instead presenting the functional workings as the formal determinates. Uniformly clad in stainless steel, each function is personified through highly functional yet descriptive forms. The resulting reading of the building is one that expresses the anthropomorphic scale of the individual functions while aggregating their repetition and collection to create a systematized composition derived from the practicalities of need. The plan reveals the density of the edge and the exterior zone of programmatic concentration as a peripheral poche that contrasts with the centralized open floor plate of the flexible office spaces.

CONTEXT

Context

06 - Context

Context is the circumstances and setting in which an architecture is produced, sited, or experienced. It refers to the physical surroundings in which a building will be built, the adjacencies of geometry, scale, material, programmatic use, and vernacular language, all of which have the opportunity to influence design.

Diverse elements define categories of contextual influence. Natural context references site, nature, and environment, for instance. Urban context references the forms, geometries, scales, and hierarchies of an urban site. Historical context references the vernacular and typological influences of a building within its historical lineage. Material context references the impact of material, process, and tectonic. Cultural context references the collective beliefs and cultural influences that can inform a design. Each of these categories relies upon the relative positioning of a building against the lineage, tradition, history, and existing conditions within which one must operate and be compared. These categories are more fully described below.

Natural Context Natural context refers to the ecological and topographical conditions in which a building exists. Physical location and position (in a valley, for instance), adjacent natural features such as rivers, lakes, and trees, and natural resources readily available to the site or region (for material- or craft-based decisions) fall into this category. Additionally, orientation, exposure, wind, water, and temperature act as influences. Often discussed as site-related issues, all of these are the natural aspects of context.

Urban Context Urban context refers to the built environment in which a building exists. Engaging all the issues of urbanity, the localities of the site's urban condition (legible at varied scales) define its context. Location, setbacks, geometries, alignments, and the engagement or disengagement with each of these existing systems determine the contextualism. The formal, geometric, scale-based, and hierarchical decisions the urban fabric provides through practical, legally mandated, or simply historical rationale become the framework within which an addition must operate and exist.

Historical Context As do other contextual categories, historical context refers to the lineage of tradition and the previous responses to a particular condition. However, historical context also refers to specific formal resultants based on geography, function, material, and construction techniques, for instance. Thus historical context engages the traditions of vernacularly, typologically, materially, and tectonically driven decisions within the broader auspices of their development and historical significance. This context is engaged not only through the physicality of site position, but also through the functional use, indigenous material, and tectonic traditions and forms relative to a broader understanding of lineage.

Material Context Material context relies on the tectonic and material expression of a building relative to the larger historical lineage of construction. Thus a brick building is evaluated against the lineage of brick buildings that came before. Similarly a brick building can be evaluated against non-brick (e.g., steel, wood, plastic) buildings of the same era or historical lineage to gain understanding through both contrasting and similar decisions made on the projects. Material context is founded in the impact of material to tectonic and assembly to form. The technological capability, tectonic expression, performative design, and formal resultant are all aspects of material context.

Relating piece to joint, unit to field, field to function, and part to form, the interrelationship of these decisions and their historical relativism are the basis for material context.

Cultural Context Cultural context is the broadest, and thus perhaps the vaguest, of contextual categories. Not focused on localized physicalities relative to spatial adjacencies, but instead relying on cultural relativism of an era, its technology, and a general social position in time, cultural contextualism refers to the larger position within the stream of humanity. Trends, attitudes, perceptions, technological abilities, and social structures all determine the intricacies of the moment. Cultural context attempts to capture this moment and manifest it in the constructed artifact of the building. Poised to venture dangerously close to fashion, an essential cultural contextualism captures the spirit of an era through a careful calibration of its decisions.

The approach to context, and thus an architecture founded in contextualism, is dependent upon the designer's response. Most importantly, it requires research and understanding of the existing conditions, both overtly physical and formal, but also hidden, historical, and cultural. Stemming from a combined understanding of the history of architecture, the geography and nature of place, and the cultural and social condition in which a building and its complex functions will be built, this knowledge establishes the baseline of that which exists. Contextualism demands a celebration or denial of existing trends. The designer can choose to deploy contextual conditions at face value, redeploy them with an ownership and intelligence, or simply ignore them. Modern contextualism comes from the idea of understanding the contextual condition and then deploying a complimentary attitude based in engagement, abstraction, and extension of the opportunities presented by that condition.

Natural Context

Natural context presents itself through the natural factors of a site, such as biological systems, indigenous plants, rock and soil types, climate, weather, orientation, water, and topography. Each of these factors presents conditions and opportunities for the collaboration of place with design. Drawn from functional responses, indigenous traditions, practical associations, position, material, and scale, for example, each of the various inputs presents the foundation of relationships that create a dialogue of form and the locality of the natural context.

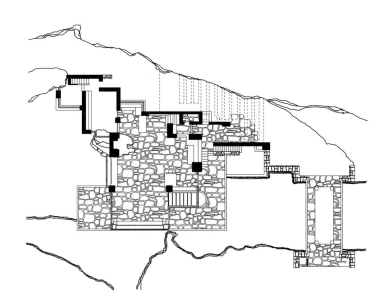

Fallingwater, Frank Lloyd Wright
Bear Run, Pennsylvania 1936
[Modern, cantilevered layered terracing floor plates, diverse]

Fallingwater, a heroic, residential modern icon, is a quintessential example of natural contextualism. Siting itself dramatically above the natural waterfall, growing like a tree out of a central stone mast with massive cantilevered concrete wings, it is based upon the engagement of site and place. Transparent, operable, glass walls and materials that spill from inside to out dissolve the barrier between interior and exterior. The continuity of space and the engagement with nature create a collaborative composition, integrating building with site. The organicism of the natural condition is woven throughout the composition, drawing its effect and inspiration from its environmental contextualism.

	CONTEXT	
	PRINCIPLE	287

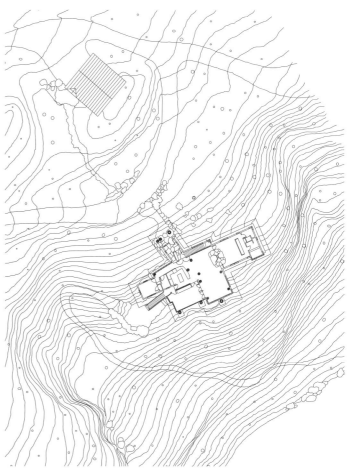

Thorncrown Chapel, Fay Jones
Eureka Springs, Arkansas 1980
[Late Modern, repetitive wood structural frame, wood and glass]

Thorncrown emerges from the site. Mimicking the verticality of the surrounding forest, the structure is built of delicate wood beams to emphasize the height and dissolve the perimeter. Transparency couples with slenderness to allow for visual and spatial continuity between inside and out. Stone emerges out of the site and aggregates to define the interior floor. Even the construction process was directed to step lightly on the land, requiring that no timber be a larger dimension than two men could carry by hand into the site, preserving a specific dimension, lightness, and sacred respect of the site. The collective composition is about a material synthesis with place; from selection to form and effect, material context dominates the design and execution.

Adirondack House, Bohlin, Cywinski Jackson
Upstate New York 1991
[Postmodern, site-based single family house, wood]

The Adirondack House adopts vernacular forms and construction techniques that emerge out of the local materials and traditions. The acceleration of the relationship of building to site through material comes through the overlay of structural columns. Located on a wooded slope, several trees had to be removed when the house was built. During construction, the trees were skinned and put back in their original location as columns. The location and pattern continue the original site condition and material through the architecture.

Urban Context

A building located in an urban area is intrinsically part of the urban context. As such, it has to respond to its local conditions. This response may be sympathetic or adversarial. The relationship formed with the neighboring buildings defines the city and is articulated by the conceptual approach to urban context. This discussion will be focused on buildings that seek to relate to their urban context and not ignore it. These relationships might come about by focusing on issues such as scale, setbacks, materiality, and programmatic function. The urban context remains one of the most powerful and recognizable design responses.

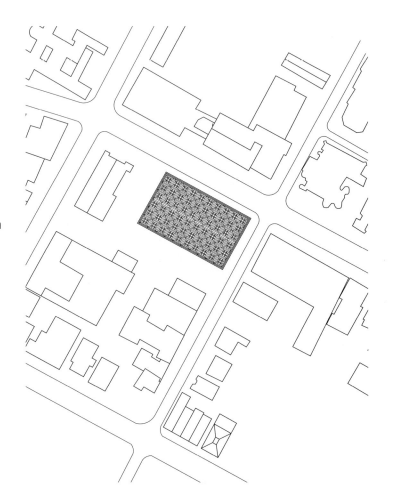

Yale Center for British Art, Louis Kahn
New Haven, Connecticut, 1974
[Modern, museum, concrete]

The Yale Center for British Art presents itself as an urban contextual building by its response to scale and form relative to site conditions. The height and scale of the building is directly drawn from its surroundings, allowing it to synthetically fit within the adjacent fabric. It similarly addresses the street, extending the methods of the surrounding context by placing shops along the street level and positioning the entry to occupy the corner condition. However, the materials of the building are in stark contrast, juxtaposing the concrete and brushed steel panels against the primarily brick buildings that surround it.

	CONTEXT
	PRINCIPLE **289**

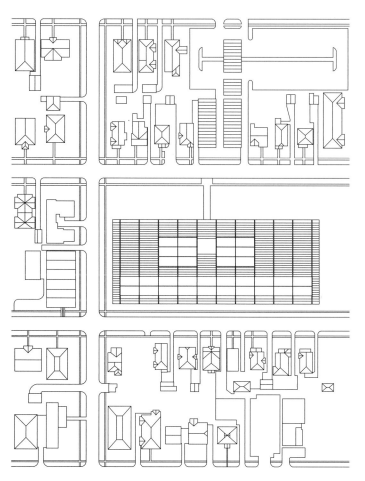
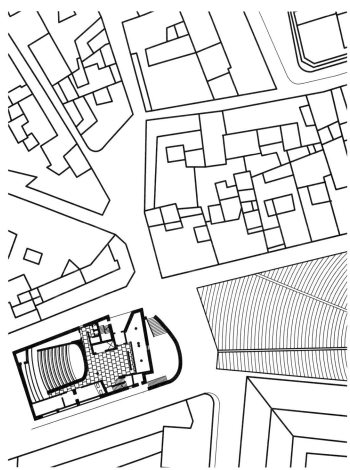

The de Menil Collection, Renzo Piano
Houston, Texas 1986
[Hi-Tech Modern, double-loaded corridor, diverse]

The de Menil Collection uses a series of formal and conceptual techniques to create a masterful urban contextualism. The museum is located in an area that is primarily residential. Piano extends the building horizontally to maintain a low scale that prevents the institutional program from overwhelming the neighborhood. He creates an arcade around the perimeter of the building that diffuses the legibility of boundary and extends the visual vocabulary of the residential porches throughout the surrounding context. He uses cypress wood siding to match the clapboard siding of the surrounding context. Finally, by purchasing the ring of surrounding houses and collectively painting them gray, Piano produces a highly coherent urban context that diffuses and synthesizes the campus with its context.

Murcia Town Hall, Rafael Moneo
Murcia, Spain 1998
[Postmodern, government building, concrete with stone facing]

Through Moneo's careful use of scale and material, the Murcia Town Hall serves as an example of urban contextualism. Concerned with the local urban scale of the piazza in which it is placed, the town hall is governed by its frontality, scale, and relationship to the urban space established by its immediate neighbors. The building incorporates a masonry material palette that matches its context, playing with the assembly system to develop an abstraction through the tectonic. Finally, the programmatic and hierarchical organization of the building, with its large piano nobile window, is derived from the governing systems reminiscent of its urban context.

Historical Context

Historical contextualism is driven by a desire to engage and an understanding of the history of a place, building type, or particular program. Historical awareness and reference, although often identified with their overt formal adaptation during the early years of postmodernism, have been essential to architecture discourse since its inception. A primary example is Rome with its continual use of the classical temple and its implied relation to the history of earlier cultures. It is quite common for an architect to consult historical models that address similar forms or programs. During late modernism, this concept began to fade as architects prioritized invention or reinvention. Postmodernism heralded the return of this methodology as a serious design concept.

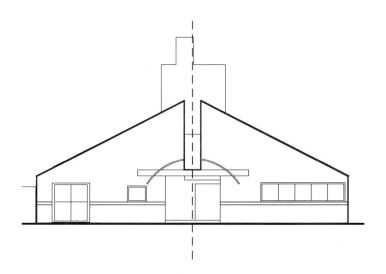

Vanna Venturi House, Robert Venturi
Chestnut Hill, Pennsylvania 1964
[Postmodern, mannerist plan house, wood]

The Vanna Venturi House uses historical contextualism through a series of subtle and major moves. Venturi deploys historical elements such as gables, pediments, and arches in such a way that they are taken out of their typical context and instead are collaged into the form and fabric of the house. The primary play of symmetry and asymmetry as governing organizational and perceptual techniques reflects the house's relationship to historical models and ideals. The front façade is layered as a way of expressing the building's relationship to Michelangelo's Porta Pia through fragmented yet recognizable historic elements. There is a delicate and subtle relationship formed in the plan, which shifts to allow for programmatic response and exterior space.

CONTEXT
PRINCIPLE | **291**

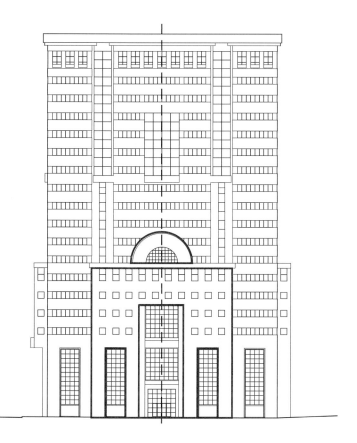

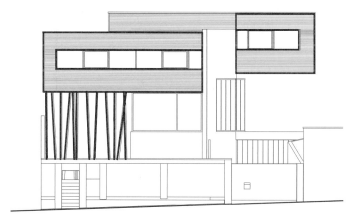

Ten Peachtree Place, Michael Graves
Atlanta, Georgia 1990
[Postmodern, layered tower, masonry and steel]

As a champion of postmodernism and historic contextualism, Michael Graves uses them as organizational and compositional devices in all of his later work. This approach is exemplified by Ten Peachtree Place where he deploys history in a vigorous and refreshing way, paying particular attention to the proportions of the various parts and their interrelationships. Each element is separated and examined almost as if it were a separate building. The forms and figures emerge as abstracted historical fragments, referencing collaged elements such as the arch, column, and portico. These elements are then recombined to form a skyscraper that follows the historically traditional model of base, shaft, and cornice. The uniformity of material collects and unifies the composition, allowing for the forms to deviate. The windows are primarily horizontal, which balances the verticality of the building, whereas the cornice caps the building with articulated abstraction and detail.

Villa Dall'Ava, Rem Koolhaas
Paris, France 1991
[Deconstructivist, free plan house, steel]

The Villa Dall'Ava is founded in historical contextualism as an homage to Le Corbusier and his five points of architecture. Located in a neighborhood that contains several of Le Corbusier's houses, the significance of the architectural legacy as a stylistic, historical, and contextual influence was adopted as a form generator. Koolhaas carefully uses the iconic ribbon window, cut and shifted; the pilotis, now distorted and tilted; the roof garden, extended and now programmed with a pool; and the free plan with formal and flexible configuration extending to interior-exterior porosity. Even the ramp, an experiential, cinematic element that Le Corbusier utilized in a number or projects, is deployed to orchestrate the sectional sequencing. Beginning with Le Corbusier's five points as historical reference, Koolhaas reinterprets them to playfully engage and adapt them, weaving together a series of episodes indebted to the original Le Corbusier projects, but cleverly inventive in their own right.

Material Context

Material context uses place, climate, ecology, geology, construction traditions, and all conditions of indigenous systems as the material that drives the architectural composition. As traditions of vernacular and cultural precedents, local building techniques, available materials, and formal and functional climatic responses, the existing natural systems provide conversational elements that frame the parameters of material engagement. Place and the existing matter of place establish the groundwork of material context.

Tristan Tzara House, Adolf Loos
Paris, France 1927
[Modern, contextual materially banded façade, stucco and stone]

The Tristan Tzara House, unlike Loos's other Raumplan-based modernist houses, uses a material contextualism to provide articulation to the elevation. The Montmartre neighborhood is defined by dramatic sectional changes and rough, indigenous-rock retaining walls. In the Tristan Tzara House, Loos continued his anti-ornament belief through the minimal white-stucco exterior, but compromised through his local response to the rock wall. In addition, he chose to deploy rich material selection on the interior to create spatial and functional effects. By extending the neighboring retaining wall height across the façade and continuing the use of rough stone material, Loos merged the purist vision of his architectural voice with a material contextualism.

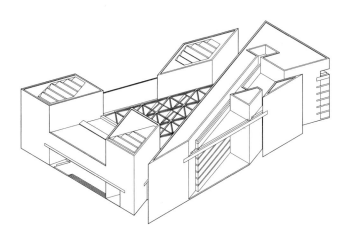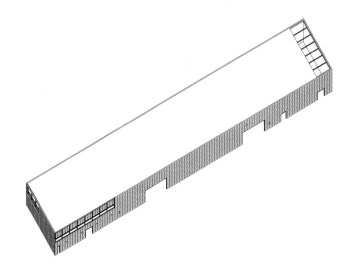

East Wing of the National Gallery, I.M. Pei
Washington, D.C. 1978
[Late Modern, central triangular atrium, concrete and stone]

As an extension of the National Gallery at the Smithsonian Institution, the East Wing had contextual responsibilities to Washington, D.C., (establishing both the triangular footprint of the site through the radial roads of L'Enfant's plan and the building height) and the existing neoclassical West Wing. With the desire for the building to reflect the modernist sensibilities of the collection, Pei needed to resolve the traditions of the old with the sensibilities of the new. To do this, Pei relied on materiality to bridge the diversity of the formal approach. He chose concrete for the structural and formal flexibility and the honesty of the expressed surface to both form and process of construction. To further the contextual bond with the existing West Wing, Pei returned to the same marble quarry to harvest the aggregate and marble dust, ensuring the same coloration and visual texture despite the dramatically contrasting material, form, and scale.

Dominus Winery, Herzog and de Meuron
Yountville, California 1998
[Postmodern, **rip rap** stone bar, glass and stone]

The Dominus Winery deploys a simple form, allowing the focus of the architecture to be on the material articulation of the skin. As a linear bar that houses the processing facilities for the vineyard, it uses a steel frame that is clad in glass and then cloaked in a rip rap perimeter. Loose stones of local rock in metal cages, typical of indigenous retaining walls, are stacked to produce a porous masonry wall. Using varying sizes of rock, the density of their size and shape determines the interlock, which in turn controls the penetration of light to the interior. The material and the collaborating tectonic determine the building's identity, formulated through a clever material contextualism.

Cultural Context

Cultural contextualism refers to a design methodology that takes into consideration the place and culture of the site. This can be achieved by including aspects of material, typology, precedent, and cultural geography. Cultural contextualism recognizes many of these aspects and is a combination of these forces synthesized into one project. This context does not rely merely on one type of context, but instead reaches out to many culturally derived aspects of site and program as a way of achieving an architecture that cannot be separated from place.

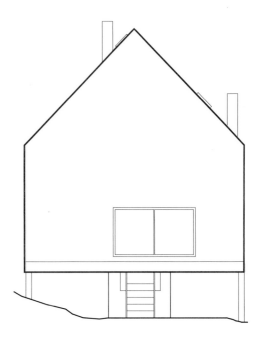

Rudin House, Herzog and de Meuron
Leymen, France, 1993
[Postmodern, house, concrete]

The Rudin House utilizes the concept of cultural contextualism in a number of ways. The primary contextualism is established through its use of the iconic house shape. The universal symbol of house is in this project elevated (literally and figuratively) to a cultural symbol that can be understood by all. Through a series of abstractions, the house does not appear as kitsch, but instead has an elegance and power that is heightened through its shape, materiality, and ability to seemingly float in space. It is only by putting the design through these various abstractions that the building becomes culturally contextual and not mere mimicry.

	CONTEXT	
	PRINCIPLE	**295**

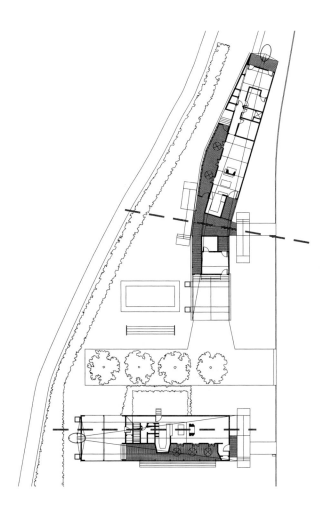

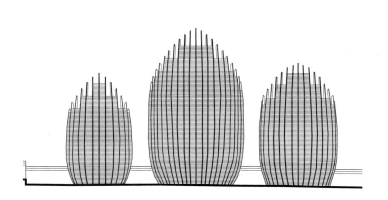

Pensacola Houses, Andrews and LeBlanc
Pensacola, Florida 1996
[Postmodern Regionalism, vernacular houses, wood]

These two houses use two separate typologies that are common in the southeast part of the United States as a way of organizing and constructing an architecture that is place specific. One house was based on the shotgun house, whereas the other house was based on the dogtrot house type. Both houses use traditional proportions, building materials, and plan organizations as catalysts for an architecture that is specific to the local culture. Ideas of sustainability, such as employing roof scuppers to catch rain water, are also a driving force in the design of the houses.

Marie Tjibaou Cultural Center, Renzo Piano
Noumea, New Caledonia 1998
[Hi-Tech, vernacular formed pods, wood]

Renzo Piano's Marie Tjibaou Cultural Center is a project that demonstrates the power of cultural contextualism. The design of the center is based on the villages in which the Kanak people lived. The complex is formed by three villages, each comprising ten Great Huts, which are linked by a long curvilinear path that is reminiscent of the ceremonial alley of the traditional Kanak village. Beyond these planametrics, Piano uses the traditional material of wood as a way of establishing and continuing the cultural context through craft and material. Finally, the shape of the buildings recalls the traditional architecture of the villages. This project truly sums up, through many different aspects, the nature of a culturally contextual architecture.

296 GEOMETRY / PROPORTION
PRINCIPLE

Geometry / Proportion

07 - Geometry / Proportion

Geometry is one of the building blocks of architectural form. It has been utilized by every civilization and continues to be a topic of intense investigation. Geometry can be used as a way of guiding design and as a catalyst in making decisions about shape and form. Historically, architects have depended on geometry as a guide in their calculations and designs. Geometry allows an architect to relate a part of a design to the whole building. It allows them to understand the shapes of fragments and components such as windows and doors. It helps them map out spatial strategies and, in general, gives the architect a ruler by which to govern local and infrastructural decisions made in the design of a building.

Point / Line / Plane These terms describe the most basic building blocks of form. The word "point" describes a single dot in a field. The point has no direction, no angle, and no mass. Moving that point in a certain direction results in a line. The line defines a border or perhaps a perimeter. The line hints at spatial definition, but stops short of achieving it. Moving the line (not simply extending it) in a certain direction produces a plane. The plane has no thickness, but it begins to define space and can produce order and geometry. These three distinct geometric elements are the building blocks of all architecture. They form the basis of any discussion concerning geometry or proportion.

Three-Dimensional Volume / Mass The next step in this evolution of form translates point, line, and plane to a three-dimensional object. Forms take on volume and mass when planes are manipulated and ordered to create volumes and three-dimensional shapes. The volume of a form refers to the amount of space bounded within a shape. Mass, in this context, refers to solidity and weight given to an object or form. Volume and mass are the primary three-dimensional building blocks of architecture. Their figurative conceptualization is foundational to the perception of architecture and, despite its fundamental beginnings, remains one of the most considered conditions in architecture.

Two-Dimensional Module in Plan The two-dimensional module in plan is primary in the design of all buildings. It refers to the idea of using a plan as a two-dimensional diagram to produce repetitive figures that have a direct effect on the subdivision of that diagram. When two-dimensional geometric and proportional systems are deployed as organizational constructs, plans take on added meaning. The square plan (a part of architecture for centuries) is a geometric system that establishes one of the most fundamental proportions of architecture. The square achieves a balance and harmony within the plan that resonates throughout the design. There is a multiplicity of modules that can be discussed and employed, but they innately stem from the square as the most used and fundamental of the two-dimensional modules.

Two-Dimensional Square Module in Elevation as a Figure / Form The idea of the two-dimensional form of the square as an elevational element has implications equal to the use of the square in plan. To apply the square module in elevation takes commitment from the architect. The square in plan, which is typically not perceived in the same manner as the elevation, seems natural on many levels. The elevation, on the other hand, provides the designer with an entirely new dimension, requiring a level of understanding and knowledge that supersedes the plan. As a Platonic form, the square establishes a benchmark of geometry and proportion that, when used as an elevational element, reinforces the architect's willful control of composition.

Two-Dimensional Square Module as a Space / Void Within the construct of two-dimensional geometric systems, the concept of space as void is more complex as it results from two separate elements (a positive figure and a negative void) that establish a distinct relationship to one other. This interrelationship of space as void (space as a subtraction from another object) is choreographed through the consideration of the two elements and their relation in proximity and form. This relationship elevates the space-as-void condition to figure in its own right. The operation requires a different intent and understanding than previous conditions.

Circle as Figure / Form The circle is an intrinsically iconic and powerful shape. Many consider it the first figure and therefore the most fundamental. Throughout the history of architecture, the circle has come to symbolize a number of ideas, including the earth, the heavens, the community, and civilization. With a purity of form it is limited in its ability to communicate proportion and instead often relies on scale as a means of interaction. The very nature of the circle or sphere is closed, thereby denying axial relationships and making entries very difficult. The circle and sphere have a primal relationship to architecture and hence continue to be widely used.

Circle-to-Drum The circle in plan is often transferred to the form of the drum. This architectural maneuver is quite common and involves the simple extrusion and extension of the perimeter of a circle upwards to form a walled, circular space of a constant height. The variables of height to width produce the power of scale and, along with materiality, govern the quality of the resultant interior space. This architectural operation of extending a two-dimensional shape into a three-dimensional form is a simple act that results in a complex and rich space. The circle-to-drum combination is perhaps the most common and powerful of these machinations.

Generically Space / Void The concept of space/void is a complex and mature idea that dictates the relationship between a solid figure and the space or void that is contained within it. There are times when the form and the void have a distinct one-to-one relationship and thus are reflections of each other. There are also conditions when the two are not so closely related and instead rely on their differences as a way of complementing each other. The issue of legibility is critical in a space /void relationship due to the fact that the success or failure of the reading is dependent upon the ability to recognize the form and, conversely, the space or the void within it.

Triangle Within the parameters a three-sided figure there are infinite possibilities; unlike other Platonic forms, there is not a singular formal solution. The equilateral triangle, where all sides are equal, is the most recognizable and often most common due to the repetition of identical pieces. Subtle variation in the connecting angles yields widely variable forms. The shape of the triangle denies the simple relationships that are common in squares and rectangles. The sides of a triangle are related to the opposing vertex rather than an equal and opposite side. These types of relationships form complexities and atypical conditions that have fascinated architects for centuries. During the Renaissance, the triangle was very popular in painting and was also used effectively, though less commonly, in architecture.

Root Rectangles (Plan, Section, Elevation) Localized and Whole The $\sqrt{2}$ square is one of the most common and ingenious of all proportional systems. It is the only shape that when doubled equals a third, similar shape. In other words, two $\sqrt{2}$ squares, placed side by side, result in a third larger $\sqrt{2}$ square. Many architects and artists (particularly during the Renaissance) were fascinated with this system and deployed it in numerous ways. The $\sqrt{2}$ square is created by starting with a square, drawing a diagonal from one corner to the opposite corner, and then rotating that line 45°. The $\sqrt{3}$ square is then formed by carrying out the same operation as a $\sqrt{2}$ square but with a differing angle of rotation and so on. This proportional system is rational, easily understood, and simple to perform. As such, it was and is used frequently.

Golden Section (Plan, Section, Elevation) Localized and Whole The golden section, or golden ratio, is a proportional system that has fascinated man for thousands of years. Elements of this system are found in architecture, mathematics, painting, and music. Architecturally, the golden ratio can be traced back to the Parthenon and to Vitruvius. Later proponents of the system include Alberti, Palladio, and Le Corbusier. The fascination with the golden ratio stems from its irrationality on one hand and on another, its ability to reproduce itself by the addition of a square. It is a system that unlike other geometric systems is found in nature and is therefore considered universal. The golden section is a ratio determined by dividing a square in half and then drawing the diagonal through the half of the square and rotating that diagonal down until it is parallel and aligned with the bottom of the original square.

Two-Dimensional and Three-Dimensional Module Form Typically the aforementioned proportional systems were used primarily in two-dimensions; however, there were times when these systems were expanded to include three-dimensional spatial applications. Though less common, they were powerful in demonstrating that proportions in architecture were different than proportions in painting. This concept of proportion in spatial terms gained credence during the Renaissance. Here proportion became a way of separating architecture from art and served as a way of creating architectural systems that replaced the role of intuition.

Distorted Geometries and Computational Complexities
Computational complexities and distorted geometries emerged during the architectural computer revolution as a way of incorporating ideas and geometries that were previously unavailable to the architect. Through warped iterations of conventional representational systems, the resulting geometric complexities create dynamic and powerfully individuated spaces. Issues relating to site, organic anthropomorphism, and parametric modeling with script controls make up some of the diverse, emergent opportunities. The complexity of these geometric mechanisms has affected both space and structure in architecture and accelerated the complexity of pattern and ornamentation in building skin design.

Point / Line / Plane

The fundamental building blocks of geometric description are point (defined as nothing but the demarcation of a location), line (the connection of two points), and plane (the extrusion of a line to define a surface). These primary elements define every resulting geometry and thus have latent significance in every composition. Their level of legibility is evident and embraced in architectural composition. The translation of these elements to architectural column and wall, frame, and surface mediate the translation of geometry from drawing to building.

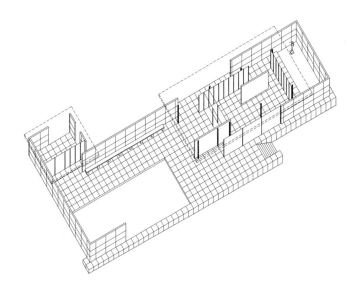

Barcelona Pavilion, Mies van der Rohe
Barcelona, Spain 1929
[Modern, free plan, masonry, steel and glass]

PLANE: The Barcelona Pavilion deploys the free plan with a clear agenda of defining a specific moment within a larger infinite space. The fluid flow of space from function to function and from inside to out allows the masterful and highly evocative ambiguity of edge. By facilitating an implied movement through the slipping figures, the use of the plane is essential to the spatial extension by facilitating an implied movement through the slipping figures. The thinness of the form and the richness of the material of the vertical planes are pinched between the dramatic extending floor and roof planes. Bridging between inside and out, the extended surfaces demarcate the form of the plane and the spatial implications of its presence.

	GEOMETRY / PROPORTION
	PRINCIPLE **303**

The Bird's Nest Stadium, Herzog and de Meuron
Beijing, China 2008
[Postmodern, irregular lattice frame stadium, concrete and steel]

LINE: The Bird's Nest stadium is defined almost entirely by the development and language of its skin. This skin is literally composed of thousands of lines that cross at seemingly random angles to ultimately create a series of web-like structures that lend the stadium its architectural morphological makeup. This ingenious collaboration of tectonics, pattern, and material all come together through the common element of the line to produce a singular composition that is graphically dense and iconic. The stadium represents how a simple architectural element, such as a line, can be reiterated to such a degree, resulting in a non-orthogonal but highly ordered system.

UK Pavilion Shanghai Expo, Heatherwick Studio
Shanghai, China 2010
[Postmodern, figurative centralized pavilion, fiber optic rods]

POINT: The UK Pavilion at the Shanghai Expo is a dramatic form that generates an intense experience. Made as an aggregation of linear sticks, the cluster of fiber optic rods has a sample collection of seeds located at the tips. The fiber rods create an amplified projection of light through the extended point, culminating at a captured node of organic seed. Each line is transferred into a nodal point, clustering to produce an atmospheric chamber of discrete points. The singular inner chamber is defined by thousands of points of light. The series of points collect to create a field that defines identity of surface and space.

Three-Dimensional Volume / Mass

The translation of the two-dimensional elements of plane, line, and point to the three-dimensional, and thus spatial, realm is accomplished through a consideration of volume and mass. Volume refers to the bounded space contained within a form. Mass refers to the solidity and weighted formal reading. Both engage the formal and spatial legibility of architecture, but their individuated terms each imply highly divergent conceptual parameters for making and articulating their intrinsic systems. The selection and articulation of a spatial conceptualization has dramatic linguistic framing that translates directly into the forms and perceptual intentions, and ultimately their reading and experience of architecture.

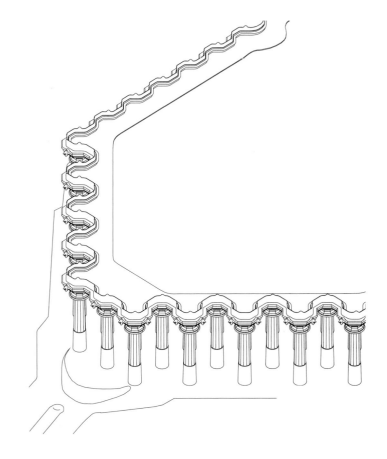

Park Guell, Antoni Gaudi
Barcelona, Spain 1914
[Expressionist Modern, structurally expressive park, masonry]

Park Guell is one of Gaudi's masterworks, fully integrating landscape and architecture. The colonnade in particular expresses the formal bridge of the site by addressing the step section of the topography. It engages the topography with the form by creating an expressive, leaning buttress that blurs between the structural optimization (a response to the requirements of physics) and the effectual sculptural agenda of his exotic and highly individuated personal forms. Structure and ornament are executed through the material, which seems to emerge from the ground naturally, both in rock type and formal unitization. Yet the material is organized and composed through the repetitive patterning, organization, and systemization of the larger deployment. The park colonnade deploys the mass of material and form to create subtractive spaces, which are embedded yet removed from the slope and mass of the site.

GEOMETRY / PROPORTION **305**
PRINCIPLE

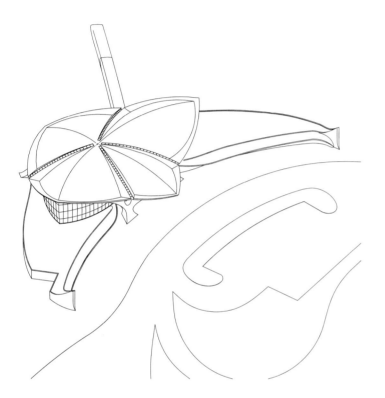

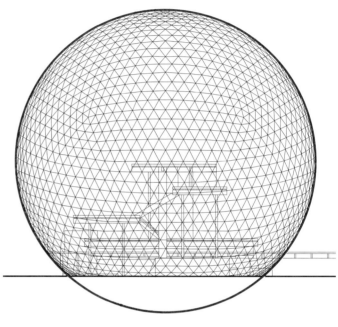

TWA Terminal, Eero Saarinen
New York, New York 1962
[Late Modern, free-flowing curved airport terminal, concrete]

The TWA Terminal is typified by the continuity of surface achieved through the deployment of mass. Solid and flowing, the organic formalism of the continuous surfaces defines the unique morphology that blurs wall, floor, and ceiling into a single continuous entity. Defined by the fluidity of aerodynamics, the organic formalism of the compositional method implies the futurism and gesture of flight. The otherworldliness of the composition prepares the visitor for the uniquely technological experience. The geometric formalism determines an unparalleled encounter with the space through the continuity of form.

Expo Pavilion, Buckminster Fuller and Shoji Shadao
Montreal, Canada 1967
[Late Modern, geodesic dome, metal]

The Expo Pavilion uses a repetitive unit with hyper-geometrically controlled deployment on both the micro and macro scale. At the scale of the building as a whole, the stability, continuity, and singularity of the sphere provide a systematized yet highly stable overall shape. At the localized dimension, the use of triangulation creates a regularized mesh that allows for equal repetitive elements to produce a stable shape. The relative relationship of piece to whole and the geometric governance of the aggregated system establish an interrelated systemization of geometry to form through the volume of space. The surface is optimized structurally, minimized in material and weight, to make this a volumetric space.

Two-Dimensional Module in Plan

The two-dimensional planametric geometric system deploys the rationale of mathematics as an organizational field. As a quintessential ordering system, it establishes the parameters of almost all architectural planning. The term "module" refers to the use of an architectural element as a "unit" that establishes an increment of measurement. Its internal organization creates the system of segmentation that is the standard for determining and regulating the proportions and order of the remainder of the building. This concept is illustrated when a building uses a dimensional system that is synthetically linked to the measurements and proportioning of a building.

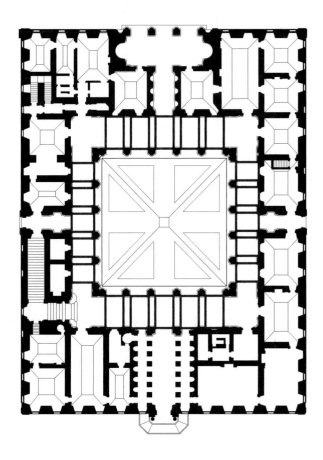

Palazzo Farnese, Antonio da Sangallo the Younger
Rome, Italy 1541
[Renaissance, courtyard building, masonry]

In the plan of the Palazzo Farnese, the module is utilized to establish the organizational geometry. The module is established by the central courtyard with each bay articulating a single unit. From this measurement all the other spaces and rooms are then derived. This concept synthetically links all the aspects of the building together. This is exhibited not only in the plan but also in the elevation and sections. There are two dimensions used in the courtyard and each is a variation on a theme. They reoccur throughout the building as abstract spatial and organizational constructs.

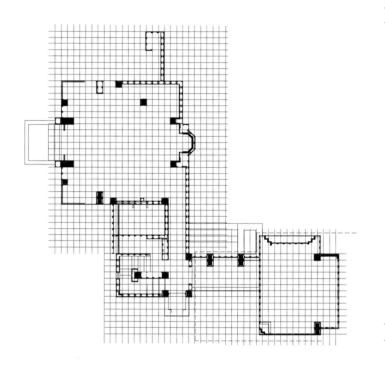
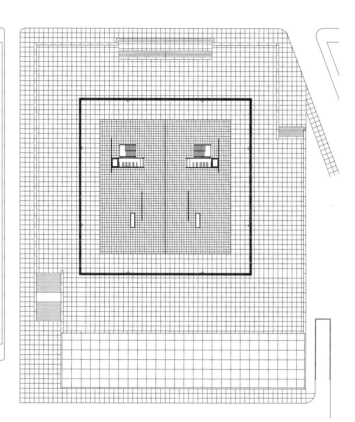

Freeman House, Frank Lloyd Wright
Los Angeles, California 1924
[Modern, textile block house, concrete masonry]

The Freeman House, as one of Wright's iconic textile block houses, uses the individual concrete masonry units to dictate the module and dimensional system of the building. All the blocks are square and iteratively identical, varying only in the quantity of surface ornament. Their formal articulation, along with their relentless deployment, created the distinctive pattern that came to be associated with these Los Angeles houses. The textile blocks go beyond pattern-making as they literally create a performative system that responds to all of the functional elements of the house. On the exterior, this is made evident by the modular coordination and alignment of the blocks with the window openings. In the interior, the floor pattern is aligned with all the internal block walls. The house offers a simple system that is an extension of the module block, creating one of Wright's complete environments.

Neue Nationalgalerie Museum, Mies van der Rohe
Berlin, Germany 1968
[Modern, museum, steel frame]

In the Neue Nationalgalerie Museum by Mies van der Rohe, the module is set by the paving pattern in the great hall. Each paving stone is square in shape and is approximately four feet in dimension. This module is then repeated and is used to generate the remainder of the building. Every aspect and measurement of the museum is based on this paving grid. This becomes clear as one walks through the building and notices that all the lines of the floor pattern bisect all the architectural elements such as columns and window **mullions**. This module is extended vertically to incorporate the elevational and sectional proportions as well.

Two-Dimensional Square Module in Elevation as a Figure / Form

Within the family of proportional systems, the square as a figure in elevation is one of the most common and most used throughout history. When deployed in elevation as in plan, this modular system can provide a sense of order and rigor. It carries with it unquestionable qualities and meanings. As a geometric system, it has crossed cultural and aesthetic boundaries. It can be found in numerous architectural movements and time frames. Its constant reoccurrence throughout architecture has given it a quintessential quality that extends beyond straightforward readings and instead posits it as a fundamental element defining the field.

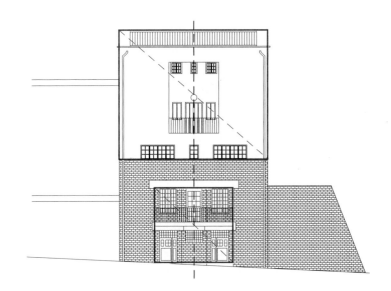

Tristan Tzara House, Adolf Loos
Paris, France 1927
[Modern, contextual materially banded façade, stucco and stone]

The Tristan Tzara House uses the square as the major organizational feature of the façade. Contextualized materially by the use of a heavy stone on the base, this stone relates the building to a preexisting stone retaining wall that is adjacent to the house. Above this wall, Loos has rendered the wall flat and white, which emphasizes the square shape of the upper portion of the façade. There are a number of voids and punched openings in the wall; however, the overall square is never lost. In this particular building, Loos used the square as a reference to Tzara's position as a Dada artist. The square represented the pure form as compared to the Dadaist beliefs of nihilism.

GEOMETRY / PROPORTION
PRINCIPLE 309

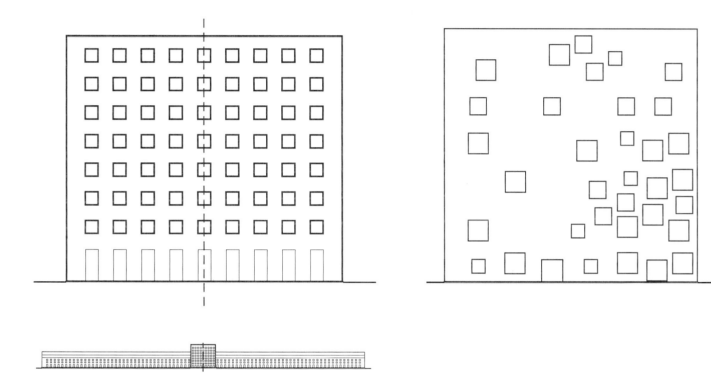

San Cataldo Cemetery, Aldo Rossi
Modena, Italy 1978
[Postmodern, grid field, stucco-clad masonry]

*The centrally located main building of this complex is a large cube with punched-square openings. The complex is a series of **columbaria** that makes up this "house of the dead." The building uses the square as its building block. The building is square in plan, elevation, and section and uses square windows to further reinforce this reading. Each urn fits into one of the squares, whereas the other columbaria are removed on a regular grid to provide windows into the building. The building's central position, red color, and exaggerated scale all reiterate the hierarchy and significance of the building within the entire composition of the cemetery.*

Zollverein School of Management and Design, SANAA
Essen, Germany 2006
[Postmodern, iteratively scaled field, concrete]

The Zollverein School uses the square as a module and catalyst. The overall shape of the building is a cube with the appropriate square plan, section, and elevations. The elevations further comprise a series of square windows that appear graphically intuitive, having little relationship with the interior other than lining up with the various floor plates. The building represents the power and purity of the square even within the context of the postmodern contemporary field. There are two square cores that penetrate the building. The roof garden culminates in a square oculus, connecting the building to the sky.

Two-Dimensional Square Module as a Space / Void

The concept of space/void in architecture is one that came about primarily during the advent of modernism. It resulted from the relationship between architecture and cubist painting. By using a figure as a space or void, the architect could explore ideas of spatial experience, chronology, and phenomenal transparency. It allowed for the breakdown between the interior and exterior to become fully manifest. The square as a two-dimensional module is exhibited as a pattern of squares that combine to form a field. This pattern is usually manifested in a structural grid or an elevational composition. Their presence governed both the positive figure and the implied subdivision of the space/void.

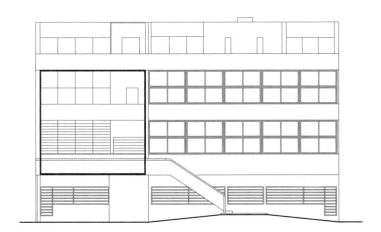

Villa Stein at Garches, Le Corbusier
Garches, France 1927
[Modern, free plan, concrete and masonry]

The Villa Stein at Garches was one of the first buildings where Le Corbusier experimented with the idea of space/void. Primarily exploited on the garden elevation, there is a large square carved out of the left side of the composition. This square serves as an exterior terrace for the villa and is connected through a sectional cut to the roof garden above. This square void establishes a sense of proportional order to the garden façade, ultimately connecting it back to the front façade.

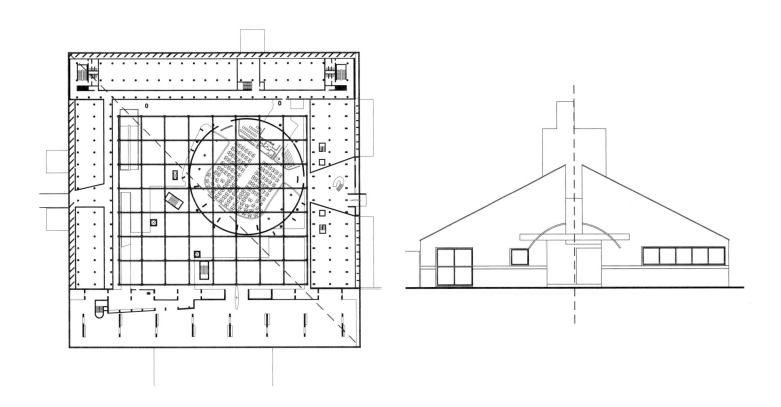

Palace of Assembly, Le Corbusier
Chandigarh, India 1963
[Modern, free plan, concrete]

The plan of the Palace of Assembly by Le Corbusier is a quintessential example of the square used as a two-dimensional module. In this building, there is a U-shaped structure with a large interior courtyard in the center. This interior space is occupied by an extensive field of columns in the form of a grid. The organizing module, expressed through the grid of columns, illustrates the concept of a three-dimensional matrix that occupies the entirety of the building. Inserted into the structural field are a number of objects that intensify the grid by challenging it spatially. The overall form of the building is a square, as are each of the modules; however, the structural field is expressed as a rectangle within the interior courtyard.

Vanna Venturi House, Robert Venturi
Chestnut Hill, Pennsylvania 1964
[Postmodern, mannerist plan house, wood]

The main elevation of the Vanna Venturi House utilizes the square as a two-dimensional composition that expresses itself as a complex system of devices. Deployed as a modular system, it is challenged with ideas of balance through asymmetry as a way of structuring the façade. Venturi uses ten equal squares as a way of achieving this. On one side, Venturi uses a large window that is composed of four equal squares; to the right of that he places a single square window. On the other side of the entry he uses a single line of five squares that are at the same height as the single window on the other side. Numerologically they balance one another; but compositionally they respond to the local functional requirements. Collectively they illustrate the opportunity for complexity and contradiction within a two-dimensional square modular system.

| 312 | GEOMETRY / PROPORTION — PRINCIPLE | PROPORTION — ORGANIZATION | CIRCULAR MODULE — GEOMETRY | ELEVATION — READING | ARCHITECTURE — SCALE |

Circle as Figure / Form

The circle as a figure or form is the simplest closed shape. Defined by one centralized and stationary point, the rotation of a radius of a fixed distance establishes the second point and creates a pure perimeter form. The intrinsically centralized nature of the geometry denies a single point of axial approach due to the equality of the figure. This equality of form similarly denies any proportionate relationships other than scale, resulting in a classical monumentality in its deployment. As drum, rotunda, or theater, the dominant nature of the geometry intrinsic to the form implies the associated practical and functional hierarchies.

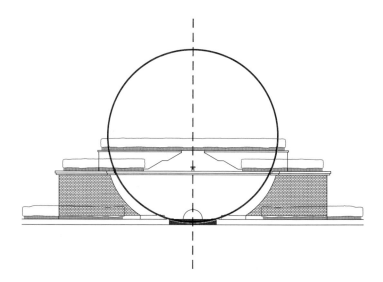

Newton's Cenotaph, Étienne-Louis Boullée
Unbuilt 1784
[Neoclassical, heroic **architecture parlante**, masonry]

Newton's Cenotaph, though an unbuilt project, is the quintessential engagement of the circle through plan, section, and elevation. As a sphere, externally exposed and internally continuous, the geometry is representative of the purity of the perceived singular space as well as the purity of the calculus that is identified with Newton. The ability of numbers to establish another language founded in the rationality of a ruled and principled system is the inspiration and method for the design. The massiveness of the scale and the simplicity and simultaneity of the form and space make the building a perfect resolution of its geometric origins.

GEOMETRY / PROPORTION **313**
PRINCIPLE

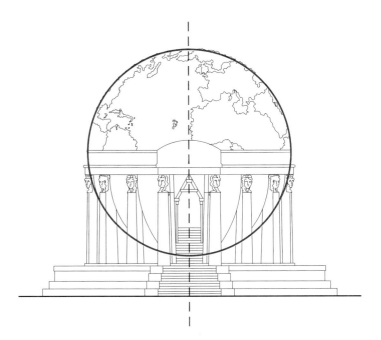

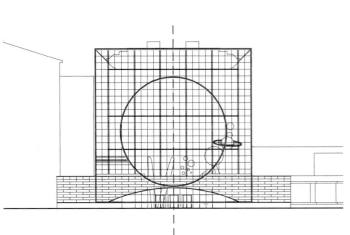

Temple of the Earth, Jean-Jacques Lequeu
Unbuilt 1790
[Neoclassical, layered monument, unbuilt masonry]

The Temple of the Earth by Lequeu has great similarities to Newton's Cenotaph, but greatly reduces the scale and encircles the sphere in a colonnade. The premise of the geometrically pure sphere is maintained; however, it is hybridized with the Greek temple typology evident in the repetitive columns that fully encircle the building. The circle as a geometry maintains dominance in plan, section, and elevation through the singular and pure deployment of the sphere as form.

The Rose Center, James Polshek
New York, New York 2000
[Postmodern, sphere of planetarium as object, glass and steel]

The Rose Center is a product of emerging visualization technologies. The shape of the sphere is used to house a planetarium, which requires the curved inner surface to simulate the arcing canopy of an artificial sky. The resulting form of the sphere is embraced in both the organization of the building (with bracketing bars cupping the figure) and the maximization of visual transparency (by means of a carefully engineered and highly dramatic glass façade). Thus the sphere is a dominant presence on the façade of the building.

Circle-to-Drum

If a simple circle is inscribed on the ground and then extended upward, a drum is formed. The drum is the simplest spatial extension of the circle. It is one of the most fundamental of architectural moves and results in a complex and rich spatial result. It remains a primary form in architecture and, due to its historical and primitive underpinnings, gives a precedent-rich reading to many buildings that use it as their parti. This particular scheme has survived for centuries, constantly changing to satisfy new programs, structures, and materials.

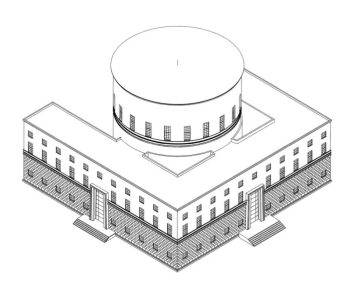

Stockholm Library, Gunnar Asplund
Stockholm, Sweden 1927
[Early Modern, centralized drum, masonry]

The plan of the Stockholm Public Library takes the form of a circle inscribed within a U-shaped building. In plan, this circle is extended upwards to form the central drum of the complex. This shape is a simple singular room that dominates the composition and speaks of the importance of the circle-to-drum as an architectural device. The most important program and complexity of geometry occur in the drum, reinforcing its identity as the hierarchical moment of the plan. In elevation, the drum extends above the surrounding buildings, again announcing its importance and distinction within the hierarchy of the building.

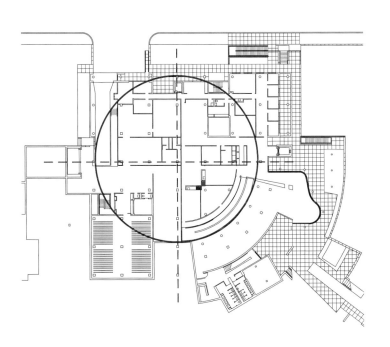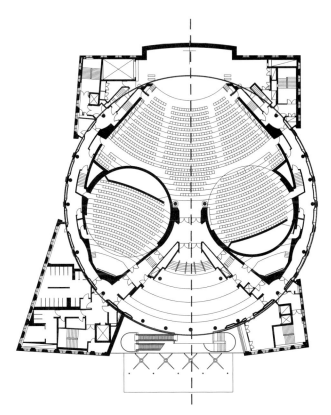

High Museum of Art, Richard Meier
Atlanta, Georgia 1983
[Late Modern, L-shaped galleries with circular atrium, metal]

The High Museum of Art uses a quarter circle in plan that is extended upwards into a quarter drum. This partial figure then contains the main vertical circulation that connects all the galleries, which are located in the L-shaped building that brackets the atrium. From the exterior, the drum has a more dominant reading as a transparent frame of glass and structure that is animated by the movement of the visitors on the ramps behind. As the main ceremonial circulation of the building, the circular ramps are formally exposed to express their position and power, again reiterating the concept of a circle-to-drum organization.

Edinburgh Conference Center, Sir Terry Ferrell
Edinburgh, Scotland 1995
[Postmodern, drum conference center, stone]

The Edinburgh International Conference Center (or EICC) is a circular drum building located at the heart of the master-planned Exchange District in the west end of the city. The large drum of the external form of the conference space mimics the contextual form of the adjacent fabric and reestablishes local hierarchy. Its geometric uniqueness establishes a nodal head to the district as a whole, announcing the significance and presence of the building and the collective complex. There was a conscious desire for an architecturally individuated reading to the building relative to the complex (for practical and political reasons) that also maintained a collective identity. Like the Pantheon in Rome, the drum-like urban figure has dominance and presence as a heroic entity within the density and orthogonality of the adjacent urban fabric.

Circle as Space / Void

The space/void relationship of the circle refers to the subtractive nature of the circle projected three-dimensionally into the sphere to create a void. The space defined by the shape as opposed to the legibility of the form becomes the focus. The void is created by the removal of either the full form (Newton's Cenotaph) or the fragmented remnants of the form (Exeter Library and the Très Grand Library). The transfer of the purity of the form to the purity of the space creates a heroic figure through the equality and simplicity of the geometry.

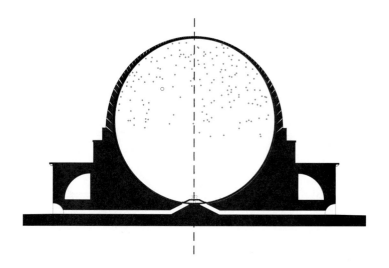

Newton's Cenotaph, Étienne-Louis Boullée
Unbuilt 1784
[Neoclassical, heroic architecture parlante, masonry]

Newton's Cenotaph epitomizes the form and idea of the circle being used as a space/void. The project's striking exterior partial-sphere shape makes it easily recognizable and readily identifies the complex and its relationship to the earth. This sphere, however, is found in the interior of the project as well, creating an abstraction of the heavens, complete with the stars and constellations. This space is pure in its continuity and in its representation of Newton and his theories. Unlike other examples such as the Pantheon in Rome, the cenotaph relies less on abstraction and more on the literal power of the sphere as a void in both its simplicity and its massive scale.

| | GEOMETRY / PROPORTION | 317 |
| | PRINCIPLE | |

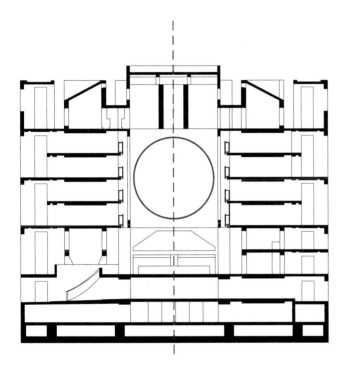 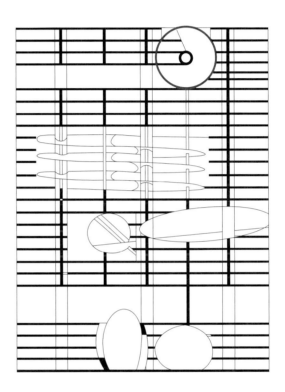

Exeter Library, Louis Kahn
Exeter, New Hampshire 1972
[Late Modern, centralized square, masonry and concrete]

In the Exeter Library, Kahn ingeniously combines the two platonic figures of the cube and the sphere. The cube takes the form of the overall exterior of the building, whereas the sphere is rendered as a void that occupies the center of that cube. There is an inner concrete square figure that organizes the central atrium; each wall of this concrete figure has a large circular cut-out that describes the central circle as a void within the complex. Created through a subtractive centralized sphere that produces the removals from the inner figure, the purity of the geometry as equally space and figure dominates the composition.

Très Grande Bibliotheque, Rem Koolhaas, OMA
Paris, France 1989
[Postmodern, cubic book field with subtracted spaces, unbuilt]

The Très Grande Bibliotheque can be described as a large cube that is full of book stacks that act as a solid poche. Carved out of this poche are series of rooms, ramps, and transitional spaces that make up the rest of the program of the library. The larger spaces are typically curvilinear and are connected by ramps, forming a type of Raumplan on a large scale. These curvilinear voids present an excellent example of circle as space/void. The library allows a subtractive methodology of design to dictate the typical and new spaces of the library in a novel way. Defined by the spatial voids, the inner figurations of the subtracted spaces determine the architectural iconography.

Triangle

The triangle as a geometric figure is the simplest closed shape possible using straight lines. The figure is intrinsically stable as a structural shape and can be deployed in two- and three-dimensional configurations to create structural trusses, diagonal bracing, geodesic domes, and space frames, for instance. The simplicity of the form allows for variability through the angles deployed. From acute to obtuse, subtle differences in nodal angles allow for massive variability of individuated pieces, permitting dramatic formal and structural variation.

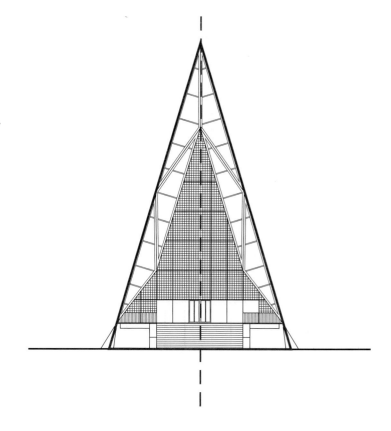

Air Force Chapel, Walter Netsch, SOM
Colorado Springs, Colorado 1962
[Late Modern, repetitive triangulated structure, steel and glass]

The Air Force Chapel uses a repetitive triangulated structure to imply flight and create a dramatic peaking sanctuary. Massed like a phalanx of fighter jets, the seventeen chapel spires create a dense, repetitive array of triangular figures. With the exaggerated scale and tight footprint, the extremity of the form is heightened and emphasized. The triangle serves as an optimized form of the apsidal arch that is typically formed with a rounded geometry. The militaristic hyper-functionalization allows for a minimalist interpretation of the traditional form, eliminating all formal transitions and favoring the drama of a single figure and the ceremonial and light effects of its composition. The stability and simplicity of the triangular geometry establishes the building's identity.

GEOMETRY / PROPORTION 319
PRINCIPLE

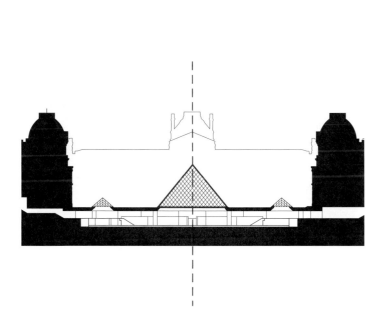

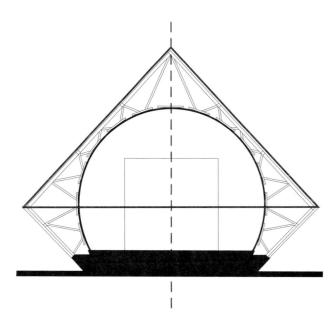

Louvre, I.M. Pei
Paris, France 1989
[Postmodern, monumental pyramidal form, steel and glass]

By deploying the pyramidal form as a formally pure geometry, but one that is contextually foreign, and then dematerializing the form, which is historically solid, through an intricately engineered and detailed glass skin, the contrast and individuation is clear. The triangular geometry was adopted to create contrast with the existing context while still subscribing to the solemnity of a pure form. The structural articulation of the glass skin through the three-dimensionally triangulated back supports produce a delicate web that allows for an optimization of the transparent surface, while creating depth and detail to the otherwise primitive shape.

Imai Daycare Center, Shigeru Ban
Odate, Akita, Japan 2001
[Postmodern, repetitive rib structure, wood and fabric]

The Imai Daycare Center by Shigeru Ban uses a circular laminated veneer lumber (LVL) beam encased in a square roof. The rotated square is truncated by the ground plane, creating a triangular extrusion. The purity of the geometry is employed to deal with the region's heavy snow load. The steep pitch allows for the speedy removal of accumulating water weight. The translation of the outer shape with the inner circular extrusion is filled with the structural lattice that allows for a pure yet differential inner and outer figure. The simplicity of the cross section is complemented by the extrusion of the figure into a self-similar, elongated tube. The triangle serves as a prismatic shape that provides the building with a distinct exterior legibility and a unique architectural identity through the formal simplicity and primitivism.

Root Rectangles (Plan, Section, Elevation)
Localized and Whole

The √2 (root-2) square is a proportional system that extends directly from the square and allows a designer to form complex combinations. The √2 square is formed by taking a square and then drawing a diagonal across it; this diagonal is then rotated down until it is parallel with the bottom of the square. The resultant rectangle is the √2 square. Executing the same operation on a √2 square (instead of a traditional square) produces a √3 (root-3) square, and so on. One of the remarkable properties of the √2 proportional systems is that when two like shapes are combined, a third similar shape is created.

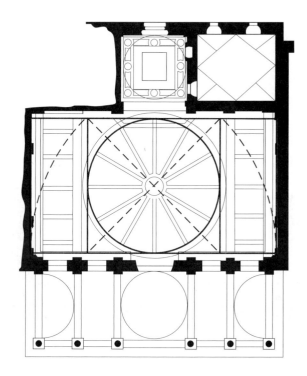

Pazzi Chapel, Filippo Brunelleschi
Florence, Italy 1461
[Renaissance, chapel, masonry]

In the Pazzi Chapel, Brunelleschi uses the √2 square as a spatial device to dictate the plan of the chapel. The center of the space is occupied by a square, which is governed by the main central dome of the ceiling. Though this square is not recognizable in the floor pattern, it is extended in both directions to form two √2 squares with the base squares of each occupying the same central square. Thus the space of the chapel is achieved by overlapping two √2 squares. In section, the space under the main dome is also proportional, using a √2 square to establish the ceiling height.

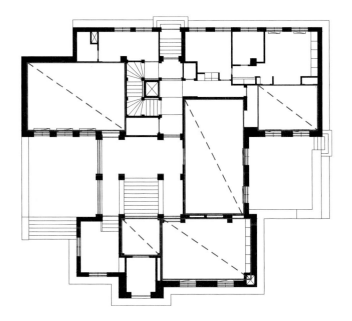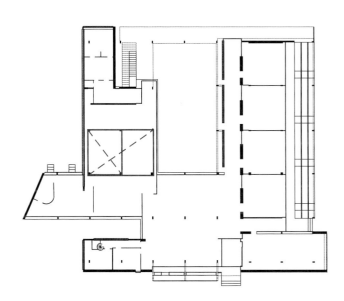

Wittgenstein House, Ludwig Wittgenstein
Vienna, Austria 1926
[Modern, house, masonry]

The Wittgenstein House, designed by the philosopher Ludwig Wittgenstein, establishes a proportional system in the plan of the ground floor. Each one of the major rooms of this plan assumes a different harmonic proportion. The entry foyer is 1:1, the salon is 1:2, the library is 2:3, the dining room is 3:4, and finally, the sitting room takes on a 4:5 proportion. By using this simple system of harmonic proportions, Wittgenstein is able to establish an organization that goes beyond mere intuition.

Asilo Sant'Elia, Giuseppe Terragni
Como, Italy 1937
[Modern, courtyard, concrete]

In the plan of the Asilo, Terragni uses the $\sqrt{2}$ square as an organizational device that dictates the placement of the structure. Each structural bay is defined by the $\sqrt{2}$ square; when two of these bays are combined, they form a third $\sqrt{2}$ square. In this way Terragni allows one system to govern the structural aspects of the building. He then utilizes a different system, that of the golden section, to dictate the spatial aspects of the building. Ultimately both systems are combined to form an overlapping system of rectangles that work together to form a complex matrix.

Golden Section (Plan, Section, Elevation)
Localized and Whole

The golden section is a proportional device that was made popular during the Renaissance and has remained relevant to the present. It is an extremely useful way of dictating sizes and proportions to various parts of architecture. The golden section has a very specific relationship to the square: if a square is marked within a golden section, the space that remains is another golden section. The golden section continually rotates on itself, creating infinite golden sections as each square is removed. This sequence is also referred to as the Fibonacci sequence, creating a spiral of connective base points.

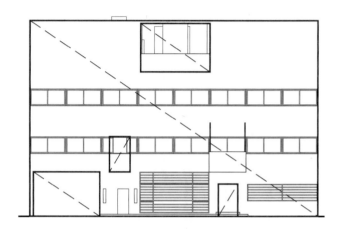

Villa Stein at Garches, Le Corbusier
Garches, France 1927
[Modern, free plan, concrete and masonry]

The main façade of Villa Stein is composed of a series of interrelated golden sections. A golden section establishes the overall form of this façade and the central top opening. Additionally, the front door and the balcony ensemble on the piano nobile are golden sections. Le Corbusier used this proportional device as a guide when composing, believing that the use of classical proportional systems illustrated a connection with the past.

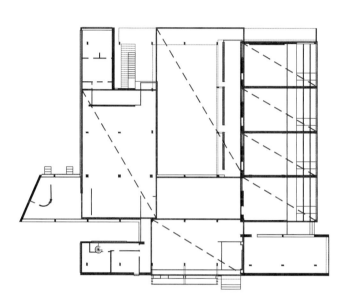 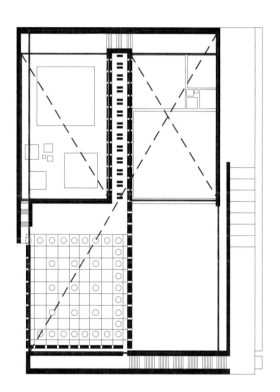

Asilo Sant'Elia, Giuseppe Terragni
Como, Italy 1937
[Modern, courtyard, concrete]

The Asilo, by Terragni, used the golden section as a way of distributing the various major spaces of the program. The indoor playroom, the refectory, and the courtyard all reflect the use of this proportional system. Each of the classrooms is a square; however, when the glass doors are opened and the classrooms extended to include the outdoor portion, they form a golden section. The golden section is also used as a proportional system in the entry façade to determine the size and shape of many of the elements, serving equally as a compositional and organizational tool.

Danteum, Giuseppe Terragni
Rome, Italy 1942
[Italian Rationalism, grid hall of glass columns]

The Danteum, an unbuilt project dedicated to Dante Aligheri, used the golden section as its primary ordering device. The project was conceived as an allegory of the "Divine Comedy" in which a series of spaces represents the different cantos of the poem including the inferno, purgatory, and paradise. Each of these is represented through an entire or partial golden section. The collective building takes its shape from the ruins of one of the bays of the Baths of Maxentius, situated across the Via dei Fori Imperiali (formerly Via dell'Impero). The Danteum is composed of two overlapping squares, which are then layered with six major golden sections. Terragni was able to connect Italian culture to this idea of proportion and mathematics through the modern application of this traditional proportional system.

Two-Dimensional and Three-Dimensional Module Form

The two- and ultimately three-dimensional modules added a complexity to the architectural equation that helped remove it from the conceptual ties to painting. The employment of geometry and mathematical principles as both surface and spatially based organizational methods allowed the understanding and conception of structures as a whole. As the architects began to use two-dimensional proportional systems to solve the graphic problems of elevations and plans, it was a natural and logical next step to introduce these proportional qualities into the space of the buildings. The three-dimensional modules gained widespread acceptance during the Renaissance and have continued to be employed and advanced into the present.

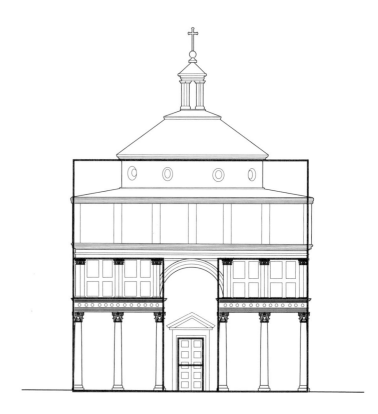

Pazzi Chapel, Filippo Brunelleschi
Florence, Italy 1461
[Renaissance, private chapel, masonry]

The Pazzi Chapel utilizes the square as a way of organizing the façade. The three columns on either side of the façade each form a square with the paneled attic story forming an additional half square. The arched opening between these two elements is formed by stacking two squares. The entry door is also two squares stacked atop one another. Finally, the width of the building and the height (up to the roofline above the small circular windows in the drum) are equal, forming a square that comprises the whole structure.

GEOMETRY / PROPORTION
PRINCIPLE
325

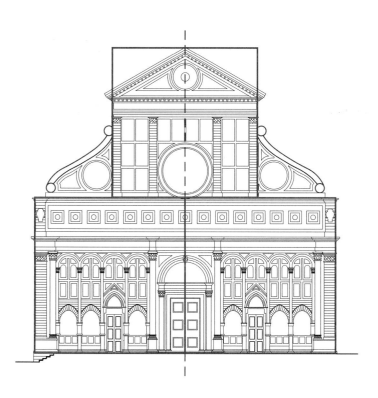

Santa Maria Novella, Leon Battista Alberti
Florence, Italy 1470
[Renaissance façade, rectangular nave with transept, masonry]

The façade of Santa Maria Novella is an example of two-dimensional modules working together to form a comprehensive whole. Alberti inherited the façade in an unfinished state and was forced to work within a specific set of contextual parameters. For Alberti, the challenge was to superimpose a façade that used classical language onto a traditional Gothic façade shape. He achieved this by using the geometry of the square. The lower portion of the façade forms two equal squares that are placed side by side. The temple front on the upper level forms the third equal square. On either side of this temple front are two scrolls that are one quarter of the original square module and are used to cover the anterior roof form.

Nakagin Capsule Tower, Kisho Kurokawa
Tokyo, Japan 1972
[Japanese Metabolist, residential tower, steel and fiberglass]

The Nakagin Capsule Tower, by Kisho Kurokawa, is a rare example of Japanese metabolism. It uses the concept of a three-dimensional module in a very clear way. The building is a series of capsules or pods that are clustered around two vertical circulation towers. Each pod is easily identified as a complete unit, both in plan and three-dimensionally. Each capsule has a circular porthole, which helps to identify it as an identical but separate unit within the complex matrix of the tower. Capsules are randomly removed as a way of providing differentiation and interest to the overall massing of the tower.

Distorted Geometries and Computational Complexities

Distorted geometries have emerged out of geometric agendas rooted in both method and form. The sophistication of digital technologies in recent decades has allowed for an acceleration of complexities, providing greater ease of representation, visualization, manipulation, and translation of ideas into fabrication. The result has been the development of varying categories of geometry. Interactive geometries allow for anthropomorphic, cultural, and site-based influences to engage the geometric system. Responsive geometries integrate technologies that allow for the translation of influential forces into either static or active form. Performative geometries functionally respond to practical or environmental necessities through a geometric response. Organic geometry, or bio-mimicry, looks towards nature for formal and functional responses. Haptic geometries engage fragmented formalism for a highly sensorial response to the geometry. The diverse systems and interest in new, more complex formalisms that emerge from geometric processes and computational systems create an ever-increasing interest in non-orthogonal systems and distorted geometries.

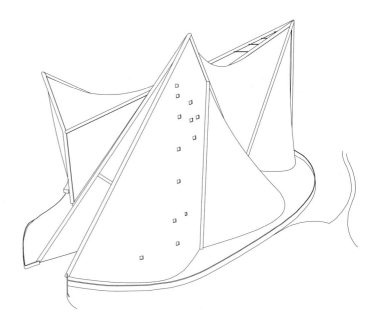

Phillips Pavilion, Le Corbusier
Brussels, Belgium 1959
[Modern, free-form exhibition pavilion, metal tensile structure]

The Phillips Pavilion is one of a series of late projects in which Le Corbusier engages a more expressionist geometric formalism. Here he adopts the structural geometric shape of the hyperbolic paraboloid shell. The ground plane is defined by the base geometry. A series of diagonals establishes the vertical masts from which the remainder of the shell is determined by the connective array of a ruled surface. Triangulated frames establish the dominant edges allowing the iterative and incremental subdivided surface lines. The heroic form and scale, in conjunction with the thinness and delicacy of the material deployment, produce a dynamic figuration that is intrinsically derivative from geometry. The forms become calculated through the regulatory systems of their distorted geometry.

GEOMETRY / PROPORTION PRINCIPLE **327**

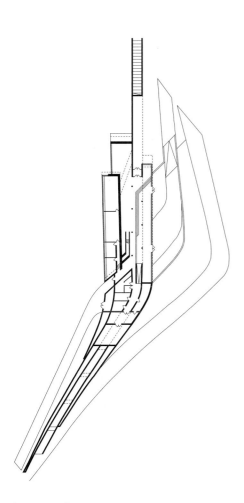

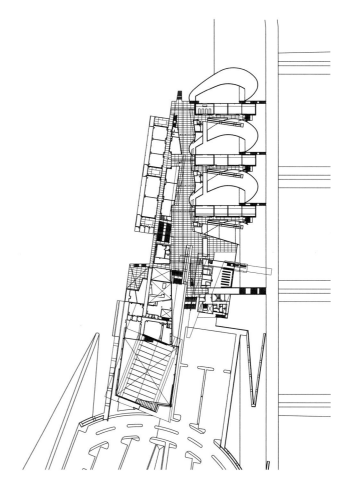

Vitra Fire Station, Zaha Hadid
Weil am Rhein, Germany 1994
[Deconstructivist, formally expressive parallel bands, concrete]

As one of Zaha Hadid's early built works, the Vitra Fire Station manifested the methodology of her projective geometric paintings. Engaging a continuity of form through a bundling of fluid shapes, the smooth formalism of the elongated surfaces are choreographed to produce accelerated perspectival vantages along a dynamic sectional promenade. Governed by expressionist formal responses, the surface is highly sculpted to produce the effectual composition. Produced through the representational methods of two-dimensional graphic techniques, the Vitra Fire Station is a three-dimensional spatial translation. Their manifestation translates from the single, fixed vantage to a mobile perspective along a sequence. The resulting smoothness and fluidity of the elongated figuration define a highly specific formalism emerging from a methodological geometric process.

Diamond Ranch High School, Morphosis
Diamond Bar, California 2000
[Expressionist Modern, double-loaded exterior corridor, metal]

This school does not use a single ordering system, but instead relies on expressionism as a means of formalizing its concept. The project contains a series of linear elements that are composed through a system of distortions meant to focus attention and highlight moments and activities within the campus. These visual and spatial focal points act as gathering spots for the students, resulting in coalescence of social, physical, and architectural activity. The overall composition is defined and rendered by the distorted composition of its geometric intention.

328 SYMMETRY

PRINCIPLE

Symmetry

08 - Symmetry

Symmetry is defined as having two equal sides that are mirror images of one another reflected across an axis. Symmetry is a foundational architectural practice rooted in the order and rationale of classicism and remains a natural ideal (the human form is intrinsically symmetrical) as well as a built ideal evident across scales and cultures. A building is most recognizably symmetrical when an axis, or a line of symmetry, can be drawn in the middle of the image and as a consequence what is on one side is conversely on the opposite side. Symmetry has been used as a way of achieving balance within an architectural composition. During modernism this compositional idea was challenged as architects began to examine the complexities of non-symmetrical, field-based compositions. Designers were now influenced by modern painting and sculpture, which eliminated the rigidity of representation and overt order; symmetry, which had been the rule until that point, now became the exception. Regardless, symmetry never seems to disappear from the architect's palette and it continues to be a powerful organizational and experiential tool even today.

Bilateral Symmetry Bilateral symmetry, or axial symmetry, explicitly describes a type of symmetry that is divided into two equal halves. This is the most common type of symmetry in architecture and is most often applied to façades and plans. This type of simple symmetry gives buildings a sense of balance and harmony. It was used across cultures and programs until the advent of modernism in the early twentieth century. It is founded in a belief that as humans are symmetrical forms, architecture should follow the same rule. Bilateral symmetry occurs not only at the scale of architecture, but also in urban design. Here it is used to create large symmetrical boulevards and blocks that produce hierarchical moments within the fabric of both ancient and modern cities.

Asymmetry Asymmetry refers to a composition that is not purely symmetrical but achieves an intentional differential balance or imbalance to the collective. It is important to note that there is a difference between a building that is asymmetrical, and an architectural endeavor that is conceived without the use of symmetry. The fundamental difference between these two is that an asymmetrical architectural composition is visually or conceptually balanced and emerges out of careful consideration, whereas a non-symmetrical composition simply does not address the issue.

Local Symmetry Local symmetry refers to a condition whereby a symmetrical condition occurs but is applied on a particular fragment or part of a composition. This allows for a collective composition to be asymmetrical, while embedding within it moments of constructed order governed by symmetry. This type of symmetry is typically more common in façades than in plans or urban design as they are more legible moments of part to whole. The use of local symmetry allows the establishment of a balanced tension between the various parts. The collective reading of part to whole then establishes a secondary tension and dialogue.

Material Symmetry Material symmetry deploys materials and their tectonic of assembly to construct the visual order and balance associated with symmetry. Typically this condition is matched with bilateral compositional symmetry; however, the material and its language can take precedent, allowing the composition to be dominated by its material articulation. The modern movement witnessed a rise in the use of material symmetry as architects searched for ways to balance compositions without relying on the strict formal confines of bilateral symmetry.

Programmatic Symmetry Programmatic symmetry is achieved by balancing functional spaces within a plan or composition. Often these spaces are similarly shaped, but this type of symmetry uniquely allows for an imbalance of form despite a balance of function. Programmatic symmetry, like material symmetry, often takes place in a bilaterally symmetrical organizational composition. Despite this, programmatic symmetry offers a criticality that can dictate the conceptual framework of a project.

Bilateral Symmetry—Architectural Plan

Bilateral symmetry is rooted in the theoretical connection to the innate symmetry of the human body. Designers, fascinated with the similarities between nature and architecture, noted that the symmetry of the human figure has a rational ordering that governs and produces balance. Through the emulation of nature, architecture translates a higher order. The overtness of this order is typified through the deployment of bilateral symmetry.

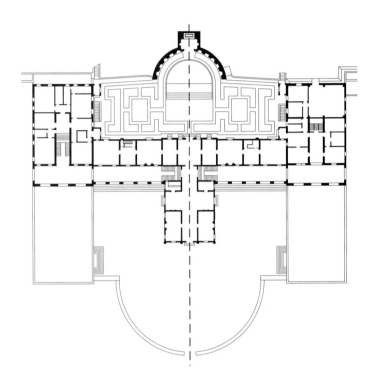

Villa Barbaro, Andrea Palladio
Maser, Italy 1560
[Renaissance, modular arcade, masonry]

The Villa Barbaro is a quintessential example of bilateral symmetry. The building is organized directly along a linear axis that runs through the center of the villa. This axis begins in the countryside, passes through the building, and terminates in the grotto that is located along the back of the villa. This axis is organizational only; due to the shifting sectional levels of the villa, it does not represent an occupiable path. There is a cross axis that similarly runs through the building; however, it does not demand the same symmetry as the center axis. The building is symmetrical around this center axis in both plan and elevation. Symmetry is a tenet of Palladio's work and his belief in its importance is seen continually throughout his career.

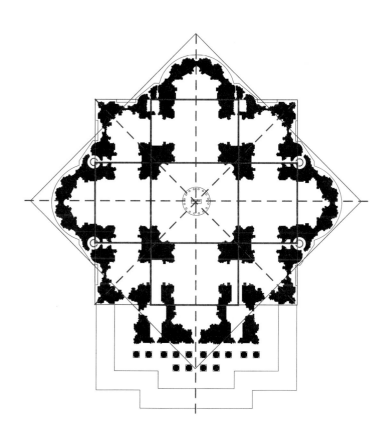 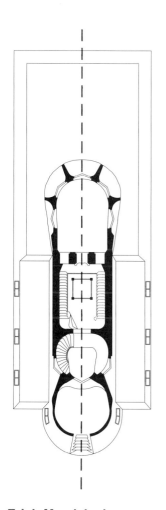

St. Peter's Basilica, Michelangelo Buonarroti
Rome, Italy 1547
[Renaissance, dome and Greek Cross plan, masonry]

Michelangelo, like Palladio, organized all of his architecture around the principle of symmetry. The plan of St. Peter's Basilica is an example of symmetry working in multiple directions. Michelangelo's plan for St. Peter's Basilica is in the shape of a Greek Cross, a plan type that is symmetrical about not only the two main axes, but also the diagonal axes. The multiaxial and symmetrical plan of the Greek Cross was considered perfection to the architects of the late Renaissance because it exploited symmetry more completely than a Latin Cross.

Einstein Tower, Erich Mendelsohn
Potsdam, Germany 1921
[Expressionist Modern, observation tower, masonry]

The Einstein Tower is one of the first examples of expressionist architecture. As such it was expected to form a break with many of the traditions of architecture that had become prevalent at that time. Despite these desires, Mendelsohn continued to rely upon the formal and compositional power of symmetry. The unique morphology of the tower explores aspects of architecture that were extremely novel. In order to balance the complexity and ingenuity of these forms, symmetry allowed a superstructural organizational order and geometric regulation.

Bilateral Symmetry—Architectural Elevation

Bilateral symmetry is most recognizable when deployed in elevation. Unlike the plan, which is only legible through a chronological mental map or a representational diagram, the elevation is readily consumable. This legibility and overt nature of the compositional organizer gives the elevation a tremendous amount of power and responsibility. Bilateral symmetry, when used in the composition of an elevation, innately imposes distinct ideals upon the building and organization as a whole. With inspiration in hierarchies of politics, nature, and religion, the architecture allows for an overt manifestation in built form and composition. The balance, legibility, and predictability of elevational symmetry established a familiarity that continues to make it a prevalent condition of usage.

San Sabastiano, Leon Battista Alberti
Mantua, Italy 1475
[Renaissance, Greek Cross votive church plan, masonry]

The elevation of San Sabastiano, by Alberti, is a classic example of bilateral symmetry in elevation. Like many of the architects of the Renaissance, Alberti was adamant about the power of symmetry and how it controlled architectural compositions. Unlike San Andrea (also in Mantua), San Sabastiano's elevation has little to do with the interior of the church. The façade was seen as a mask that depends on its symmetry to synthesize the entire project into a balanced and honorific whole. Unfortunately, Alberti was unable to finish the façade to his specifications, and only portions of the original remain.

	SYMMETRY
	PRINCIPLE **335**

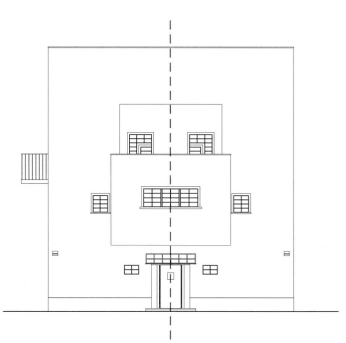

Altes Museum, Karl Friedrich Schinkel
Berlin, Germany 1830
[Neoclassical, museum, masonry]

The Altes Museum, by Schinkel, is a symmetrical building with a façade that responsively addresses the scale of its context. Located adjacent to a very large urban space, the façade accordingly needed to present an appropriate face. The urban space lacked a dominant axis, leaving Schinkel free to compose the order of the space through the articulation of his elevation. Ultimately the elevation takes its cue from its urban context, appropriately scaling itself to symmetrically engage the scale of the adjacent space. Schinkel's façade is a series of large columns organized in a repetitive colonnade that perceptually continues indefinitely. The translation of this scale and order to the internal ordering is entirely masked in the composition.

Villa Moller, Adolf Loos
Vienna, Austria 1927
[Modern, Raumplan, sectional arrangement of spaces, masonry]

The Villa Moller, by Adolf Loos, illustrates his ideas about the private façade. Loos believed that the house should not offend anyone, and to maintain a traditional reference to external order, most of his residential structures are rendered with a symmetrical façade. Loos understood that the symmetrical façade was culturally something that everyone understood and found familiar. Each house presented itself as a good neighbor. The order of the exterior was then contrasted with the spatial hierarchies of the interior and governed by the sectional organization of the raumplan. For Loos, the façade was really a face that presented a correct and comfortable image to the world, allowing only the interior of the house to break from symmetry and reflect the complex inner workings of the owners' lives.

Bilateral Symmetry—Urban

When deployed on the urban scale, bilateral symmetry emphasizes a hierarchical and grandiose experience on the scale of the city. Establishing axial conditions with highly ordered duplications across the line of symmetry makes for important urban pathways and heroic spatial rooms. The balance of repeated figuration intrinsic to symmetrical spaces, when taken to the massive urban scale, results in powerful effects formed by the residuals of a simple move employed on a massive scale. The balance of form, produced through the mirroring action, creates highly ordered compositional spaces that result in urban monuments of the fabric itself.

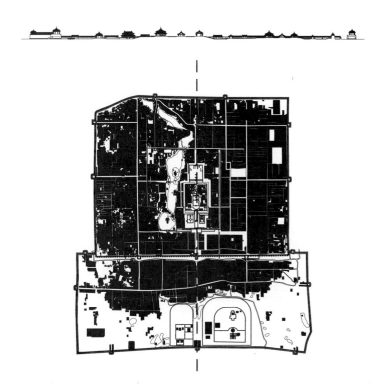

The Forbidden City
Beijing, China 1420
[Chinese, central axis with nodes, masonry and wood]

Built as the Imperial Palace of the Ming Dynasty, the Forbidden City has served as the residence of the emperor and his household as well as the ceremonial and political hub of the Chinese government. Organized to establish levels of hierarchy and ceremony, the extensive compound consists of nearly one thousand buildings. Positioned with strict hierarchies along a dramatic central axis, the buildings are governed by a relentless symmetry that repeats and maps the geometric purity of the balanced plan. The resulting composition is a ceremonial complex that is heroic in both scale and effect. The organization allows for the hierarchies of social and political interaction to be exquisitely orchestrated.

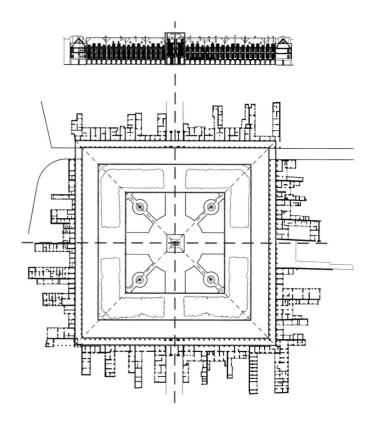
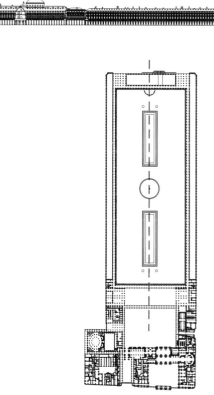

Place des Vosges, Baptiste du Cerceau
Paris, France 1612
[Henry IV style, repetitive bayed symmetrical square, masonry]

As the oldest planned square in Paris, the Place des Vosges is one of the first examples of royal city planning. It is a true geometric square measuring 140 m x 140 m. The square is defined by the uniformity of its repetitive bays of red brick interlaced with strips of stone **quoins** *over vaulted arcades. The north face, interrupted by the Pavilion of the King, and south face, similarly interrupted by the Pavilion of the Queen, are given hierarchy through the rise in the otherwise unified roofline. These pavilions offer access through their three archways and reinforce the axial line of organizational symmetry. The harmony of the whole is accomplished through the uniformity of the repetitive façade, the purity of the geometric figure, and the axial symmetry of the surface and space alike.*

Palais-Royal, Jacques Lemercier
Paris, France 1629
[French Baroque, axial rectangular courtyard, stone]

Originally known as the Palais-Cardinal, the Palais-Royal was built as the residence of Cardinal Richelieu, the chief minister to King Louis XIII. Its organization is defined by a narrow perimeter of residential spaces surrounding a rectangular central courtyard encircled with a colonnade. Announced with a dramatic gateway, a central axis is established with the heroic main entry. The repetitive bay and the tripartite banding of the elevation make a pure and regularized space that imparts order and power while creating an intimate room within the Parisian fabric.

Asymmetry

As a concept, asymmetry is not merely the opposite of symmetry, but refers to a compositional balance that occurs without a direct mirroring of form. If a building is not symmetrical and this condition is accidental, then it is obviously very different from a building that is purposefully designed in an asymmetrical fashion. For architects before the art nouveau or modern movement, the default was to compose façades in a symmetrical fashion. These new movements introduced complex themes of balance that moved beyond simply mirroring. It is only when there is an intent behind the asymmetry that it can be discussed within this particular architectural context. Here a larger sophistication of balance is expected to accomplish a similar regulatory balance traditionally provided by symmetry.

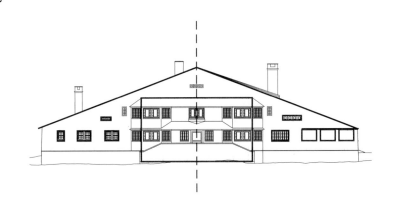

W. G. Low House, McKim Mead and White
Bristol, Rhode Island 1887
[Shingle Style, house, wood frame]

The Low House has an overall symmetrical form with asymmetry occurring through the orchestration of the details. The main façade is dominated by the large triangular form of the roof, which blankets the form of the entire house. The two sets of bay windows are identical, establishing a sense of symmetry. In contrast, the lower right side of the house is developed as a large exterior porch, forcing the building into an asymmetrical but completely balanced composition. The non-symmetrical position of the chimneys prioritizes functional need over the absolute purity of the composition.

SYMMETRY
PRINCIPLE **339**

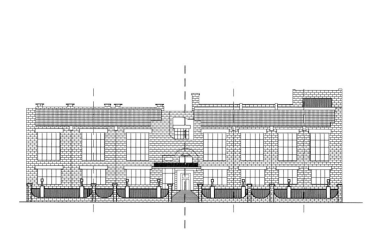 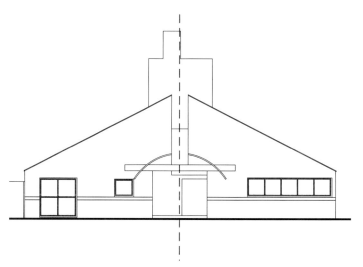

Glasgow School of Art, Charles Rennie Mackintosh
Glasgow, Scotland 1909
[Art Nouveau, linear plan school, masonry]

The Glasgow School of Art serves as a quintessential example of asymmetry in Mackintosh's work. Though the plan is essentially symmetrical in overall form, Mackintosh makes small but critical adjustments to the building's façade (and particularly the entrance) to create slightly different compositional layouts to the two identically sized wings of the building. The left side has three bays of windows, with each window containing five panes. The right side has four bays, with two windows that have five panes and two that have four panes. The entrance door and the director's window above it are shifted slightly to the right to compensate for this variation. The fence situated in front of the building marches along, establishing an irreverent datum to this phenomenon. The building is subtly asymmetrical, finding balance through composition.

Vanna Venturi House, Robert Venturi
Chestnut Hill, Pennsylvania 1964
[Postmodern, mannerist plan house, wood]

The Vanna Venturi House has symmetry to its overall form, but deploys asymmetry in the window composition. The right side has five windows in a line, referencing the Corbusian strip window. The left side employs the same five windows; however, here Venturi has gathered four of them into a giant four-square-window door with the fifth window placed to its right, recalling the strip windows on the other side of the entrance. Venturi has achieved complete balance within this asymmetrical scheme.

Local Symmetry

The use of symmetry does not need to occur exclusively on the superstructural level. Often there are conditions within a composition when a number of objects or elements form or have local symmetry. This condition can occur as a module within a symmetrical façade or as an element within an asymmetrical façade that acts as a focal point within the overall composition. It is a fairly sophisticated operation, requiring the architect to be aware of the asymmetries of the composition, how this one moment of symmetry will affect the façade, and what it means in terms of the building itself.

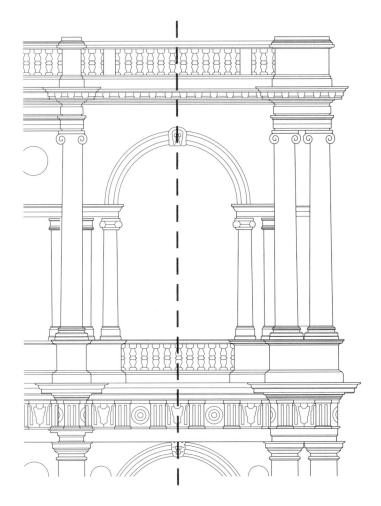

The Basilica, Palladio
Vicenza, Italy 1549
[Late Renaissance, basilica, stone façade]

The Palladian windows of the Basilica in Vicenza are an example of local symmetry occurring in the detail of a repetitive module. This window type is described as an arch with two accompanying columns on either side. Serlio was responsible for the origins of the motif, though it was certainly popularized by Palladio, thus deriving its name. The element, no matter its location on a façade, demands attention due in part to the compositional power of its internal symmetry, which relies upon the central arch and its consequential relationship to the matching columns on either side. This motif is intrinsically symmetrical, regardless of its location or context within a larger composition.

	SYMMETRY	
	PRINCIPLE	**341**

 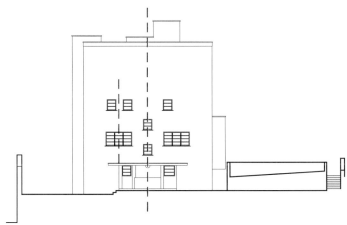

Casa Mila, Antonio Gaudi
Barcelona, Spain 1910
[Art Nouveau, apartment building, stone façade]

The Casa Mila's primary form is defined by an undulating stone skin. Within this organic formalism, local symmetries are deployed to provide datums of order. Fragments of the building, such as the elevational cluster to the left of the main corner entrance and the most public portion of the building, are in fact completely symmetrical to provide geometric regulation and hierarchy. Gaudi allows the building to begin as symmetrical; as it travels away from the entry, the elevation undulates and becomes more irregular. The local symmetry of this portion acts as a base that anchors the remainder of the façade.

Villa Müller, Adolf Loos
Prague, Czechoslovakia 1930
[Modern, Raumplan house, masonry]

Likely due to its remote location and legibility as a free-standing object, the Villa Müller broke the tradition of Loos's earlier residential projects in its use of symmetry. Typically, only the public faces of Loos's houses exhibited symmetry. At Villa Müller, local symmetry is used on all four elevations, each of which has a line of symmetry running through it. Each elevation is balanced and yet contradicted by the isolated placement of select windows. Consequently, each elevation has a local symmetry that is first set up and then broken by Loos. Villa Müller remains one of the most sophisticated series of elevations to come from the modern movement and clearly demonstrates Loos's belief in the organizational power of exterior composition.

Material Symmetry

Material symmetry refers to the reliance upon material selection and use to produce visually effectual organizational geometries. Still deploying the conventions of symmetry, with axial lines of mirroring and visually balanced compositions, material symmetry is different in that the form is subordinate to the matter of which it is made. Though the two often coincide, legibility occurs at the material and tectonic scale. Thus the symmetry relates to the dimension of the piece, the module, or the overall material composition of the building as a whole.

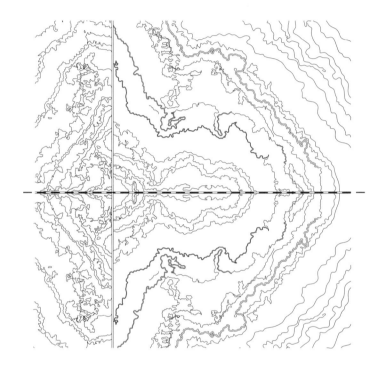

Barcelona Pavilion, Mies van der Rohe
Barcelona, Spain 1929
[Modern, free plan, steel, glass, and masonry]

The Barcelona Pavilion deploys the free plan with a clear agenda of defining a specific moment within an infinite space. The fluid flow of space from function to function and from inside to out allows the masterful and highly evocative ambiguity of edge. The hyper-articulated material palette introduces a richness and regality to the formally simplified and minimally detailed components. The celebrated marble walls are book- matched, a process through which the stone is cut in parallel slabs and positioned as though unfolded like the pages of a book. The result is the production of a line of symmetry at the seam. Spatially, the marble is the only material that has an illustrated grain, scale, and detail. As a result, the dominance of this mirroring presents itself as an organizing reading of the entire building as a sectionally symmetrical space, equal up and down. This bold, formal organization is reminiscent of the order and systemization of monumental classical architecture, but is adapted as highly modern through the reorientation of the axis.

SYMMETRY
PRINCIPLE 343

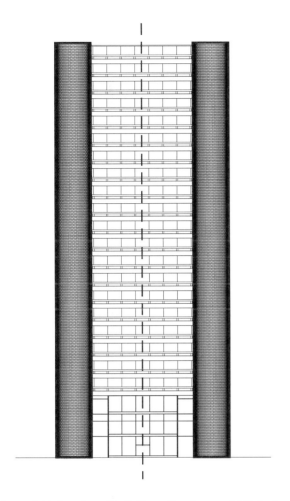

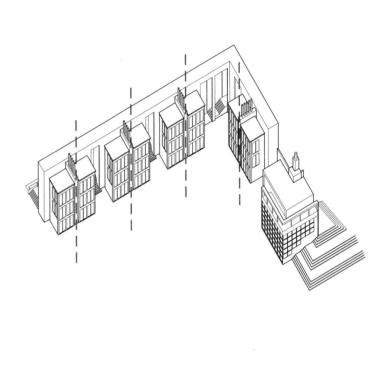

Knights of Columbus Building, Roche and Dinkeloo
New Haven, Connecticut 1969
[Late Modern, symmetrical corner pillars, concrete and masonry]

The Knights of Columbus Building is a simple office tower organized with a central elevator core. The remainder of the floor plate is left open and flexible with the four corners highlighted with brick-clad concrete service towers containing the fire stairs and restrooms. The resulting elevational effect of this functional decision produces massive symmetrical towers. Symmetrically positioned swatches of opaque clay masonry produce a collectively simple and uniform figure that is equal in all directions and anonymous in presence. The figurative and material symmetries occur at the scale of the whole building's composition, producing a heroic effect and scale.

The Inn at Middleton Place, Clark and Menefee
Charleston, South Carolina 1986
[Late Modern, symmetrical room towers, wood and masonry]

The Inn at Middleton Place uses localized material symmetry to create bays of residential rooms that are stacked three high and separated with a thickened wall that serves as the mirror line. The localized material symmetry produces repetitive modules that reoccur rhythmically across the long length of the L-shaped plan. The local order based in material creates a heightened compositional effect through the provisional regulation of the repetitive hotel room units, while permitting an irregularity to the overall composition that is governed by site, scale, orientation, and formal composition. The repetitive wood units are anchored by a variant node that houses the common shared spaces and furthers the asymmetry and localized hierarchies of the collective organizational composition. To highlight this internal order, the material shifts from stuccoed masonry to a boldly painted, black clapboard siding that draws and holds the eye with its stark contrast.

Programmatic Symmetry

Programmatic symmetry is a powerful organizational system. Inherently, it can only be enjoyed by the direct users of a building. It is a type of symmetry that is intensely connected to the plan and the functional workings of a building and therefore cannot be easily consumed by a study of an elevation or by a third-person user. Typically there exists some reference to a bilateral symmetry that is part of the project, though programmatic symmetry typically takes precedent over other forms of symmetry due to its dominance of function and typological reference to history.

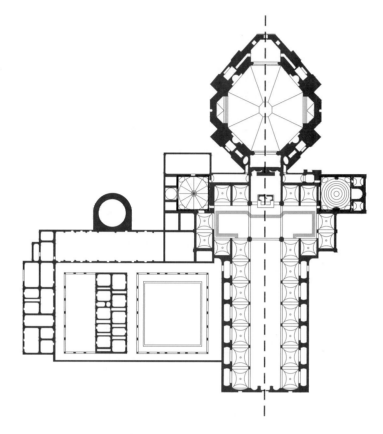

San Lorenzo, Brunelleschi and Michelangelo
Florence, Italy 1459
[Renaissance, Latin Cross church, masonry]

The Old Sacristy, by Filippo Brunelleschi, and the New Sacristy, by Michelangelo, form a magnificent programmatic symmetry within the Church of San Lorenzo. The Old Sacristy is a square room that opens off of the north transept of San Lorenzo. Brunelleschi designed it as a perfect square; in section it is a cube intersected by a sphere. It was an early example of Renaissance architecture that had a profound effect on later architects. Almost one hundred years later, Michelangelo designed the New Sacristy and had it placed exactly opposite the Old Sacristy, producing a seminal example of programmatic symmetry. Though Michelangelo's sacristy was more complex in every way, the functional shape, and ultimately the location of the plan, remained equal to the original by Brunelleschi.

SYMMETRY PRINCIPLE 345

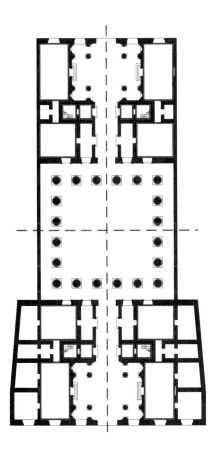

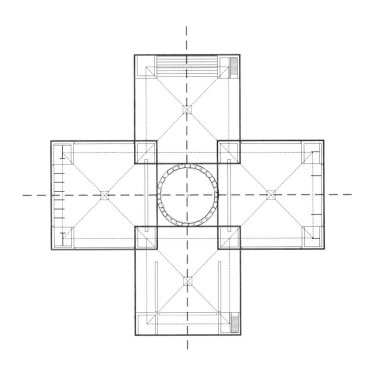

Palazzo Porto, Andrea Palladio
Vicenza, Italy 1544
[Late Renaissance, Courtyard Palace, masonry]

The Palazzo Porto, though unfinished, was originally designed as two identical buildings joined (or separated) by a large courtyard. Palladio conceived of this building as a rather conventional palace type, but instead of positioning the court in the center of the palace, as was the tradition, he doubled the building and placed its twin on the other side of the courtyard. Palladio likely made this decision due to the fact that the shape of the site was very long and not very wide and hence evoked programmatic symmetry as a solution.

Trenton Bath House, Louis Kahn
Trenton, New Jersey 1959
[Modern, Greek Cross bath house, masonry and wood]

The Trenton Bath House exemplifies programmatic symmetry through its organizational division of the building by gender. Each of the two balanced changing rooms occupies an opposing side of the Greek Cross, accommodating male and female users respectively. This symmetry is made more manifest by Kahn's positioning of pyramidal roofs over each of the pods, thus making them more separate and identically linked at the same time. The project is so intent on the governance of programmatic symmetry that other ideas such as hierarchy and entrance seem to fade away.

346 HIERARCHY

PRINCIPLE

Hierarchy

09 - Hierarchy

Hierarchy remains one of the most complex aspects of architectural principles. Its complexity arises from the fact that unlike many aspects of architecture, hierarchy is not completely self-evident and instead must often be interpreted. This interpretation allows for multiplicity and innate uncertainty. It is, however, perhaps the nature of this uncertainty that allows hierarchy to remain a relevant and evolving architectural principle. By its very nature, hierarchy will never be abandoned by architects; its descriptive nature, founded in the interrelationship of diverse forms, remains a principle tenet of the profession. It remains a generative mechanism for nearly any aspect of any project, be it a plan, façade, room, or city.

Formal / Geometric Hierarchy Formal, or geometric, hierarchy is the most overtly recognizable type of hierarchy. It occurs when a building develops contrast through shape and form. This does not imply that a building must be foreign or chaotic; it refers simply to the use of contrast and differentiation of shape from compositional elements within or surrounding it.

Urban formal hierarchy occurs when there is a clear break in an urban system that allows for a different organization to take place (for instance, Frank Lloyd Wright's Guggenheim Museum in New York City). Again, there are no clear repetitive models in terms of this type. The only clear and important aspect is that the urbanism does not synthesize with the surrounding fabric. For example, one could imagine a medieval block within a Renaissance framework or vice versa.

Axial Hierarchy Axial hierarchy describes a system in which an axis dominates a condition, placing hierarchical emphasis on the terminus. There are countless buildings that exercise this rule, such as the church typology founded on the axis that exists between the entrance and the altar. Axial hierarchy can occur at a multitude of scales, spanning from the dimensions of a single room to the scope of an entire building complex.

Urban axial hierarchy is a condition in which an axis, or a series of axes, organizes and controls entire sections of a city. This typically only happens when politically it is possible. To have control over the streets of an entire city is something that in today's society seems nearly impossible. In the past, however, popes, emperors, and kings were able to redesign entire sections of cities and exercise any number of urban axial hierarchies. These conditions were typified by a series of axes that connected hierarchical nodes in the fabric, cutting through the city and crossing to form critical moments and spaces.

Visual / Perceptual Hierarchy Visual, or perceptual, hierarchy is intrinsically one of the more complex types of hierarchy. It relies largely on perception, which varies from person to person. Visual hierarchy occurs when one is confronted by a view that is critical to understanding an architectural, landscape, or urban project. The orchestration of this view can be aided by an axis or collective organizational system, or it can be locally orchestrated and discovered through a series of discrete moves that culminate in a visual revelation.

Hierarchy of Scale Hierarchy of scale relies upon the gradient of size as a way of conveying importance. Hierarchy of scale is highly dependent on site as a way of comprehending and understanding the relationships between the elements in question. It is typical to explode or increase scale as a way of communicating significance; however, this reading can be reversed and instead utilize smaller elements as a way of communicating importance. This principle applies not only to architecture, but also to how buildings are perceived within an urban context and how scale effects their perception. In many urban conditions there exist moments where scale plays a tremendous role in the understanding and comprehension of the city as a whole. Typically these moments are recognized by an increase in scale; however, it can and often does work in reverse.

Hierarchy of Monument Monuments occupy a unique role in terms of hierarchy. Their very nature ensures them of a certain hierarchical consideration. The hierarchy of monuments is often increased by their location, scale, materiality and form. All of these aspects contribute to one of the most powerful and recognizable types of hierarchy.

Hierarchy of Visual Control Hierarchy of visual control refers to a condition in which the designer limits and orchestrates the user's perception as a way of expressing an architectural idea. This can occur on multiple levels; the designer may conceal, or reveal, elements of interest or may provide visual cues that lead to the discovery of a space. This type of hierarchy can also be utilized as a method of control through cropping and revelation, orchestrating the sequence, quantity, and content of an experience.

Programmatic / Functional Hierarchy Often architects will separate the functions of a building and articulate them in such a way that their relationships take on a hierarchy within the building itself. This type of hierarchy, for example, can manifest itself through form, material, and axis. The functional parts orchestrate their own articulation and form a hierarchical relationship within the architecture itself.

Color Hierarchy Hierarchy through the use of color is a common and yet powerful way to express relativity of significance in an architectural project. This effect is achieved by contrasting one color with one other and establishing a relativistic hierarchy between the parts, or by camouflaging elements through the use of color to cause a varied reading within an architectural context. Colors carry with them certain connotations that architects have attempted to exploit for centuries.

Material Hierarchy Material hierarchy has always been a critical aspect of the architectural profession. Using materiality as a way of imparting importance to an object is a fundamental concept across cultures. The exploitation of material value to relate to architectural significance is a direct translation of materials to the importance of an architectural moment.

Formal / Geometric Hierarchy—Architecture

Formal or geometric hierarchy is achieved by contrasting a system with a dissimilar form. Operating with a relativism to ones surroundings, the eye naturally identifies variation within a pattern. This transitional quality of individuation is fundamental to the principles of hierarchy. Differentiation alone, however, does not imply a hierarchy—it is only a difference. It is the complexity of the relativism and the interrelationship that establishes a dominance or hierarchy.

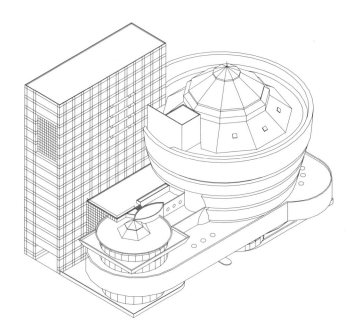

The Guggenheim Museum, Frank Lloyd Wright
New York City, New York 1959
[Modern, ramping spiral gallery around atrium, concrete]

The Guggenheim Museum in New York City derives its form from its organizing interior spiral ramp. This shape is expressed throughout the structure and the exterior form, distinguishing the museum as a singular object of uniquely variable geometry amongst the other orthogonal buildings along the avenue. Sited directly opposite Central Park, this urban spatial condition allows it to be seen from a distance, heightening the form of the building. It is a simple concept to contrast an unusual shape against regular shapes; however, Wright weaves the building into the urban context in such a way that the transition is well conceived and not as formally jarring as it could have been.

HIERARCHY PRINCIPLE 351

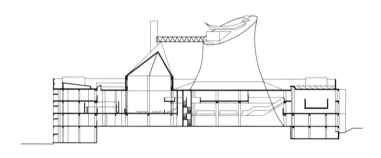

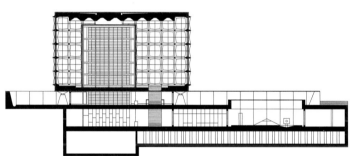

Chandigarh, Le Corbusier
Chandigarh, India 1963
[Modern, parliament house, concrete]

The Palace of Assembly by Le Corbusier is the central figural composition of Chandigarh. The interior of this building is unusual in that it contains a relentless field of columns. Placed just off center within this field of columns is the main meeting hall that is articulated with a dramatic curvilinear form. The contrast of this circular form against the orthogonal perimeter boundary establishes a quintessential example of formal or geometric hierarchy. The meeting hall, with a complex, curvilinear three-dimensional shape, is reminiscent of a nuclear cooling tower. The juxtaposition of this looming figure against the field of stick-like columns is extremely powerful. Moving through the open rows of columns, the user is constantly aware of the large, curvilinear shape hovering nearby. The form of the meeting hall extends through the roof and is illuminated by a massive skylight, creating a dramatic space that announces its hierarchical governmental purpose throughout the complex.

Beinecke Rare Book Library, Gordon Bunshaft, SOM
New Haven, Connecticut 1963
[Modern, centralized glass stack, concrete and masonry]

The Beinecke Rare Book and Manuscript Library is a severe rectangular box with no windows. From the exterior it seems to hover in space. Surrounded by self-similar neo-Gothic buildings on the Yale University campus, contrast is achieved through the juxtaposition of the large, field-based, rectangle with the traditional ornament and detail of its context. The library employs the difference of form as a way of achieving formal hierarchy. Here this is accomplished through a removal of detail, which juxtaposes style, ornament, and simplicity of form.

Formal / Geometric Hierarchy—Urban

Formal or geometric hierarchy within the urban condition operates as it does in architecture. Hierarchical urban moments are achieved by inserting urban elements that are in contrast with their counterparts. Typically these elements take the form of an open space or void situated against a denser field. They can also take the form of axial subtractions from the fabric of the city, connecting and ultimately focusing on one hierarchical moment. When there is a local change in the quality of the urban fabric, through density, scale, or form, an urban formal hierarchy has been achieved.

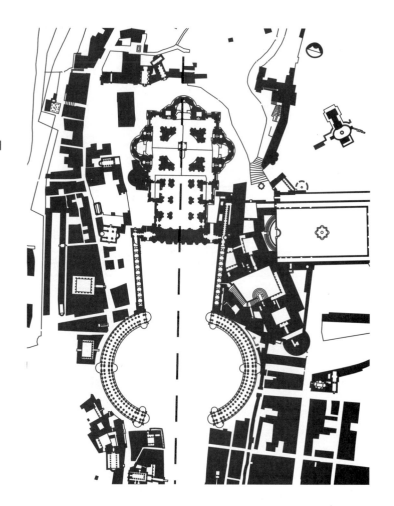

St. Peter's Basilica, various
Rome, Italy 1667
[Renaissance, dome and Greek Cross plan, masonry]

The Vatican complex establishes an urban formal hierarchy through the governing axis of St. Peter's Basilica, which formally connects the Vatican to the city of Rome. Completed by Piacentini during the 1940s, the axis runs from the façade of St. Peter's Basilica, through the obelisk, and continues on to the Castel Sant'Angelo. It dominates the surrounding urban fabric and creates an important moment of urban formal hierarchy. The size and shape of Bernini's piazza, on the other hand, exemplifies a formal hierarchy through its ovaloid geometry and repetitious perimeter colonnade. The hierarchy of the ovaloid geometry creates an invisible barrier that demarcates one's entrance into a different spatial and hierarchical condition.

HIERARCHY PRINCIPLE 353

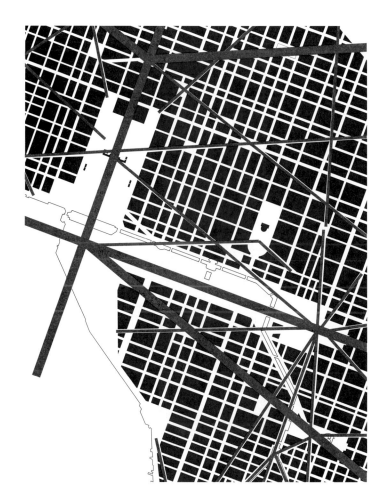

The National Mall, Pierre L'Enfant
Washington, D.C. 1791
[Neo-Baroque, large, open, green space, vegetation]

The National Mall is a large, flat, open lawn that occupies the center of L'Enfant's plan for Washington, D.C. The urban plan was modeled after European baroque plans that emphasized the ability of grand boulevards to create a connection between monuments. Within the urban fabric of the District of Columbia, which has numerous circles and diagonals, this orthogonal strip establishes a central-dominant formal urban hierarchy. This hierarchy is strengthened by the placement of politically, culturally, and commemoratively prominent buildings in its vicinity. The U.S. Capitol, backed by the Library of Congress and the Supreme Court, initiates the mall; the axis continues through the Washington Monument to the Lincoln Memorial. A cross axis is formed at the White House. The edge buildings house the Smithsonian Institution. Through its neoclassical style, consistent marble material, and formal geometric urban planning, the collective composition overtly identifies the hierarchy of place and function.

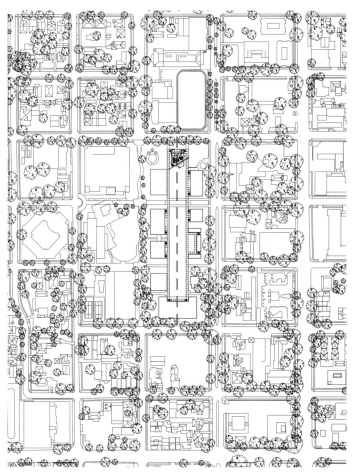

University of St. Thomas, Philip Johnson
Houston, Texas 1959
[Modern, axial university, masonry and steel]

Philip Johnson's design for the University of St. Thomas is a hybrid of the organization of Jefferson's University of Virginia and the form and tectonic of Mies van der Rohe's Illinois Institute of Technology. Johnson uses an axial condition in combination with the scale and openness of the suburban landscape to establish the formal hierarchy of the central quadrangle. The main space of St. Thomas is surrounded by the most important buildings, with the chapel and library facing each other on the main axis. They are linked by a delicate steel colonnade that creates a uniform connecting perimeter and assigns a singularity and hierarchy to the resulting urban void.

Axial Hierarchy—Architecture

Axial hierarchy is simple and overtly recognizable. It has been used for thousands of years across diverse cultures to express importance and rank. Axial conditions, like symmetry, are a natural consequence of man's desire to create order out of chaos. Visually, this alignment of openings or objects carries with it ideas of authority and power.

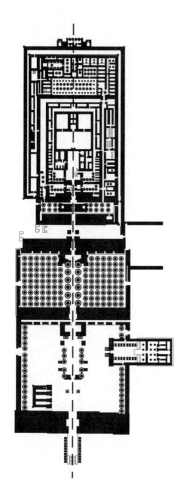

Temple Complex at Karnak, Pharaoh Ramses II
El Karnak, Egypt 1351 BCE
[Egyptian, temple, masonry]

The temple complex at Karnak is an archetypal example of axial hierarchy. The placement of six pylons that mark the major transitions along the axis, from the Processional Way to the Courtyard of Middle Kingdom, also denotes the importance of the various sequential spaces. Axial hierarchy is intensified by the constriction of space, which becomes progressively smaller with each transition into the depths of the temple.

HIERARCHY

PRINCIPLE 355

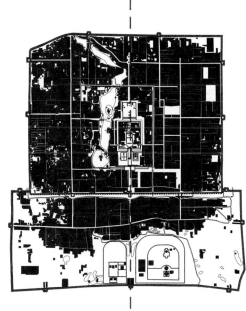

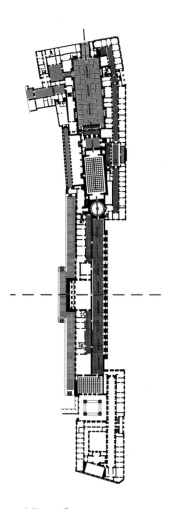

The Forbidden City
Beijing, China 1420
[Chinese, central axis with nodes, masonry and wood]

The Forbidden City is defined by its axial hierarchy. Surrounded by fortress walls, the city is organized along a central axis running from the Meridian Gate, passing through a number of halls, and terminating in the Imperial Garden. The axis of the Forbidden City was designed for the Emperor. Only he was allowed to travel along the axis, thereby making the organizational aspect of the complex more powerful through its combined spatial, political, and religious agendas. The form prevalent in the architecture was similarly reinforced through its use, governing the structure of ceremonies and events.

The Chancellery, Albert Speer
Berlin, Germany 1938
[Neoclassical, linear building with multiple courtyards, masonry]

The Chancellery by Albert Speer utilized the axis as an organizing method of leading a visitor through numerous ceremonial halls, rotundas, and courtyards, and finally into Adolf Hitler's private office. Here, axial hierarchy was used to establish control. The visitor was never allowed off the axis, but instead forced to follow it to the final destination. Despite this controlling aspect of the plan, the building's use of the axis was spatially sophisticated as it allowed the axis to turn and fold back on itself, creating a longer journey than the site would allow.

Axial Hierarchy—Urban

Axial hierarchy within the urban context is one of the most recognizable and powerful attributes of urban design. Typically axial hierarchy occurred when an older city, formulated as a result of medieval construction, would be redirected and based on the axis as its principle design methodology. Cities such as London, Rome, and Paris all experienced that transformation to some degree. There were also cities that were planned from the beginning to recall the great cities of Europe and hence utilized aspects of axial hierarchy as their main design concept.

Rome
17th Century
[Baroque, linear streets connecting piazzas]

Rome, which existed as a medieval city until the sixteenth century, exhibits a compelling example of urban axial hierarchy. Pope Sixtus V redesigned the city based on the location of the seven pilgrimage churches of Rome. He carved straight streets out of the medieval fabric, creating a series of axes that connected these important religious sites. He also created a number of enclosed urban spaces that typically contained a church. Rome served as the model of urban axial hierarchy for centuries and still illustrates the fundamental potential of this approach.

| | HIERARCHY | 357 |
| PRINCIPLE | |

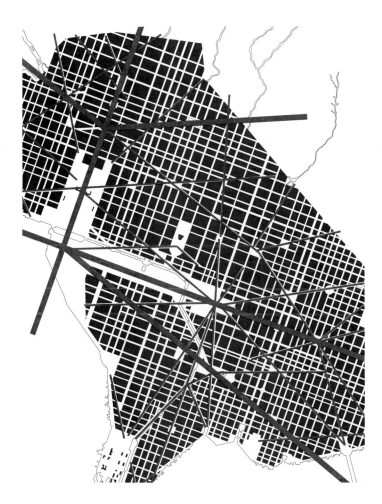 

Washington, D.C., Pierre L'Enfant
1791
[Neo-Baroque, broad avenues radiating from circles and squares]

When L'Enfant designed Washington, D.C., he visualized a series of axes connecting the important buildings and the monuments within a baroque framework. The visual connection of these moments, crossing the democratic grid, has come to represent the city and place it in the canon of great cities. Because Washington, D.C., was designed in totality as a new city, it has none of the medieval fabric; thus the axes are more numerous and visually more comprehensible. These divergent gestures are contrasted against a regular gridded fabric that allows the important axes and radial boulevards to be dramatically recognized and hence form the axial hierarchy.

Paris, Georges-Eugène Haussmann
1870
[Second Empire, connecting boulevards]

Paris was a very dense and somewhat chaotic city when Haussmann was asked to renovate it by Napoleon III. Haussmann followed the same pattern set by Rome in that he carved out long streets that served as axes between the important buildings and monuments. Haussmann went much further in his designs. He dictated open spaces, parks, the width of the streets, and even the heights of buildings, along with details such as cornice heights, material, and setbacks. Haussmann's plan has come to symbolize the grandeur of Paris with its wide boulevards and consistent building fabric.

Visual / Perceptual Hierarchy - Architecture

Visual or perceptual hierarchy relies upon the five senses to determine the relativity of significance of the architectural composition. Relying largely upon the ocular senses as the most dominant perceptual interfaces with architecture and space, it visually establishes the legibility of the formal interrelationships. The formal composition of scale, position, material, distance, form, or a combination of these, produces visual relationships that allow for a hierarchical language to emerge. Presenting dominant and submissive figures, forms, and functions, the piece-to-piece relationships emerge, transferring a universally legible language that establishes architectural readings. The perception of this visual hierarchy and relationship of pieces, spaces, and positions determines architectural experience.

Santa Maria della Pace, Pietro Cortona
Rome, Italy 1667
[Baroque, theatrical façade, masonry]

Santa Maria della Pace, located near the Piazza Navona, is approached by a long medieval street that winds through the fabric of Rome and offers tantalizing though obscured views of the magnificent baroque façade. The church is increasingly revealed as one approaches. The visual and perceptual hierarchy is extremely effective within this context as it holds the attention of the traveler until their arrival in the small piazza in front of the façade. The form of the façade is a convex curve that projects into the space, creating even more tension and perceptual hierarchy. This building was so popular in Rome that a number of buildings were removed to create a slightly larger space so that the façade could be more fully appreciated.

HIERARCHY PRINCIPLE 359

Salk Institute, Louis Kahn
La Jolla, California 1965
[Late Modern, symmetrical laboratories, concrete and wood]

Salk Institute is an iconic project that establishes hierarchy through the establishment of a central void. Symmetrical parallel wings of layered office, laboratory, and service spaces flank a central court. Defined by its central line of water, which flows to the Pacific Ocean, the localized space establishes a framed microcosm of the ocean beyond, capturing the shifting movement of light and weather across the dramatic La Jolla site. Employing individuated form, reduced scale, and material detail of wood infill, the offices draw specific articulation in front of the monolithic and uniform laboratory bars, followed by another layer of smaller service and utility functions. The clear power of the space and the interrelationship and organization of the pieces provide an overtly legible visual and perceptual hierarchy.

The Getty Villa, Machado and Silvetti
Malibu, California 2006
[Postmodern, museum addition and renovation, masonry]

The Getty Villa addition and restoration project by Machado and Silvetti engages a ceremonial circulation sequence from the parking garage nestled into the Malibu hillside, through a series of nodal outdoor rooms, and ultimately arriving at an amphitheater built into the natural topography that descends downward to the existing replica of the Roman villa. This promenade extends the thread of the historical sequence of the Roman house and garden. It moves through a dramatic topographic landscape, ascending up the hillside through lavishly articulated earthen rooms that are banded in diverse material layers and treatments to simulate the geologic strata and texture typical to the archaeology of Roman antiquity. Articulated by form and position, the visual and perceptual hierarchy relative to the ground plane and position along the sequence create a constant and legible relativism of place and perception.

Hierarchy of Scale—Architecture

The hierarchy of scale in architecture is a proven method that architects have used within the construct of a single building to manipulate various aspects of the design to achieve a sense of contrast and order through the lens of scale. This can be implemented in numerous capacities, engaging entire portions of the building or operating on the smallest of details. In three dimensions, any building that is composed of parts might exercise this concept of using scale as a way of dictating and prescribing relative relationships. This hierarchy may be accomplished through variations and contrast in size, but can extend to engage ideas of proportion and placement.

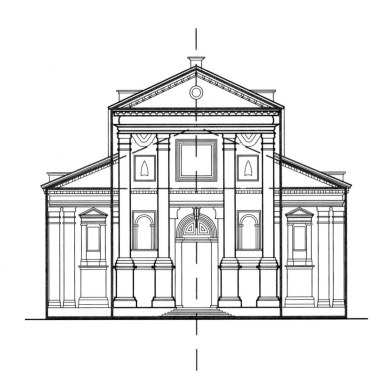

San Giorgio Maggiore, Andrea Palladio
Venice, Italy 1580
[Renaissance, transept and head additions, masonry]

The façade of San Giorgio Maggiore exhibits a masterful display of architectural hierarchy of scale through the use of three different scales of orders on the façade. There is a large Corinthian order that is used for the main temple front that dominates the façade. The broader and lower temple front forming the next layer of the façade is composed of a series of smaller Corinthian pilasters. Finally, the two niches on either side of the façade make the third scale by using a much smaller Corinthian order. The application of the Corinthian order in three different scales, united through a layering system, produces a collective composition that remains whole and unified.

HIERARCHY
PRINCIPLE 361

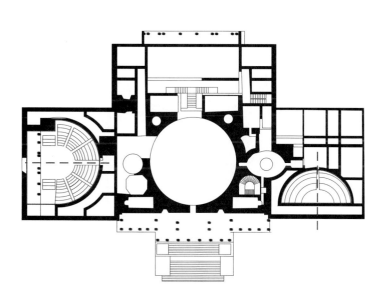

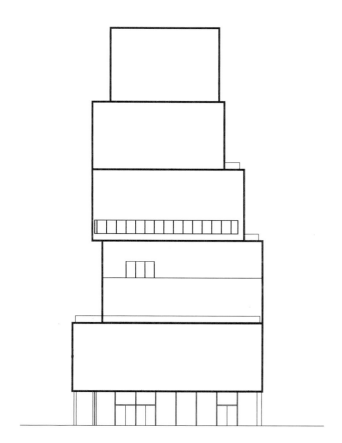

U.S. Capitol Building, Thornton, Latrobe, and Bullfinch
Washington, D.C. 1811
[Neoclassicism, government building, masonry]

The U.S. Capitol Building in Washington, D.C., exhibits scalar hierarchy primarily in the plan. While the elevation of the Capitol Building remains completely balanced and symmetrical, the plan is significantly different. A semi-circular amphitheater that has a very direct relationship with the exterior façade forms the Senate and occupies the right side of the building. The House of Representatives, which is on the opposing end of the building, is much larger and connected to the façade in a completely different orientation. The two scales of rooms operate and are oriented in different ways, yet are still contained within a completely symmetrical building.

New Museum of Contemporary Art, SANAA
New York, New York 2007
[Postmodern, museum, diverse]

The New Museum of Contemporary Art in New York takes the form of six metal-screened boxes stacked slightly askew of each other. Each box has a slightly different proportion and loses width while gaining height as the building ascends. The spectrum of hierarchy is established by the lowest and the uppermost boxes. The middle boxes mediate between these two extremes. The boxes, each shifted off of a central core, form a sculptural whole that allows sameness as well as a subtle sense of hierarchy.

Hierarchy of Scale—Urban

One way of clearly establishing a hierarchy within an urban fabric is through scale. By building something large enough, it will "automatically" achieve a hierarchy. Architecture that successfully utilizes this method is more than simply "big," but is complemented by other aspects such as materiality, placement, and meaning.

Florence Duomo, Filippo Brunelleschi
Florence, Italy 1462
[Late Medieval, Latin Cross church, masonry]

As the most important building in Florence, the scale of the Duomo is so large that it dominates the cityscape. It can be seen from almost any street in Florence and one can locate a bearing by one's relative position to the red dome of the Duomo. The Duomo establishes its primary hierarchy through its contrast in scale to the urban fabric. The drama of the massive scale of the Duomo works not only from a distance but equally when one is near. The building, which was the largest in the world for centuries, continues to receive admiration and awe due to its size and design.

HIERARCHY
PRINCIPLE 363

Monument to Vittorio Emmanuele II, Giuseppe Sacconi
Rome, Italy 1935
[Neoclassical, national monument, white marble]

The scale of the National Monument to Vittorio Emmanuele II is comparable to that of the Coliseum. Often referred to as "the wedding cake," its pure white marble gleams in the sunlight, creating an unforgettable moment of urban contrast and hierarchy of scale. The monument has always been controversial due to its location, size, and color. A large portion of the medieval part of the Capitoline Hill was demolished in order to position the monument at the end of one of the city's most important axes. Many see the immense scale as a problem aggravated by the jarring contrast of white marble against the more contextually scaled brown and red neighboring buildings.

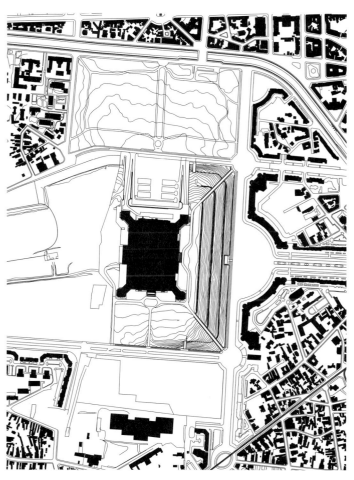

Palace of Parliament, Ceaușescu Regime
Bucharest, Romania 1989
[Neoclassical, government palace, concrete w/ marble]

Few buildings compete with the urban scale of the Palace of Parliament, the world's second largest building. Its lavish materials and placement at the end of a long axis that runs through Bucharest make it a seminal example of an urban hierarchy of scale. For the Romanian people, the massive structure remains a scar—a reminder that as its construction was carried out, the country itself spiraled into debt and deprivation.

364	HIERARCHY	MONUMENT	OBJECT / FABRIC	PLAN ELEVATION	URBAN
	PRINCIPLE	ORGANIZATION	GEOMETRY	READING	SCALE

Hierarchy of Monument—Urban

Monumental hierarchy refers to the position and significance of monuments within an urban fabric. Constructed to commemorate a person or event, monuments assume unique forms, dramatic scales, and honorific positions. The result of their presence draws the eye, from near or far, clearly establishing significance and intrigue. The hierarchy can occur relative to the object itself or the relationship to its surrounding fabric.

Washington, D.C., Pierre L'Enfant
1791
[Neo-Baroque, broad avenues radiating from circles and squares]

As the nation's capital, Washington, D.C., is defined by its collection of monuments, museums, and governmental buildings. The city's urban design was laid out using a baroque methodology that identifies hierarchical nodes within the city fabric and connects them with grand radial boulevards. The intensity of the point-to-point visual and physical connection of the city's civic infrastructure is the juxtaposition of the formal, linear, and nodal condition of these distinct moments against the regularized and democratically gridded urban fabric. This condition is further emphasized through the standard height of the fabric buildings, which is uniformly capped to give dominance to the U.S. Capitol Building. As a result, an evenly backfilled density allows for the clear articulation of formal subtractions that do not subscribe to the orthogonal geometry of the fabric.

	HIERARCHY	
	PRINCIPLE	**365**

 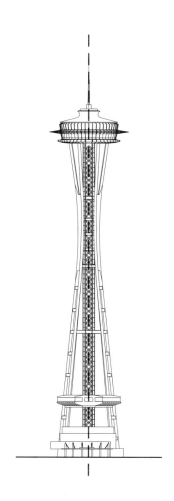

Eiffel Tower, Gustave Eiffel
Paris, France 1889
[Early Modern, vertical monument for World's Fair, iron]

The Eiffel Tower is the most iconic monument in Paris. Built for the World's Fair, the structurally expressive monument is massive in both scale and presence within the uniform height and material of the Parisian fabric. Deriving its form from the combination of structural necessity and material selection and limitations, the monument is expressively honest. Functionally, it is largely ineffective as it is only a restaurant and viewing platform, but as an icon it is highly successful as one of the world's most identifiable monuments. The diversity of its form and the dominance of its scale make its presence hierarchically monumental.

Space Needle, John Graham
Seattle, Washington 1962
[Late Modern, vertical monument for World's Fair, concrete]

The Space Needle follows the same formula as the Eiffel Tower to accomplish its monumental hierarchy. Using height and uniqueness of form, the Space Needle is both structurally and representationally determined. The futuristic form, dynamic presence, and height make it a uniquely recognizable icon. Its position relative to the urban fabric of Seattle and the dramatic topography of the region, in concert with the large scale of the urban fabric with diverse building types and competing formal and dimensional buildings, leaves the Space Needle less potent than the Eiffel Tower in contrasting its geography and context.

Hierarchy of Visual Control—Architecture

The dominance of vision and the hierarchy of power established through the relationship of viewer to the viewed introduce the engagement of control in architecture. Position, vantage, and sequence of revelation establish the parameters for a visual relationship. The interaction between the viewer and the viewed is a conversation that through architectural techniques can introduce a distinct dialogue about power and hierarchy.

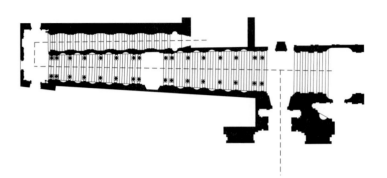

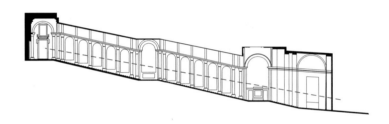

Scala Regia, Vatican, Gian Lorenzo Bernini
Vatican City, Rome, Italy 1666
[Baroque, papal pathway staircase, masonry]

The Vatican Scala Regia is a unique architectural moment designed as a point of entry and departure to the Vatican. As the papal pathway connecting the pope's private residence through the Sistine Chapel to St. Peter's Basilica, this heroic staircase is accessible only by the pope and his esteemed visitors. Deploying baroque techniques, the stairway is a barrel-vaulted colonnade. The combination of site constraints with desires for a choreographed elongation of the space requires the walls to angle and taper the space, creating an extended visual perspective. Bernini employed the statuary's ornament and position to complement the space, placing a statue of Constantine at the base of the stairs, adjacent to a window that dramatically illuminates the inscription "In hoc signo vinces" meaning "In this sign you will conquer." This serves to remind the viewer of Constantine's vision and their own individual religious responsibility.

HIERARCHY
PRINCIPLE 367

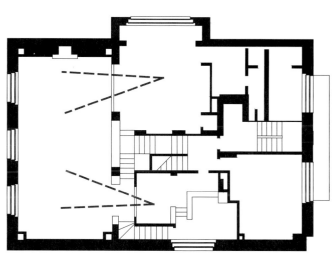

Panopticon, Jeremy Bentham
Unbuilt 1785
[Neoclassical, radial plan for visual surveillance, masonry]

The Panopticon Prison was conceived of as a surveillance device that allowed one individual to view many others. The Panopticon was a large circular building that had at its center a guard tower from which one guard could exercise visual control over hundreds of prisoners that were arrayed in a circular, radial fashion against the exterior of the prison's wall. This concept of a few watching many, especially within this context, illustrates an example of the hierarchy of visual control. This control was geometrically reinforced by the shape and location of the various parts. Despite the clear control and hierarchy of this example, the type was relatively short-lived due to its inhumane system of forced surveillance.

Villa Müller, Adolf Loos
Prague, Czechoslovakia 1930
[Modern, Raumplan house, masonry]

Villa Müller uniquely develops its domestic organization based upon the Raumplan. Three-dimensionally extending the house into discretely organized rooms that sectionally sequence and interpenetrate, the hierarchy of space is constantly questioned. The sectional arrangement of spaces allows for visual overlaps. The diagonal sightlines present hierarchical, positional relativisms that place the lower viewer in a position of full view but unable to see the upper viewer; whereas the upper viewer is able to see the lower room without being seen. The drama of this voyeuristic condition establishes a visual hierarchy that reinforces the positional, programmatic, and material articulation.

| 368 | HIERARCHY / PRINCIPLE | FUNCTION / ORGANIZATION | | SECTION ELEVATION/AXO / READING | ARCHITECTURE / SCALE |

Programmatic / Functional Hierarchy—Architecture

Programmatic or functional hierarchy is achieved through the relativism of use. The specific functional activities establish a relational hierarchy based in the sequences of practical or ceremonial rituals. Distinctly articulating the individual uses and their relative significance in terms of duration of use, level of privacy, scale, or social significance allows the functional practicalities to be embodied with a relative hierarchy. Service and served, public and private, day and night, messy and clean—all become dialectical categories for diagramming and discerning the hierarchical interrelationships of function.

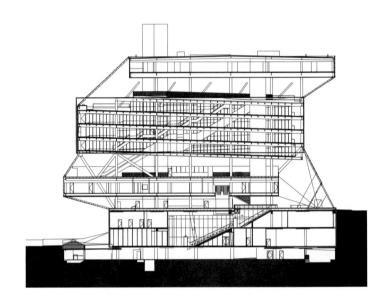

Seattle Public Library, Rem Koolhaas OMA
Seattle, Washington 2004
[Postmodern, continuous free plan, concrete and glass]

The Seattle Public Library utilizes program, or function, as a catalyst for its planning and form. In this project, the programs of a typical library are housed in novel ways, producing a series of interconnected paths and spaces, resulting in a programmatic or functional hierarchy. The book stacks are realized as a ramp that negotiates many levels of the building, essentially becoming the poche of the entire project. Placed along this book stack ramp are other programs that are architecturally situated to produce a linear architectural promenade that constitutes the programmatic aspect of the library. Innovative functional ideas are exploited in different areas to produce unique environments that produce a functionally driven architectural form. The collective and individual organization methodologies collaborate in their diversity to produce an architecture rich in hierarchy of form and program.

HIERARCHY PRINCIPLE 369

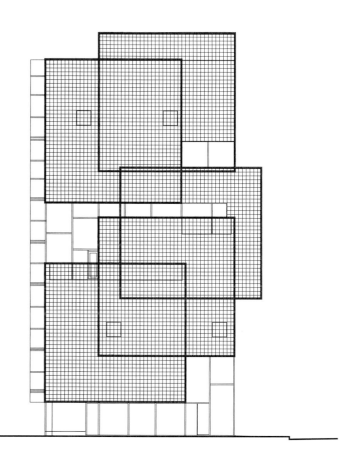

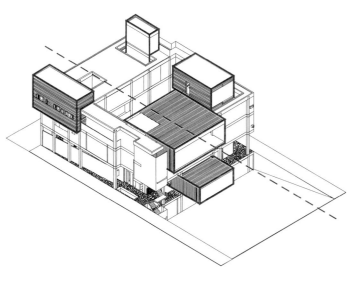

Ftown Building, Atelier Hitoshi Abe
Sendai, Japan 2008
[Postmodern, description, material]

Built as a speculative office building, the Ftown Building uniquely aggregates the individual functions of each user into formally identifiable pavilions. As a tower, the stacked and somewhat interlocked forms allow for a visual discreteness of the sublet zones of floor space within the continuity of the high-rise typology. Typically the vertical repetition of a serial floor plate provides a single form to create architectural continuity, but provides no individuation of space or recognizability of specific user. The Ftown Building creates tailored zonal suites that produce the form from the disparate tenants and their functional spaces. This mechanism is extended as the governing factor in determining the exterior form.

Chengdu Building, Borden Partnership
Chengdu, China 2010
[Ultramodern, courtyards and object volumes, diverse]

Built as a speculative office building with distinct functional and ceremonial requirements, the Chengdu Building defines its primary form of a rectangular opaque stone mass. Through a series of subtractive moves, interior courtyards are created to subdivide the formal figure and introduce light and ventilation into the building. Every interior space has a reciprocal exterior space. To this voided volume an overlay of functionally specific and hierarchically significant pavilions are added, cantilevering and protruding in the X, Y, and Z direction from the main figure. Housing banquet halls, conference rooms, cinemas, guest quarters, and fitness/spa facilities, the hierarchy of the functional spaces is both materially and formally read through the organizing principle of the architecture.

Color Hierarchy—Architecture

The use of color in architecture is an easy and powerful method of expressing hierarchical relationships (either codified or simply relativistic), amplifying light, or creating atmosphere. It can be used to reinforce color theory, contrast with material, or contextual tones; it can also be deployed to camouflage, blend, and dissipate. As a sensibility differentiated from materials and their natural coloration, its use and effect are closely aligned in sensibility, though uniquely different as the application of color is a conscious secondary overlay. The intention of its presence determines effect, mood, and perception.

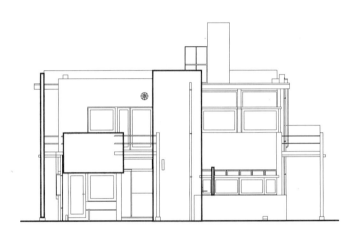

Schröder House, Gerrit Rietveld
Utrecht, Netherlands 1924
[De Stijl, open plan of shifting planar walls, masonry]

The Schröder House extends the compositional and spatial concepts of the De Stijl movement into the realm of architecture. The use of color in primary tones (blacks, whites, and grays) allows for the establishment of compositionally dominant and recessive planes. The coloration allows for the individuated articulation of pieces, implying fragmentation despite the continuity of the surfaces. The fractured breakage and slippage of the surfaces allows for visual movement. At times this visual movement becomes physical movement through telescoping walls as movable partitions. The discrete color complements the form and defines the architecture.

HIERARCHY | 371
PRINCIPLE

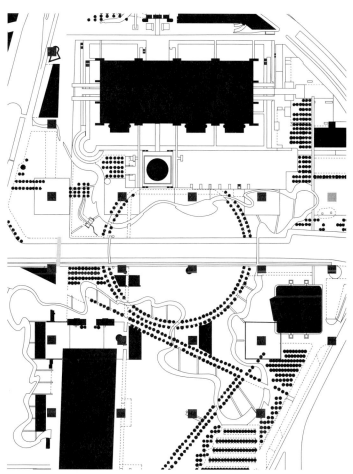

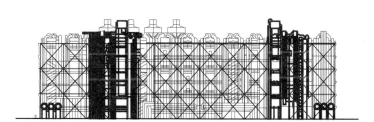

Centre Pompidou, Renzo Piano and Richard Rogers
Paris, France 1974
[Hi-Tech Modern, free plan museum, steel, glass, concrete]

The Centre Pompidou is singular in its use of color as an architectural element. The building reveals itself through the act of essentially turning itself inside out, thus exposing many of its systems and functions. Each of these elements is then colored as a way of providing programmatic or mechanical identification. The museum presents itself as a brightly colored object set against the typical gray stone of Paris. Blue is used for air, green for water, gray for secondary structure, and red for circulation. The building becomes a series of high-tech, colored symbols that communicate the various functional and mechanical aspects of the complex.

Parc de la Villette, Bernard Tschumi
Paris, France 1987
[Deconstructivist, iterative nodal grid field, metal]

Parc de la Villette uses red powder-coated metal to define and connect the array of buildings. Designed as a series of pavilions arranged on a grid, each pavilion is functionally and compositionally different. Color is used to maintain continuity, collecting and unifying through the datum of pigment. The foreign nature of the red coloration furthers the abstraction of the composition, referencing the color and forms of Russian constructivism and graphically highlighting and unifying the dispersed collection of pavilions into a unified architectural composition.

Material Hierarchy—Architecture

Material hierarchy refers to the compositional and associatively referential nature of materials to demarcate interrelationships. Engaging qualities such as the rarity, cost, weight, density, source, and geography of a material to describe a sensibility, the articulation through position and tectonic produces the dialogue of a design's intent. The presence of material in all architecture inevitably forces its legibility and reference into the conversation. The material's application as an inevitable consideration has the opportunity to establish these interrelationships. They are not intrinsic by nature or affiliation, but provide opportunity for implementation and expression. Material and the visual relationships of its choreographed use have the opportunity to define a formal language.

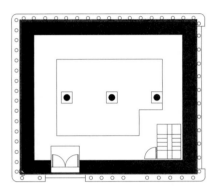 

Kaaba, Abraham
Mecca, Saudi Arabia 5th century BCE
[Islamic, cubic religious structure, masonry and silk]

The Kaaba in Mecca holds a position of cultural and religious significance in the Muslim faith. As the pinnacle of the pilgrimage every Muslim is expected to take during their lifetime as an outward sign of their commitment to their beliefs and religious faith, the hierarchy of the architecture is firmly established. Materially, the Kaaba is expressed as a black, stone-clad, cubic building further wrapped in a fabric skin. The grade of material, color and rarity of its contrast to the local and vernacular stone, and its hand-hewn nature express the intricacy, quality, care, and craft of the making. The visual, physical, and material descriptions denote the hierarchy and significance of the architecture.

	HIERARCHY	
	PRINCIPLE	**373**

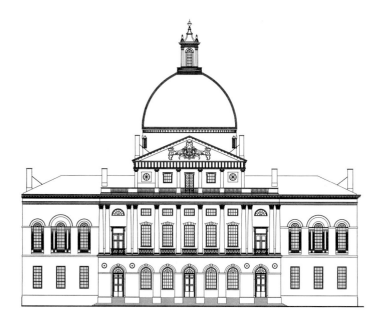

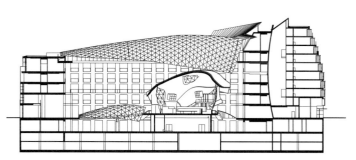

Massachusetts State House Dome, Charles Bullfinch
Boston, Massachusetts 1789
[Neoclassical, gilded dome, masonry and wood with gold leaf]

The Massachusetts State House Dome is a unique figure within the building and the surrounding urban fabric. Dramatic in its hemispherical shape and striking position, the composition is furthered by the use of gold leaf. The preciousness of the material, in combination with the scale and reflectivity, presents a visible material hierarchy through the intensely special nature of the topping. The effect denotes a cultural and visual effect that is legible through contrast and composition.

D.G. Bank, Frank Gehry
Berlin, Germany 2001
[Postmodern, expressionist figure in internal courtyard, diverse]

The D.G. Bank, like much of Gehry's work, contrasts expressive curvilinear forms with orthogonal and nondescript elements that frame and recede to establish hierarchy and compositional focus. In the D.G. Bank, the figure of the primary building holds the geometry, scale, and materiality of its context. The minimal simplicity of the building is established to frame a central atrium. Covered with a delicate glass ceiling, the central space becomes the focal point of the composition. Here the expressionistic figure of the horse-head-shaped conference room, with varied geometry of the custom-formed metal panels, clearly announces itself. It becomes the jewel inside the box. The objectification creates a focus of all ancillary spaces, connecting them through their dialogue with the hierarchy of the central figure.

374 | MATERIAL
        | PRINCIPLE

10

Material

MATERIAL PRINCIPLE 375

10 - Material

The primary categories discussed and diagrammed here focus on the use of materials as formal and tectonic generators. Materials generate form through their intrinsic geometry of process, physical properties, systems of construction, and the resulting associated patterns and ornamentation possible, given their specific natures. In essential dialogue with these are the intrinsic structure of the material and the innate relationship of static forces to physical practicalities. Finally, material detail and the way in which tectonic language of assembly provides a conceptual syntax of material thinking form a basis for expressing the relationship of assembly and making through the connection.

Form—Geometry / System / Pattern / Ornament

The primary governing factor of form is geometry, which provides descriptive mathematical rules. Geometric specificity, as established through rule systems that are either parameter controlled or instinctually and expressively determined, creates the visual legibility of the three-dimensional figure. The governing geometric and formal figure establishes a rule set that translates to the material, allowing an interweaving of overarching formal intent, construction, and assembly systems. These governances may be site-driven, type-driven (formal or functional), geometric system-driven, material, or performative, for instance. The control of these systems regulates the segmentation of unitized construction and introduces pattern, and the visual control of pattern, to embody ornamentation into the inevitable systemization. The integration of material and geometry, through systematized pattern, establishes the premise of material form.

Structure Structure as a physics-driven performative factor is an essential collaborator in design. Though collectively determined by the programmatic and functional needs and the formal and spatial requirements, the material properties of the matter from which the system is made determine the physical and thus specific figurative form. The intrinsic strengths of masonry and concrete in compression and the dual tensile and compressive capabilities of steel and wood establish the application and associated form of both the localized member, or element, and the collective form. The depth of a beam, the nature of the system as a wall or cage typology, and the material ability to join in these systems determine the form and nature of tectonic connection and establish the relationship of matter with physical forces through the performative responsibility of structure.

Detail The detail of material is a resultant of the raw nature of a material, the process of manufacturing that determines its "produced" form, and the systemization of the tectonic joinery. These decisions aggregate and determine the assembly technology, massively impacting the collective of the system. The detail, as the marker of the synthetic thinking of part to whole, establishes the sensibility and physicality of the material. The nature of the connection and the localized decision determine the legibility and effect. The detail is the expression of the material process.

In each of these thematic categories, the primary material types—wood, masonry, concrete, metal, glass, and plastic—each develop a distinct and individuated response and significance depending upon their material and processes.

In addition to these aforementioned topics, the following categories represent other lenses through which once can view materiality. As materials are fundamental to the physical presence of architecture, they are latent in every other theme discussed in this book. Examples can be found throughout the various chapters, but are discussed below to illustrate the significance of their thinking as conceptual methods and classification systems in their own right.

Surface / Skin In contemporary construction systems, the segmentation of craft allows for an efficiency and independence for construction and tradesmen. As a result, this separation produces an overlay of varied but integrated construction systems that are discrete in tools, forms, scales, and geometry. One key system in this composite construction is the surface itself. As an applied skin, much like the human body, the external layer of the architecture provides the final visual reading. The materiality of this outer surface dramatically impacts the tectonics of the assembly, the form of the segmentation and surface, and the visual and cultural reading of the building as an object.

Tectonic / Technology The tectonic (as the expressive control of the conceptual and visual articulation of the assembly system) and its associated level of technology are established both within a material and across the systems of construction and assembly. The diversity of technology and complexity within the material system relate to the desired form and the level of continuity or individuated articulation of components within the larger field and figure. Whether building on traditions of construction and craft, or engaging emerging technologies and advanced building construction technologies, the system of tectonic construction and the associated level of technology speak towards the significance of detail in relation to the conceptual and formal legibility of material.

Perception The perception of material is dependent upon macro and micro decisions that affect the scale and vantage of the viewer. Beginning at the scale of the unit, which is dependent upon the specific material and its intrinsic scale, the dimension of the material in both unit and grain relative to the anthropomorphic dimension contributes to the perceptual dialogue. The joinery system, scale of aggregation, and formal decisions on the local and collective scale create the visual environment. The play of light, color, and texture collaborate to produce space and atmosphere, establishing the experiential factors of perception.

Ecology The ecology of materials refers not only to the natural conditions from which the raw materials emerge, but also the life and process of the material as it is worked through manufacturing and assembly. This ecology also includes the performance of materials within the building and the energy expended throughout these processes. All physical matter has intrinsic properties, responses to external forces, durability, toxicities, and capabilities. Sustainability of a system is embedded in the material itself and the associated processes of manufacturing, fabrication and assembly. This sustainability can be examined in terms of environmental impact issues such as lifecycle, energy, carbon, water, and waste, for instance. . The ecology of the material is a collective view of all of these factors that weighs them against human costs, functional necessities, and formal and effectual capabilities.

Program Material program refers to the relationships of specific functional responsibilities of varied building typologies and the associative material capabilities that can practically and functionally respond to these requirements. Ceramic glazing in hygienic applications, concrete structure that reduces vibration, and sand-filled concrete block that deadens acoustics are representative examples of programmatic materials. They are specific applications of distinct physical and material properties for specific performative, functional or practical purposes. This relationship is key in the selection of materials as the performative and programmatic associations of material establish a powerful rationale for its use, detailing, and application.

Meaning—Color / Association / History The meaning of materials through color, association, and history refer to the cultural relationships that have evolved over time. Emerging from vernacular traditions rooted in technology and available local resources, but honed through the transference of cultural associations distinct to the object, materials are laden with meaning. Much of this meaning is localized and personal. Its perspective is often produced by the culture and experience of the viewer; but within this specific vantage there are associated trends and styles that are identifiable. The meaning and reference of the material is essential to understand its perception.

Form—Geometry / System / Pattern / Ornament
Wood

Wood, due to its ease of workability, has a dramatic ability to be ornamentally articulated. Repetition of individuated pieces (whether ornamental or performative) allows for the production of serial fields of geometric systems that produce ornamentalism through visual effect. Early decorative carving, elaborate shingling and scrollwork of the Victorian gingerbread houses, and contemporary layered open-lattice louvers and screens, are just a few examples. In the establishment of pattern, the exploitation of form is most direct when iterating position and scale to produce a formal effect. This is then further complemented and augmented by the addition of color, grain, and orientation. These discrete elements allow for the patterning of the individuated systems to assemble and engage one another to produce their compositional effect. The diversity of the formal effects possible through wood construction illustrates the flexibility and ease of workability of the material and results in the specific derivative systems and forms.

Pope-Leighey House, Frank Lloyd Wright
Alexandria, Virginia 1941
[Modern, Usonian L-shaped house, wood]

The Pope-Leighey House is a quintessential example of Wright's Usonian House. Intended as an affordable single-family house solution for the American middle class, the house uses many of Wright's basic design principles of: streamlined ornamentation, open floor plan, bifurcated public and private zoning, and a pinwheeling plan configuration anchored by the hearth. Using a unique wood-laminated wall, the tectonics of the assembly build a structural laminated sandwich that integrates siding, structure, and interior surface. Incorporating a ribbon of narrow operable windows, an ornamental mullion framing establishes a visual and iconographic motif that is repeated throughout the house. The innovation of the system as derived from the material tectonic and performance establishes the form and perception of the project as a whole. The clear sealant of the exposed wood-grain siding celebrates the material premise and visual quality.

MATERIAL PRINCIPLE 379

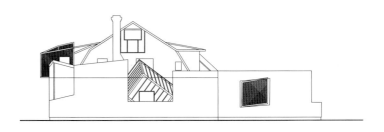

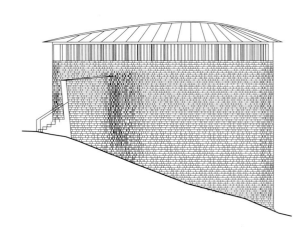

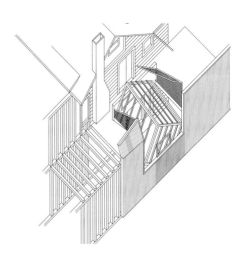

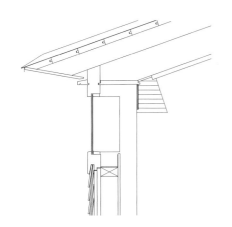

Frank Gehry House, Frank Gehry
Santa Monica, California, 1978
[Deconstructivist, reworking of existing suburban house, diverse]

The Frank Gehry House is an experimental addition and renovation that establishes a deconstructed composition. It reveals the wood construction system of the existing house as performative ornament and then adds compositional overlays of a second surface that create a thickened shell casing that performs both functionally and visually. Using standard 2 x 4 stud wall construction, the systems that are typically covered and masked are revealed and celebrated. Dissolving the sandwich system to reveal the internal structure and the delamination, the tectonic is deployed compositionally. The rawness of the structural lumber, corrugated siding, and chain link addresses the industrialized materiality with a compositional flexibility that whimsically integrates form and function.

St. Benedict Chapel, Peter Zumthor
Sumvtig, Graubünden, Switzerland 1989
[Postmodern, ovaloid wood chapel, wood]

St. Benedict Chapel uses wood in diverse applications, illustrating the breadth of its capability. It employs an exposed, skeletal, heavy-timber perimeter of structural columns and arrayed roof beams. Each system is founded in the process by which it was made, its intrinsic material properties, and its visual qualities. The field of shingles that were allowed to naturally weather, plank flooring and ceiling, and even the furniture are all materially expressive of their individuated system. The collective result is an encompassing environment that celebrates the serenity and simplicity of the form and material, establishing a pious yet spiritual environment. The repetition of simple, singular pieces, from the shingle to the structural column, creates a pattern and form through collective systemization.

Form—Geometry / System / Pattern / Ornament
Masonry

Masonry, an innately unitized system, is dependent upon the module and dimension of the individual unit. The singular base unit establishes the field effect through the aggregation of the piece. The construction system of the rigid unit, in conjunction with the plasticity of the mortar, introduces a visual pattern throughout the system. Variation of the coursing and patterning impacts not only the aesthetic legibility, but also the structural and practical aspects of the system. As either a load-bearing or structural system, masonry results in a planar system that permits variation of color, shape, and position (and thus ultimately form and pattern) through the variation of the piece and its joinery. The simplicity of masonry and the durability of its performance have historically established masonry as a construction system with deep historical traditions. The formal relationship of the piece to the whole establishes an intrinsic orthogonal formalism that scales and extends the geometry and form of the piece to the collective whole.

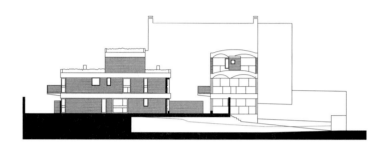

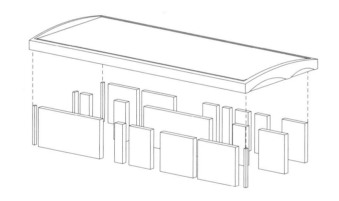

Maisons Jaoul, Le Corbusier
Neuilly-sur-Seine, France 1956
[Modern, concrete frame with textured infill, masonry]

*Maisons Jaoul expresses the masonry through the exposed parallel planes of the side walls. Using load–bearing, concrete–filled, reinforced-clay units, the unitization and coloration of the vertical planes establish a visual contrast with the concrete floor plates. The brutalism of the materiality is representative of a broader trend in Le Corbusier's work, which includes an acceptance of the roughness of the material and the crudeness of the construction as representative characteristics that give the texture from the process of casting the surface. The masonry wall was built without centering to introduce a rawness to the laying that allows for subtle shifting of the planar surfaces. These walls serve as the formwork for the site-cast, rough–formed, **béton brut** of the vaults. Defined by interlacing, stacked levels of alternating load-bearing masonry planes and concrete, shallow, arched, floor plates, the building is organized by its striated layering. The material and its process of construction define the form and the spatial texture of the house.*

MATERIAL PRINCIPLE **381**

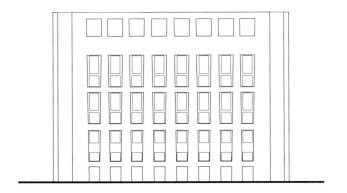
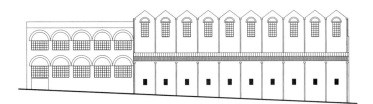
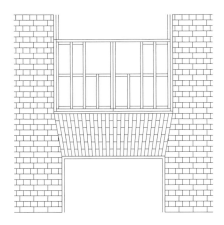
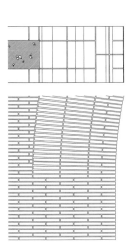

Exeter Library, Louis Kahn
Exeter, New Hampshire 1972
[Late Modern, centralized square, masonry and concrete]

Exeter Library is defined by its load-bearing masonry skin. Using a repetitive grid façade, graduated vertically to reduce the structural weight of the load-bearing walls by one brick per level, the effect is an increase of the aperture size to accommodate the transition and diffuse the mass of the building. The simple modularity of the material, defined by the manufacturing process and the anthropomorphic dimension of the mason's hand, the building is overtly and honestly expressive of the material, the process of fabrication, the process of construction, and the structural physics that govern them all. The clarity of its expression provides the minimalism of its architectural expression and the infallibility of its formal regularity relative to the material from which it is made.

The National Museum of Roman Art, Rafael Moneo
Merida, Spain 1986
[Postmodern, series of parallel masonry walls, masonry]

At the National Museum of Roman Art, Moneo references the content of the museum's collection through the form and material. Adopting a roman brick (which is longer, broader, and flatter than standard brick) as a load-bearing system, the shape of the unit defines the overall form of the building. Repetitive layered walls, configured as parallel planes with arched openings, replicate the visual quality of Roman construction and thread between the ruins of the archaeological site the museum sits upon and houses. The simple expression of the material tectonic establishes the visual and formal quality of the building. The structural layering of the massive walls, the coloration of the individual units, and the clarity of the forms that translate the physical forces into vaults and piers all establish the impact of the material and form.

Form—Geometry / System / Pattern / Ornament
Concrete

Concrete as a fluid material is dependent upon the vessel into which it is cast. The **formwork** thus becomes an integral participant in the determination of form. The plasticity of the possible shapes and the intrinsic continuity of the material joining at the molecular level permit formalism unlike any other material. Amorphic forms, curvilinear figures, and continuous surfaces are readily possible. As a combination of raw materials (water, aggregate, sand, and Portland cement), the chemical process of hydration is an exothermic reaction that occurs during curing. Thus concrete does not dry but cures and can be set underwater. These dynamic properties create a material that performs extremely well in compression but lacks tensile strength. This property dictated the early Roman forms that transfer loads through the geometry of the form. With the improvement of the Bessemer furnace and the ability to mass produce steel, the development of reinforcing rod allowed for a composite system to emerge. Structural cast-in-place, reinforced concrete permitted the development of a system that performs equally in compression (concrete) and tension (steel). Variable aggregates, finishes, and surface treatments, as well as casting techniques, create a broad field of concrete types and visual and performative properties. The following examples engage casting technique, form, and finish to establish their material identity.

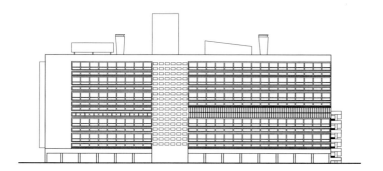

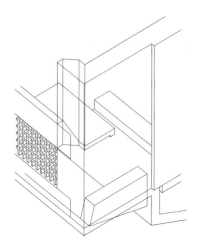

Unite d'Habitation, Le Corbusier
Marseille, France 1952
[Modern, double-loaded corridor, concrete]

The Unite d'Habitation is primarily constructed of the singular material of cast-in-place concrete. This complex was built using a limited availability of technical craftsmanship, thus making this material choice appropriate and feasible. The flexibility of the concrete construction allowed Corbusier to incorporate the material as structure, finish, enclosure, columns, floor, brise-soleil, and furniture. The brise-soleil particularly gives a language to the building: as the system of overhangs and panels repeats, a pattern is composed, resulting in a sophisticated type of ornament.

MATERIAL PRINCIPLE 383

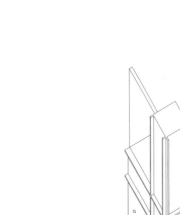

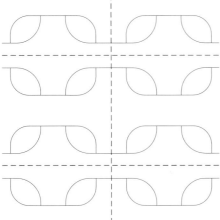

Design 3, Erwin Hauer
Various 1952
[Late Modern, geometric screen wall, concrete]

Design 3 is one of a series of Erwin Hauer's modular investigations into repetitive unitization. Engaging three-dimensional depth through curvilinear transitions in the cross section of the unit, Hauer developed complex geometries that allow for repetitive fields when aggregated. The intelligence of a serial piece creates a multi-directional field effect. The removal of the unit, intended to erode enough to allow literal transparency and openness across the unit, allows for the production of screen walls that are legible from both faces. The materiality is deployed for its continuity and smoothness of surface. Using a repetitive mould process, the intricacy of a piece can be carefully crafted and established and then repeatedly cast to create serial units. The process of construction is optimized to create the repeatable unit that, when serially aggregated, allows the complexity of the field-effect geometry to eliminate the legibility of the piece and instead establish a collective surface. The form and process is one that is dependent upon the materiality of concrete and its process of formation.

Church on the Water, Tadao Ando
Hokkaido, Japan 1988
[Postmodern, double square sanctuary, concrete]

The Church on the Water, as a representative project of nearly all of Tadao Ando's work, establishes its identity through the precision and modularity of the formwork system. The careful control of the color, surface, and detail establishes an astonishing level of craft through the material itself. The formwork, carefully planned in dimension and position, scribes into the surface a three-dimensional cage. Through the panel joints of the modular system, the conceptual frame of the space is projected from material to perceptual order. The control of this systemization is even further articulated through the form ties. As lateral braces that tie the two sides of the formwork together (to prevent shifting or blowout from the tremendous pressure of the poured concrete), they are capped and left in the surface. These nodal points establish an anthropomorphic dimension to the continuous surfaces. This careful rigidity then allows for the dynamism of natural systems such as light, vegetation, and human activity to be framed and foregrounded by the serene calm of the material architecture of concrete form.

Form—Geometry / System / Pattern / Ornament
Metal

Metal has been formally and ornamentally used for centuries for both jewelry and weaponry. The limited quantities of its availability established an intrinsic preciousness that was furthered through the draftsmanship of its workability and making. Cast, forged, or otherwise formed, the metallurgic properties of the specific metal type determine the scale of availability, method of workability and the resulting formalism. Despite massive technological advancements, the primary properties of availability, strength, malleability, corrosion and coloration all determine the formal capabilities of its application. Mass production results in categories of sheet material, structural shapes, tubing, and other special shapes; each creates a standardization of raw material "types" available for translation into building form. The inter-relation of the raw material types are determinate governors of the final formal types in dimension and fabrication technique. Rolling, extruding, bending, punching, expanding (to name a few), are the operable processes that determine and effect the possible forms. Shapes of components, in combination with assembly techniques such as welding, riveting, and bolting establish methods to aggregate both structures and surfaces into the frames and surfaces that establish the articulation of form. The intricacy of these systems is most recently accelerated through the introduction of computer numerically controlled (CNC) fabrication techniques including laser cutting, CNC milling, water jet cutting and plasma cutting. These techniques in conjunction with advanced methods of forming and shaping, allow for the production of intricate forms including the fully amorphic geometries of compound curvatures. The workability of metal under heat and pressure in combination with its intense strength to size ratio make it one of the most dominant and flexible building materials as both structure and enclosure. The following projects represent a performative, a formal and an effectual application of metal to produce architectural form.

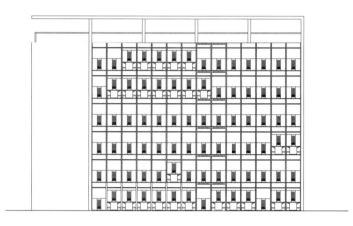

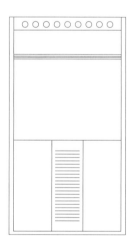 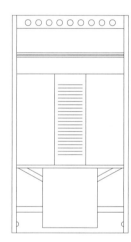

Mozart Place, Jean Prouve
Paris, France 1953
[Modern, operable façade of apartment building, metal]

Defined by the highly articulated operable skin, Mozart Place uses an intricately conceived aluminum façade system in serialized segments. Each module is designed to create a thin performative surface. These modules serve as a shutter system (for light and security), ventilation system (with varied scales and positions for quantities and control of natural breezes), insulative layer (to deal with thermal conductivity), and ultimately the composition of the **fenestration** *that composes the view and light. The thin aluminum material, in conjunction with the use of cross sectional hollow shapes, introduces strength and rigidity, while maintaining the lightness and durability. The hyper-engineered nature of the skin and its celebratory expression of practical functionality embed an honesty and overt technical expressionism that derive from a collaboration of the functional response with the mastery of the material and its processes of fabrication.*

MATERIAL PRINCIPLE | 385

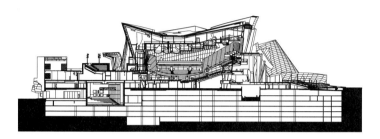

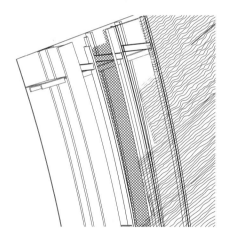
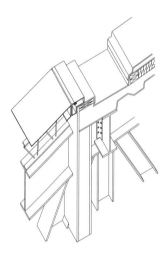

Disney Concert Hall, Frank Gehry
Los Angeles, California 2003
[Postmodern, free formed skin over orthogonal box, metal]

The Disney Concert Hall uses both surface sheet geometries to define form and stick structural systems to create the skeletal system. Using a panelized shingling system, the complexity of the overarching shape is segmented, divided, and locally formed. By using metal, the thin gauge of the surface and the malleable metallurgy allows for the fabrication of a compound curvature. The system and surface interrelate to allow for a segmental continuity of form. The sublayer of structural steel uses a variety of orthogonal and super-structurally systematized grids with localized adjustments. The stick assembly provides the necessary structural performance as needed, and then fully masks this organizational and formal sublayer with the dominance of the surface skin. The postmodern application of form independent of function and the subordination of structure to emphasize the legibility of exterior surface and formal figuration over space are extended through the material and its application.

The de Young Museum, Herzog and de Meuron
San Francisco, California 2005
[Postmodern, free plan with courtyards, metal]

The de Young Museum's innovation comes through the carefully articulated development of surface through the variably perforated and dimpled skin. The three-dimensional dynamism of the surface comes through a method of introducing diversity and locality into the surface. Using a patterned vocabulary of perforation (utilizing variation in scale and the position of the perforation) and dimpled deformations (utilizing both recessive and projective depths), the simplicity of the formal surface derives its visual complexity through the superimposition and effect of the techniques. The two-dimensional optical reading of the variable repetitive fields permits the systemization of the pattern to produce identity. The referential imagery of the grove of trees previously on the site, in conjunction with an understanding of the multiplicity of readings that depend upon the distance from the surface, the translation of the image into patterned field creates a balance between referential image and abstract geometry of the system.

Form—Geometry / System / Pattern / Ornament
Glass

The ephemerality of glass as a material of variable perceptual presence becomes uniquely dependent upon the detailing of its assembly system. The articulation of the joint establishes the formal dimension of segmentation and establishes the relative visual reveal of the joinery system. This tectonic system builds on the intrinsic qualities and effectual and performative characteristics of the glass itself. Reflectivity, opacity, color, surface texture, embedded imagery, and emissivity determine the visual characteristics of the resulting surface. The performative responsibilities, in conjunction with these effectual decisions, mediated by the practicalities and engineering of the assembly system, determine the form and effect of the glass surface and its reading as an architectural system.

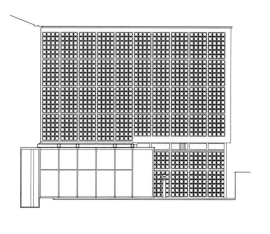

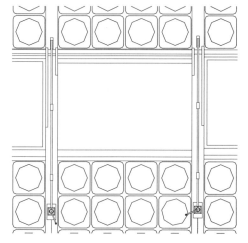

Maison de Verre, Pierre Charreau
Paris, France 1931
[Modern, additive and subtractive spaces, glass]

The Maison de Verre uses glass block as a translucent infill material that defines the main façade. As a two-story wall in the main living space of the home, the modular, unitized, masonry system opens the space through the luminous nature of its translucent surface while providing a dimensional subdivision to the surface. Breaking the planar surface into a unitized field, established by the anthropomorphic dimension of the mason's hand, the wall embeds pattern of joint as well a pattern of the textured and figured surface of the repetitive unit itself. The simplicity of the surface field juxtaposed with the enigmatic nature of the material itself establishes the effectual qualities of the architecture. Allowing light and shadow to dapple and move through the repetitive surface, the visual complexity comes through the revelation of the surface itself. The result is a dynamic yet regimented surface that relies upon material performance to orchestrate the perceptual impact and the iconography of the project as a whole.

MATERIAL
PRINCIPLE 387

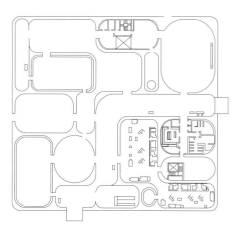

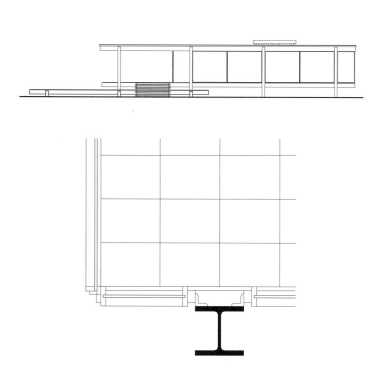

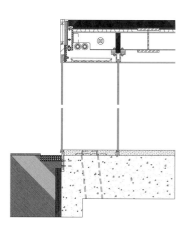

Farnsworth House, Ludwig Mies van der Rohe
Plano, Illinois 1951
[Modern, free plan house, glass]

Glass Pavilion Toledo Museum of Art, SANAA
Toledo, Ohio 2006
[Postmodern, nested transparent galleries, glass]

Using materiality to evaporate the presence of the building, the Farnsworth House is the quintessential glass project. The hyper-regimented and considered order of the architectural systems collaborates with the ethereal qualities of the materials. The result is a reductive architecture that defaults to the surrounding natural context to complete and enclose the composition. The house dissolves the boundaries and presence of any conventional compartmentalization of function, or even space, by deploying the materiality of glass to integrate the landscape with the activities of occupation.

The Glass Pavilion at the Toledo Museum of Art uses a basic streamlined perimeter figure to bound and gather a collection of individual interior rooms. Articulated through mullionless glass walls, the ephemerality of the material is further blurred and softened by the gently arcing surfaces. The resulting palimpsest of the layered glass walls accentuates the delicately reflective and refractive surfaces. This effect dissolves the perception of the traditional wall and instead emphasized the ethereal nature of light as an evaporative surface. The layered effect of the transparent surface creates a blurred reflectivity that absorbs the objects in the visual field and disperses their presence into an abstract continuity.

Form—Geometry / System / Pattern / Ornament
Plastic

Plastic, as the most recent material category, has the most complexity in its processing and formation, but equally the most flexibility of form and capability of surface. With varied chemical compounds that offer varied manufacturing processes, distinct material properties and associated formal capabilities, the material shifts from rigid cast, milled, or bent forms to flexible fabric-like surfaces. This diverse spectrum allows for massive variation within the material category. It has been a material typically specified for its practical material capabilities as a hidden collaborator. In recent decades, a series of technologies and new celebratory projects have deployed plastic as the primary material of their architecture, collaborating with it to define both performance and form simultaneously. From the early experimental Monsanto House of the Future (founded in the continuous surface and associated formal curvatures) and the Water Cube (with an **ETFE** surface figuration of segmented air pillows) to the high performance performative and figurative surface of the L House the resulting forms are derived from the varied material properties of each specific iteration of plastic.

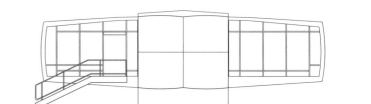

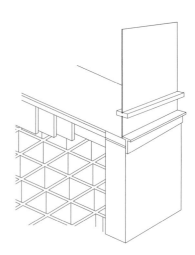

The Monsanto House of Future, Goody and Hamilton
Anaheim, California 1957
[Late Modern, cantilevered cruciform house, plastic]

*The Monsanto House of the Future is defined by its centralized cruciform figure with four identical cantilevered plastic loops. Goody and Hamilton employ a repetitively formed shape that repeats in each cardinal orientation and is mirrored in elevation. The specific curvilinear surface of the molded form is hyper-flexible in its customization, but made more efficient through a repetition of its primary figuration. The form of the house transposes the material qualities into the distinct and seamless anonymity of the localized surface, emphasizing the overall shell. The simultaneous continuity of form, enclosure, and structure evolves into a **monocoque** skin that allows for a single surface to assume the responsibility of typically aggregated systems.*

MATERIAL PRINCIPLE 389

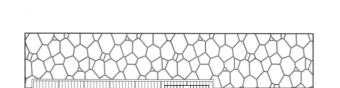

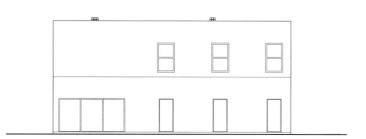

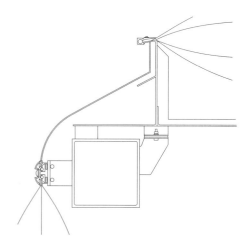

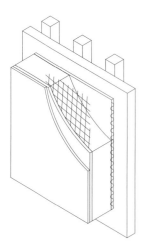

Water Cube, PTW Architects
Beijing, China 2007
[Postmodern, irregular ETFE-skinned rectangle, plastic]

The Water Cube uses ETFE with an irregular segmentation to create a field of individually inflated air pillows. Referencing the natural form-finding of soap bubbles, the double-walled surface allows for an internal lighting system that produces a lantern effect to the building as a whole. As a materially light surface, the delicacy of the skin allows for the singular massive space of the interior natatorium to be spanned, while the double surface allows for a continuous unbroken surface to be read from both the interior and the exterior. The insulative efficiency of the air pillows maintains a transparency to the skin and provides highly identifiable formal characteristics while allowing functional performance.

L House, Moo Moo
Lodz, Poland 2010
[Postmodern, materially innovative residence, plastic]

The L House begins with a conventional house form, but abstracts it through a formal cutting and peeling of the skin, an effect that is furthered through the seamless, enigmatic material application of its surface. Using Thermopian, a plastic insulating material available in any color, the monolithic surface has good thermal, acoustical, and insulating properties. The deep bluish color of the skin and the smoothness of the unbroken surface make the prismatic form feel taught. Sloping from one house-shape figure up to a larger house-shape figure, the perspectival bending is further complicated by a delaminating angled wall. The tautness of the surface allows for a formal legibility of the overarching form.

Structure
Wood

Wood as a structural material is one of the most affordable and thus ubiquitous construction types (particularly in the United States). Spanning diverse technologies and varied levels of material efficiency, each system has evolved with its diverse associated tools. Beginning with log-cabin construction that uses simply hewn and largely whole trees that are stacked and locked, it requires little working but tremendous amounts of material. Thinning the material mass, heavy-timber construction mills the wood into large elements that are assembled into frames and cages. This material efficiency requires structural calculation and a more specific understanding of material physics. The final system is stick framing and exists in two primary categories: balloon framing and, more commonly, platform framing. In balloon framing, the walls carry the full height of the building and the floors are hung off the vertical walls. In platform framing, each level is framed and then the floor plate is stacked and then the level is framed, and so on and so on. This construction type uses repetitive mass-produced elements that are modular and easily field cut to formulate any system through their flexibility and repetition. As a fully concealed system (both on the interior and exterior) that effectively works in vertical, horizontal, and angled surfaces, the density of the framing system allows for the ad hoc flexibility. Below are examples of contemporary wood systems that use combinations of traditional construction techniques hybridized with engineered lumber. Their formal articulation evolves from the material tectonic and the manufacturing system that establishes the specific material properties, capabilities, and forms.

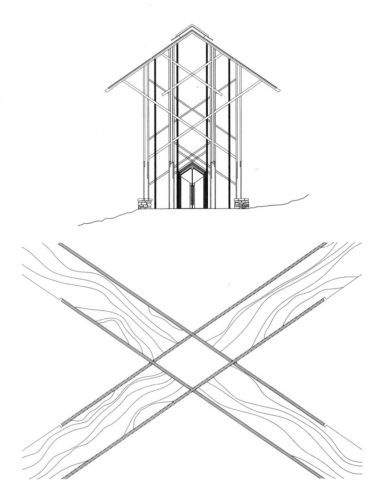

Thorncrown Chapel, Fay Jones
Eureka Springs, Arkansas 1980
[Late Modern, repetitive wood structural frame, wood and glass]

The Thorncrown Chapel is an essay on the use of wood as a structural element. Through its extreme vertical structural emphasis, the chapel mimics the surrounding forest. The building uses a palette of natural materials common to the site. The wood columns and beams create a dominant figure that repeats itself continually, forming a pattern or matrix that blends into the hovering tree limbs surrounding the chapel. This mimicry gives way to clarity as one enters the space and recognizes the hierarchy of the wood structure as a form-giving language.

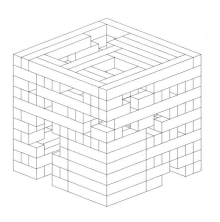

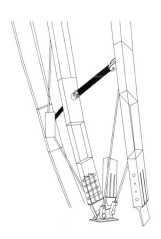

Marie Tjibaou Cultural Center, Renzo Piano
Noumea, New Caledonia 1998
[Hi-Tech Postmodern, vernacular formed pods, wood]

The Marie Tjibaou Cultural Center is defined by its tectonic articulation and evolutionary expression of vernacular and site-responsive forms. The Center is built out of indigenous woods in expressive, cupping, wind-scooping forms. These forms are defined by massive, tapering, glue-laminate beams that are joined by a gradient of lateral wood beams that diffuse as the figure ascends. Defined to optimize and respond to both passive and active lighting and ventilation needs, the forms, their expression, and their materiality all respond to the localities of climate and culture. The resulting building is a synthesis of place, site, material, and highly articulated performative responses to the building's systems.

Final Wooden House, Sou Fujimoto
Kumamura, Japan 2008
[Postmodern, stacked heavy timber pavilion, wood]

The Final Wooden House uses a combination of heavy timber materials (large, orthogonally milled, solid, natural members) with log-cabin assembly systems (stacked and pegged using gravity connections). The result is a modular, multi-directional environment that has precise orthogonal edges that form a Platonic shape, whereas the interior has a variably stepped surface in all six axes. These varied surfaces allow for performative interpretations and associative programmatic occupations. The simplicity of the form and tectonic is coupled with the diversity and abstraction of the form and functional engagement.

Structure
Masonry

Masonry, as a structural material, has undergone significant evolutions in its application despite minimal technical evolutions. There are primary categories of clay masonry, concrete masonry, stone masonry, and glass block. They are all premised on the same simple process: material is taken from the ground and fired under heat to create hand-sized units. In concrete masonry units, the fired soil is replaced with concrete that is cured through hydration. The resulting individual pieces are then aggregated and assembled to created planar masses. Originally these masonry systems were load-bearing, resulting in linear wall systems and vaulted spanning techniques. In contemporary practices, the wall has thickened to include airspaces for thermal and moisture efficiency, layers of insulation, more performative flexibility, cost-efficient skeletal structural systems, moisture barriers, and diverse infrastructural systems like electrical, phone, and data. As a result, contemporary clay masonry is largely used as a veneer system. Only concrete block fundamentally remains a load-bearing system yet with limited applications as when it is used in a single **wythe** configuration, it lacks space for the aforementioned secondary systems. As a result, the legibility of masonry as a single structural system is a rarity; instead it is deployed in combination with other material systems in a hybrid capacity. The examples illustrated are case studies where the qualities and capabilities of their masonry construction are fully deployed and engaged in the formal and functional conceptualization of the building.

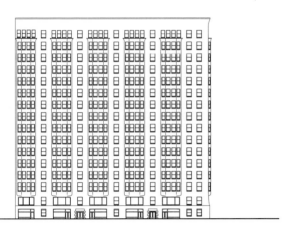

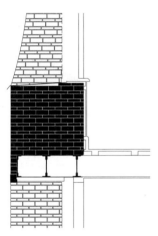

Monadnock Building, Burnham and Root
Chicago, Illinois 1891
[Richardsonian, double-loaded linear office building, masonry]

The Monadnock Building is often considered the first skyscraper. It was sixteen stories high but was built using traditional load-bearing masonry construction methods. The combination of the desired building height with the natural limitations of the material dictated six-foot-thick walls at the base. The thickened wall dimension in relation to the height optimized the material capability, stressing the clay masonry to just below its maximum point of failure. The addition of another floor would have caused an actual crushing of the material. Thus the structural responsibility and the construction method resulted in the graceful, curved, vertical lines that identify this building and distinguish it in the lineage of high-rise construction.

MATERIAL PRINCIPLE 393

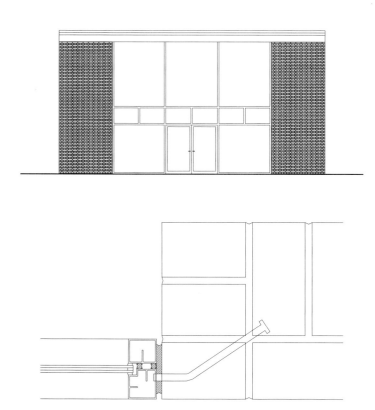

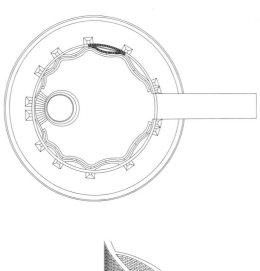

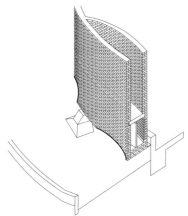

Chapel of Saint Savior, Mies van der Rohe
IIT Campus, Chicago, Illinois 1952
[Modern, orthogonal load-bearing brick chapel, masonry]

The Chapel of Saint Savior uses standard brick as a modular bearing wall. Dividing the building into five equal bays, the masonry is deployed as opaque, unbroken surfaces that flank the three central transparent bays. The essentialism of the brick is displayed through the simple celebration of the unit, its repetitive unbroken field, and the intrinsic module of the aggregated field. The same material philosophy that governs Mies's iconic approach to glass and steel is here taken to masonry, intensely celebrating the precision and resolution of the material.

MIT Chapel, Eero Saarinen
Cambridge, Massachusetts 1955
[Modern, circular sanctuary, masonry]

The MIT Chapel by Eero Saarinen derives its overall form from the plastic assembly of clay masonry units. Made possible through the incremental rotation of the individual units at the head joints, the collective rotation creates a cylindrical form to the overall building. The surface is given subtle formal, color, and reflective variation through Saarinen's choice to use clinker bricks (bricks that were deformed in the firing process and are typically discarded). The load-bearing double wythe maintains a continuous exterior plane, whereas the interior wall oscillates in a serpentine form. This second shell uses an open lattice of alternately removed units that allow for acoustical pockets to contrast the otherwise hard and thus reflective surfaces. The use of masonry extends beyond the wall plane to include the floors and the altar. The material is a constant collaborator throughout the building, being understood and collaborated with for diverse performative and aesthetic purposes.

Structure
Concrete

Concrete as a structural system is defined by the two major classifications of cast-in-place and precast. This distinction of the location of forming and setting the concrete results in dramatically different techniques of craft and subsequent formalisms. For structural applications, cast-in-place concrete allows for more monolithic forms that are specific to the constraints and particulars of the site, whereas pre-cast is typified by its precision, predictability, and modular nature (governed by the need for transport). The dependency upon the formwork, the natural compressive characteristics of concrete (that can be easily augmented with steel reinforcing that provides the necessary tensile resistance), the natural continuity and the hyper-flexibility of the material's intrinsic nature allow for diverse formalisms from column-to-frame, plane, and cage-to-shell. In each configuration, the material capabilities remain largely intact.

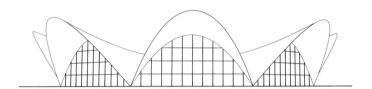

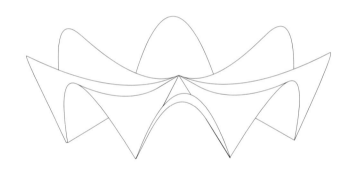

Los Manantiales, Felix Candela
Xochimilco, Mexico 1958
[Late Modern, radial parabolic shell, concrete]

The use of concrete as a structural shell is exaggerated in Los Manantiales through a maximization of the span to the thinness of the surface. The structural development of a continuous shell system allows for even distribution across the entire surface. This approach requires a collective consideration of the overarching form. Developed as a series of parabolic arches joined as a continuous surface, the merger of the forms creates a fluid surface shell. The singularity of the surface allows for a minimal purity to the overall form, optimized through the delicacy of its thinness. The remaining infill is largely transparent. The interior is a singular unbroken space that visually and effectually expands outward to the surrounding lake and landscape. The singularity of the surface, required for structural continuity, and the plastic curvilinear surface that introduces rigidity and formal depth to the form are only made possible through the intrinsic material qualities of cast-in-place concrete.

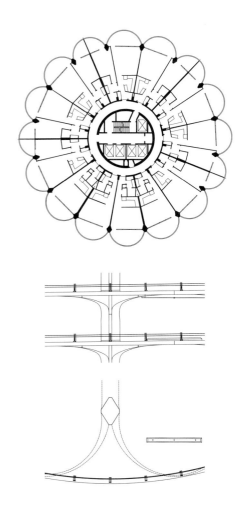
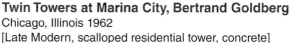
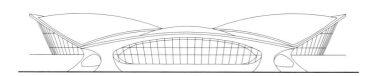
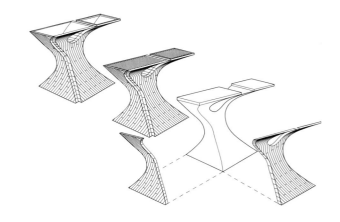

Twin Towers at Marina City, Bertrand Goldberg
Chicago, Illinois 1962
[Late Modern, scalloped residential tower, concrete]

The Twin Towers at Marina City are characterized by the formal "corn-cob" scalloping of their cylindrical shape. Having sixteen scallops per floor plate, each articulated in a balcony off the main living spaces of each apartment, the curvilinear form is directly derivative of its plastic materiality. The divergent formalism responds in contrast to the glass and steel orthogonality that dictated the Miesian formalism that dominated the high-rise building typology of the era (as typified by the similar double-tower configuration of Mies van der Rohe's 860-880 Lakeside Drive Buildings). Similarly exposing the material and celebrating the simplicity and purity of its formal expression, the plasticity of the concrete permits the curvilinear formalism of the local module and the collective cylindrical form of the building as a whole.

TWA Terminal, Eero Saarinen
New York, New York 1962
[Late Modern, free-flowing curved airport terminal, concrete]

Saarinen's TWA Terminal typifies the plasticity and fluidity intrinsic to concrete. Developing a formalism of continuous surface, the transition of floor to wall to ceiling as a single surface is a direct result of the material. Made possible at the microscopic scale of joinery, the singularity of the material and the resulting singularity of the form create a dynamic spatial effect. The fluidity of the forms and the sculptural carving of the surfaces create a dialogue between the subtractive spaces and the massive aerodynamic figuration. The form evolves from an interrelated understanding of how the material is made into a spatial agenda.

**Structure
Metal**

Structural metal, though evolved from cast iron technologies, typically refers to steel in contemporary construction. As extruded or rolled shapes, the material exists in standard dimensional length and limited cross-sectional shapes. The resulting forms are thus derivative frame-based geometries. These frames can be standardized, rationalized, or serialized to establish an economy of means, or more irregularly tailored to divergent localized forms. The connection of the systems (for example, bolted, welded, or riveted) and the introduction of lateral stability into the systems (for instance, using a moment frame, braced frame, or shear wall) establish the articulation of the joint and the visual, formal, and performative identity of the system as a whole. The collective system and these local articulations derived from the material and its process of forming, detailing, and systemization compose the formal resultants.

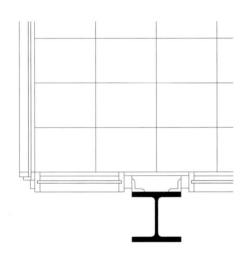

Farnsworth House, Ludwig Mies van der Rohe
Plano, Illinois 1951
[Modern, free plan house, steel and glass]

The order of the Farnsworth House is created and established by the metal structural framework deployed at multiple levels. The white-steel columns form a grid system that organizes an implicit system throughout the building. The floor pattern and ceiling beams are extensions of this overt structural methodology. These industrial and mechanical sensibilities allow the interior space to explode beyond the plenum of the ceiling and continue out into nature. The transparency of the glass walls allows and encourages the dominant reading of the steel structure against the distant trees.

MATERIAL
PRINCIPLE 397

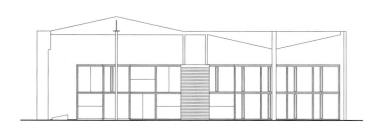

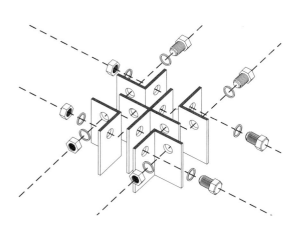

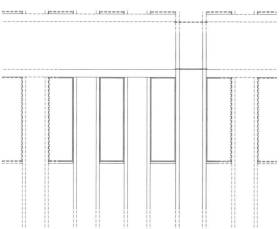

Heidi Weber Pavilion, Le Corbusier
Zurich, Switzerland, 1965
[Modern, rectilinear cage under independent faceted roof, metal]

The Heidi Weber Pavilion has a tripartite composition consisting of a modular bolted steel frame, a cast-in-place concrete ramp, and a superstructural, welded, sheet-formed, independent roof that hovers above the entire complex. The diversity of the parts is furthered by their independent and individuated tectonic systems. The orthogonality of the base building is formed through the regular repetition of a steel "L" section, radially aggregated and bolted to create a cruciform column. The array of the system carries into a three-dimensional frame that is further enforced through the infill and enameled panels that are brightly colored and variably patterned. This bolted system is juxtaposed against the monolithic sheet-welded and facet-formed roof. Dynamically alternately in its oscillating facets, the expressive figure of the roof contrasts the regularity of the cage system below. Each compositional element results from the celebration of the individuated tectonic system.

U.S. Steel Building, Harrison, Abramovitz and Abbe
Pittsburgh, Pennsylvania 1970
[Late Modern, crenellated triangular office tower, metal]

The surfaces of the U.S. Steel Building use corten steel (a steel typified by its rust-brown, oxidized surface that seals and prevents further penetration). This distinct material choice combines the rigidity of its metal form with an organic texture and coloration. Further, the building is uniquely designed with massive, exterior, water-filled columns to combat the susceptibility of metal to fire. Under high temperatures, metal deforms or even liquefies. Encasing a steel building in concrete, spraying on fireproofing, using fire-retardant paints, or protecting with other materials are common solutions, but obscure the material from visual expression. The use of the water-filled column is an innovative solution that permits the expression of the exposed metal form and material while maintaining the necessary fire resistance. In ownership, visual material qualities, and transparently expressive form, the building is rooted in the expression and celebration of its material.

Detail
Wood

Wood is a comparatively soft and malleable material. As such it can be manipulated into almost any form. From the time that man first had the appropriate tools, wood has been used for its ease of detailing. These details can be a result of its use as a structural material, as ornament, as connection, or as a cladding. Wood is probably the most common material used in detail work because of its ability to solve numerous problems within the context of architecture as well as the fact that it is a renewable resource cultivated throughout large portions of the globe. Over time, the production of lumber and use of wood has developed to a point where there are standard sizes and shapes, making it easier to use in construction and detailing. This broad availability, when combined with the ease of manipulation, which allows customization, makes for a proliferation of use around the world.

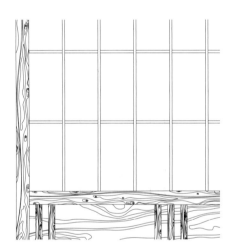

Katsura Imperial Villa, Prince Toshihito
Kyoto, Japan 1624
[Traditional Japanese, large palace, wood]

This villa, which is often described as the quintessence of Japanese taste, is a series of exquisitely detailed wooden buildings that had a tremendous impact on modernist architects. The austerity that is associated with many of the buildings originates from the careful and skillful use of wood as the primary building material. Every aspect of the Villa seems considered and rendered through this use of wood. The beams, the joints, the floors, and the numerous screens combine to produce what is definitively one of the most highly considered wooden structures of the world.

	MATERIAL	
	PRINCIPLE	399

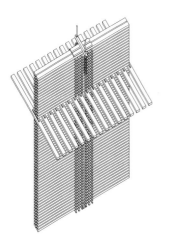

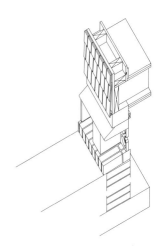

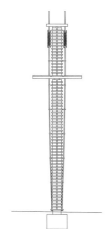

Gamble House, Greene and Greene
Pasadena, California 1909
[Arts and Crafts, asymmetrical central hall house, wood]

The Gamble House in Pasadena is a seminal work of Greene and Greene, exemplifying an architecture dominated by wood detailing. The house was constructed in one year by 12 men using only hand tools. Every inch of the building boasts some type of wood detailing. Celebrated elements such as the sumptuous stair case, the mantels, and the paneling are clearly recognized by their craft of construction and detailing. Other aspects, such as the siding shingles (which are all three-feet long), are stained and tectonically hidden, but represent the overall consideration of wood as the primary material. The structure of the roofs is exposed, clearly defined in the large overhangs, articulating a tectonic of organic ornament.

Swiss Pavilion at Expo 2000, Peter Zumthor
Hanover, Germany 2000
[Postmodern, stacked temporary pavilion, wood]

Zumthor conceived of the Swiss Pavilion at the Expo 2000 as somewhat of a respite from all the virtual worlds offered to the visitors in the other pavilions. The Swiss Pavilion was constructed only from wood boards that were all the same size (144 cm x 20 cm x 10 cm). These boards were stacked and connected without using glue, nails, or screws; they were instead kept in place using tension cables and rods. The resultant language of the building was intrinsically related to the very nature of the material itself. The details of this building came not from the ability to manipulate wood, but rather as a celebration of the standardization of sizes, and obviously the wood itself. After the Expo, the building was dismantled and all the wood was sold as seasoned lumber.

400	MATERIAL	DETAIL	MASONRY	ELEVATION / DETAIL	ARCHITECTURE
	PRINCIPLE	ORGANIZATION	TYPE	READING	SCALE

Detail
Masonry

Masonry, which is typically thought of as a structural material, has also been used in other applications to provide a sense of materiality, texture, and detail to buildings. Certainly there are details that arise out of pure masonry construction, resulting in a performative ornament of sorts, however masonry has also been utilized to go beyond mere structure and instead take on attributes that are usually reserved for more honorific materials. Cladding, made popular by the Romans, is certainly one of these methods. Other systems that satisfy this genre are decorative masonry, paving, and the manipulation of the individual units. Including both the unit and the mortar joint, the detail emerges from each system and their collaboration.

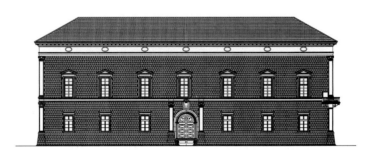

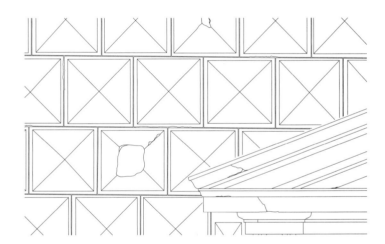

Palazzo dei Diamanti, Biagio Rossetti
Ferrara, Italy 1493
[Renaissance, Palace with marble cladding, masonry]

The Palazzo Diamanti in Ferrara is one of the most famous palaces in Italy. It is named for its expressive and unusual white-marble cladding. This cladding consists of approximately 8,500 white-marble blocks, each carved as a pyramid to represent diamonds. The texture and visual excitement of this building is truly one of the highlights of the Renaissance and continues to inspire architects in terms of the conceptualization of masonry as a detail.

MATERIAL | PRINCIPLE | 401

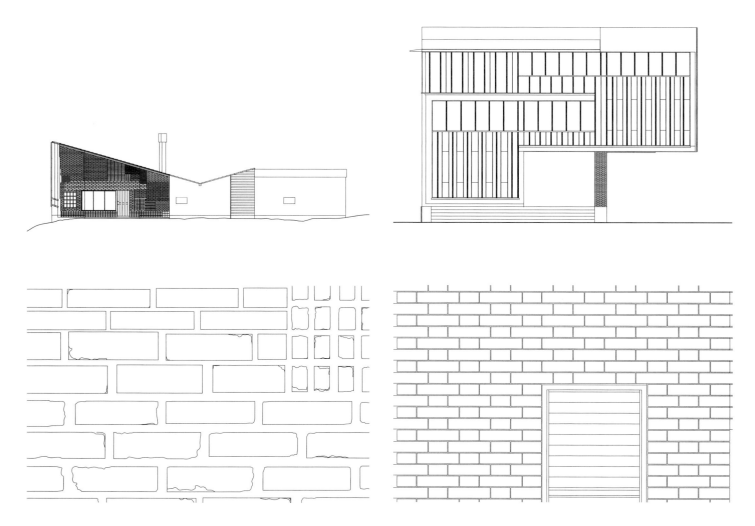

Aalto Summer House, Alvar Aalto
Muuratsalo, Finland 1953
[Modern, experimental courtyard house, wood and masonry]

Aalto used his Summer House as an experimental palette whereby he could research the properties of different architectural solutions of space and materiality. His investigations into masonry are most notable in the brick walls of the house that face into the small courtyard. Aalto used these walls almost as a sampler, exploring and experimenting with the patterning of brick and the compositional implications he could later deploy in his larger public buildings. Applications of numerous bonds and brick sizes resulted in the wall becoming a compositional brick painting. This rich use of masonry is contrasted in the outside walls of the house, which are constructed in a white-painted, straightforward, clay masonry with regularized coursing.

Tongxian Gatehouse, Office dA
Beijing, China 2003
[Postmodern, gatehouse, masonry and wood]

In the Tongxian Gatehouse, Office dA used a local gray brick in ways that were expressive and novel. They were able to combine running and stacked bonds and remove certain bricks to create textual patterns and produce formally dynamic and dematerialized walls. These operations created a broad variety in the otherwise blank surfaces. The material emphasis was translated to a detailing system that allowed for the celebration of the beauty of brick as a simultaneous structural material and decorative and ornamental element.

Detail
Concrete

Concrete is a material with great plasticity. As such it is dependent upon the structure and design of the vessel into which it is poured. The formwork thus determines the relationship of material and form. Created from raw site-delivered materials of sand, rock aggregate, water, and Portland cement (as the chemical binder), it is mixed into a liquid that is solidified in form through the chemical process of hydration. The detail of connection occurs on the microscopic scale. The architectural articulation of the material occurs on the scale of the formwork. Its materiality and scale, along with the tectonics of its assembly relative to the necessary technical components of the construction process, are relative to the overall form and design intention. The detail becomes a collaborative material articulation of associated crafts and formal techniques that are temporarily imposed then removed. They remain latent in the surface as a formal organizational residue in the final composition and artifact.

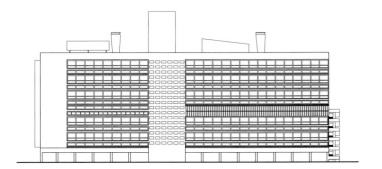

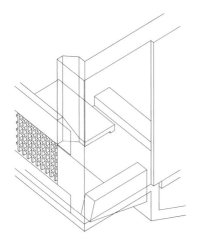

Unite d'Habitation, Le Corbusier
Marseille, France 1952
[Modern, double-loaded corridor, concrete]

The Unite d'Habitation is an accurate and self-revealing building in terms of its concrete details. The entire complex is structured using cast-in-place concrete elements that essentially tell the story of their construction. Walls, columns, floors, and elements such as the brise-soleil illustrate the detail-oriented aspect of this particular construction system. Each part of the Unite d'Habitation adds up to a complete and collective whole. The functional parts and the more honorific portions are constructed with the same precision, or lack thereof, culminating in a tour de force of concrete-construction and detailed execution.

MATERIAL PRINCIPLE 403

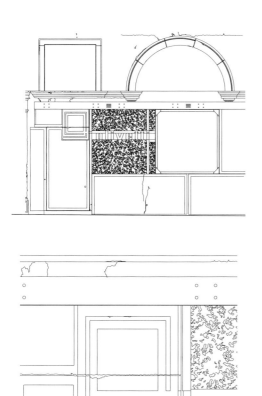

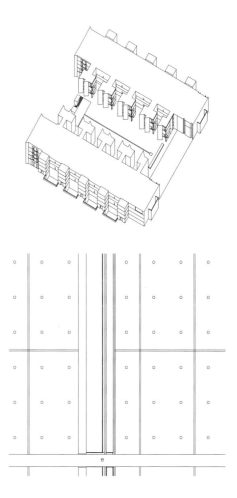

Olivetti Showroom, Carlo Scarpa
Venice, Italy 1958
[Modern, materially articulated sign and door, concrete]

The Olivettii Showroom, in typical Scarpa style, uses exquisite and intricate detailing to create an interrelation of the localized piece and the collective whole. As an essay in concrete, the subtle two-dimensional working of the surface is accomplished through playing with scales, textures, subtle depths, and varied formworks. The intricate surface is facilitated through the careful crafting of the formwork. Shifting depths, varying textures, detailed integration of text, careful joinery, and articulation of both construction and expansion joints produce a dynamic surface out of a single material surface. The craft and intricacy is furthered by the presence and precision of the operable concrete door. Balancing the solidity and material mass of concrete with the balanced mechanization of the hinged door, the concrete is dematerialized and ephemerally detailed.

Salk Institute, Louis Kahn
La Jolla, California 1965
[Late Modern, symmetrical laboratories, concrete and wood]

Salk Institute is an iconic project that establishes a pinnacle of cast-in-place concrete construction. The essential role of its materiality and construction dominate the form, effect, and presence of the final building. Using cabinet makers to craft the highly precise formwork, Kahn deployed the concept of the open joint (a condition that doesn't allow materials to touch, but rather articulates itself through a delicate reveal in which the shadow creates the connection). Within the concrete surface itself, the edges of the formwork are mitered to accentuate the seams with a protruded line that creates a surface tracery, whereas the lateral form ties are plugged with lead and left as process marks. The articulation of these joints and construction processes create an overlain anthropomorphic scale that subdivides the continuity of the surface, material, and system, creating a legible formal and geometric overlay. The result is a material celebration that works on all scales defining the project through its material and detail.

Detail
Metal

Metal as a detail category broadly refers to the collective families of ferrous and non-ferrous metals and their diverse characteristics, scales, properties, and applications. Each specific metal type has varied coloring, weathering, malleability, thermal movement, and melting points that affect the formal articulation and the performance. The specific form and application are determined through the detail. As mechanical articulations of the connection, the detail becomes a focused and revealing expression of the designed and engineered conceptualization of a material and its system.

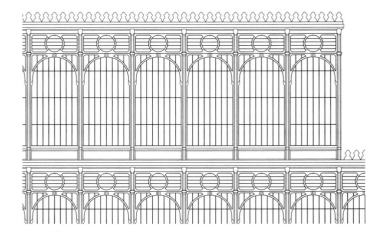

Crystal Palace, John Paxton
Hyde Park, London, England 1851
[Industrialism, prefabricated greenhouse, metal and glass]

The Crystal Palace is materially significant for its innovation in both metal and glass technologies, repetitive use of structural elements and spatial modules as design techniques, use of prefabrication, and industrialized production techniques. These systems manifest themselves in the detail. Still highly ornamental, the segmented structural system of cast iron creates small elements that aggregate and assemble to create localized frames to the space. This intense segmentation breaks down the mass of the building to a structural lattice. This allows for a prioritization of the transparent surface while providing an anthropomorphic scale to the building through the relationship of body to piece.

| MATERIAL | 405 |
| PRINCIPLE | |

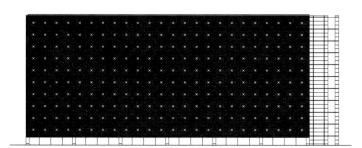

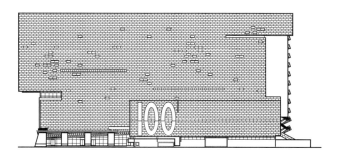

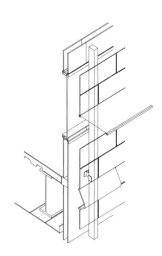

L'Institute du Monde Arab, Jean Nouvel
Paris, France 1988
[Postmodern, sunscreen with operable iris, metal and glass]

Caltrans District 7 Headquarters, Morphosis
Los Angeles, California 2005
[Postmodern, dynamic double-skinned office building, metal]

L'Institute du Monde Arab is defined by its mechanized skin. As a grid of accumulated panels, each panel contains a patterned field of intricately articulated individual **irises**. *Their collective visual pattern references traditional Islamic motifs, updated through the industrialized, high tech, performative celebration of the mechanics and their machined expression. The entire composition is dynamically furthered through the responsive solar movement of the panels. The optical effect of the metallic surfaces, under a glazed, taught skin of transparent glass, objectifies the material and pattern. The effect is drawn through the rest of the building by the expressive articulation of all systems. The emphatic transparency of both vertical and horizontal surfaces allows for visual continuity and a dominance of piece, part, and the detail of connection.*

The Caltrans District 7 Headquarters finds its identity through the varied expression and articulation of a single material. Playing with varied optical densities of metal mesh, the articulation of surface (plane), and the frame that holds and supports (line), the galvanized steel and the exposed connection create a visual vocabulary of blunt functional ornament. The expression of a regularized system and the divergent variation of or within that system, dominate the linear forms. Their resulting impact is one of efficiency that evolves into expressive formalism. The surface is more subtly evocative, engaging the variable densities of overlapping meshes and perforations. The optical ephemerality to the skin and its visually indecipherable depth evolve as the light changes. Moments of mechanized oscillation (the flipping of the operable panels) create a dynamic effect through the careful systemization of the material reflection, surface, and detailing.

406	MATERIAL	DETAIL	GLASS	ELEVATION / SECTION / DETAIL	ARCHITECTURE
	PRINCIPLE	ORGANIZATION	TYPE	READING	SCALE

Detail
Glass

The ephemerality of glass is made possible through its evaporative lack of presence. The delicacy of its articulation becomes essential to the reading of the frame and the revelation of the surface. The detail of glass comes through the material properties themselves (for instance, the clarity, coloration, reflectivity, and emissivity) in conjunction with frame composition and thickness. The relationship of the border to surface, the transparency of framed interior and exterior, and the reflection and refraction of the surface itself define the material reading and effect.

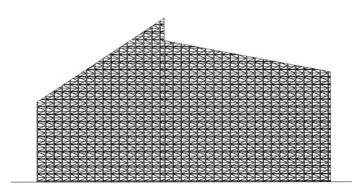

Crystal Cathedral, Philip Johnson
Garden Grove, California 1980
[Postmodern, all glass symmetrical mega-church, glass]

The Crystal Cathedral translates the material typology of the cathedral by inverting the legacy of mass with a transparency of glass. The metaphor of the televised mega-church as open and connected is materially reinforced. The detailing of the surface comes through an economy of dimension and standardization. The thickness of the mullion as a somewhat clumsy but affordable articulation is made up for through the uniformity and massive scale of its deployment. The collective effect dominates the rather conventional detail.

MATERIAL
PRINCIPLE   407

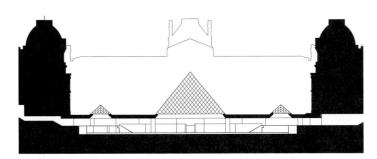

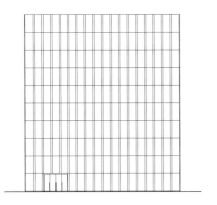

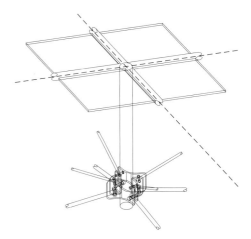

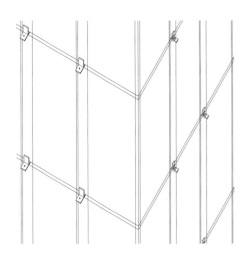

Louvre, I.M. Pei
Paris, France 1989
[Postmodern, monumental pyramidal form, steel and glass]

When confronted with the daunting task of adding onto the Louvre, a historical and museological icon on both the architectural and urban scales, Pei relied upon the Platonic form of the pyramid to add to the dominant axis. Using the materiality of glass to create a juxtaposing technological figure, he allows the traditionally solid form to visually and effectually evaporate. By creating a new entry located in the center of the courtyard, Pei was able to create a single point that reorganized the circulation and hierarchy of the entire complex. By dematerializing the form through an intricately engineered and detailed glass skin, the opening becomes a light icon during the daylight, allowing for a continuity of space, and serves as an illuminated beacon at night. The continuity of the glass surface is maintained through an elaborate rear-positioned tension cable network. The result is a thinness to the line, providing the functional depth. The localized geometry of the panel reiterates the figuration of the whole.

Kunsthaus Bregenz, Peter Zumthor
Bregenz, Vorarlberg, Austria 1997
[Modern materialism, stacked square, glass and concrete]

The Kunsthaus Bregenz utilizes a singular material application as a way of establishing the building's language and form. Glass dominates the exterior of the project, resulting in a sophisticated appliqué of material form. On one level, the building appears almost without detail in its simplicity and relentless character. Closer observation illustrates the carefully considered glass detailing system that gives the structure its architectural appeal. Each piece of glass in the exterior wall is slightly angled to produce an enigmatic, faceted whole that appears to change with the direction and conditions in which the building is observed.

**Detail
Plastic**

Plastic, as the newest genre of architectural materials, is diverse in the collection of types within its family. Hyper-flexible in color, size, malleability, and durability, the diversity of applications and fabrication and molding processes allows for a dynamic array of formal characteristics. In the included examples, three diverse configurations and applications are described: one addresses a rigid but complexly formed, glossy, panelized surface as shell; one uses a thin, ETFE air-inflated pillow as a pocketed and segmented skin; and one uses the repetition of an injection mould to produce repetitive, modular, stackable units in the traditions of masonry. The diversity of these systems and technologies represents the broad flexibility of plastic as a material and the emergence of an equally diverse array of tectonic and formal responses.

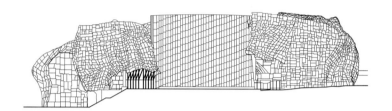

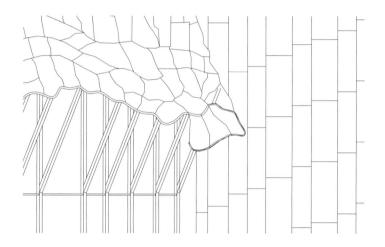

Experience Music Project, Frank Gehry
Seattle, Washington 2000
[Postmodern, four merged organic forms, diverse]

At the Experience Music Project, Gehry deploys his signature external formalism but modifies it through the collage of four juxtaposed materials and their individual figures. Variably treated metals give dynamic and referential figurations, juxtaposed with one segment clad in blue plastic. Despite the variation of material and the architect's focus on the holistic form, the necessary segmentation of the individually formed and customized panels results from the application of material such as paint, requiring a forcing and taming of the material into a predetermined shape. Representative of the flexibility and innate materiality but little else, the nature of plastic is used as a cultural reference as opposed to a material form generator.

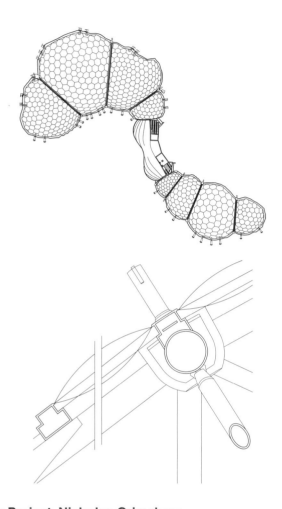
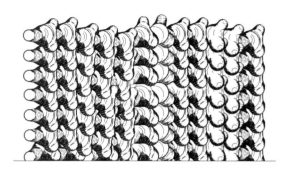
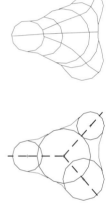
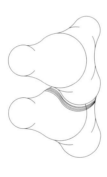

Eden Project, Nicholas Grimshaw
Cornwall, United Kingdom 2001
[Hi-Tech, truncated spherical greenhouse bubbles, plastic]

The Eden Project uses a repetitive hexagonal module that holds an inflated ETFE air pillow. The serial shape allows for the production of repetitively joinable sections that aggregate to form portions of overlapping spheres that nestle into the hillside. The intricate material skin and air-inflated pillows allow for a heroic scale of delicate pieces that structurally span and engulf massive interiors. The thermal barrier of the plastic skin allows for precise control of the enclosure to regulate the necessary internal environment for the greenhouse climates. Here, the skin as a performative surface is equally an effectual generator. Both are fully facilitated by the material.

Blobwall, Greg Lynn
Los Angeles, California 2008
[Postmodern, organic unit, plastic]

The Blobwall builds on the long-standing traditions of masonry unit construction. Hybridizing the conventional premise of aggregated units with digital design techniques, contemporary rotational injection- mould manufacturing techniques, complex geometries, and modern plastic material, Lynn creates a new tectonic. Beginning with the anthropomorphic dimension of the hand, the organic shape allows for a stackable assembly system in a slightly varied running-bond configuration. The visual effect in both color and shape produces a highly organic form juxtaposed with the intensely inorganic materiality. The pop product-like nature, though suffering from many impracticalities and limited application, suggests a material system that is uniquely dependent upon plastic for its system. It is definitively unique to the lineage of plastic construction and tectonics.

410 ORNAMENT

PRINCIPLE

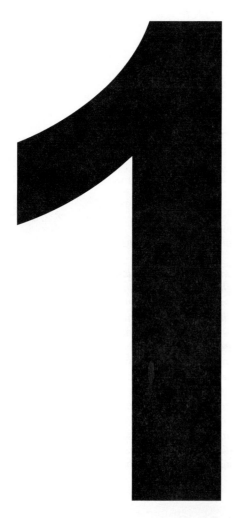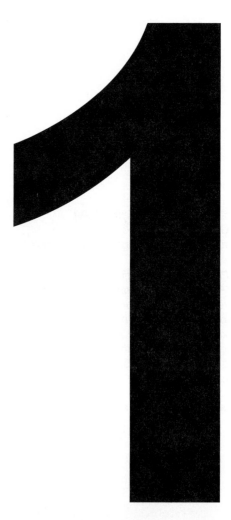

Ornament

11 - Ornament

Ornament is articulation and embellishment intended to lend grace and beauty. It has been defined as an element not belonging to the essential harmony or melody of a project. Herein lies the dichotomy of ornament: It is comprehended as a positive principle in one light, but seen as nearly superfluous in another. As an architectural subject, ornament has seen many phases. There were movements in which architects were judged by their ability to use ornament and yet conversely, there were times when architects were judged by their ability to deny the use of ornament. It has been said that ornament is, in fact, all that separates architecture from building. We have divided ornament into a number of different types, each focused on a specific thematic and formal aspect of ornament and its application.

Material Construction—Ornament Material construction as ornament is derived from the actual fabrication and assembly processes of the building itself. Architecture, by its very nature, leaves a record of the construction systems that underlie the creation of a building. These systems typically take the form of joints, connections or other construction applications (such as stacking, layering, or joining) inherent to segmental construction. The revelation of the inevitable hand-of-making produces a typology of ornament that is based on construction. This type of ornament is actually the most fundamental and has been in use since the first architecture. It is so closely linked to all architecture that any separation is impossible.

Applied Ornament—Religious Applied ornament is the most visually dominant classification of ornament. Within the types of applied ornament, religious ornament is the most historically dominant and widely recognizable. There is a tremendous history to this type of ornament due to the fact that past civilizations typically bestowed great importance and significance to their religious buildings. These buildings were often equally as critical as institutions of information or propaganda as they were pieces of architecture. The ornament on religious buildings was responsible primarily for the communication of the ideals, history, and stories of religion to an often largely illiterate population. Historically, because these buildings often formed the basis of the society, their ornament and scale were typically unrivaled.

Applied Ornament—Performative—Structural Within the classification of applied ornament, performative, or articulated, elements that derive from functional responsibilities, emerge as a sign of technology and building systems. The structural responsibility of performative systems allows for the use of a structural system as an exposed and articulated composition to produce a type of ornament. The functional rationale of the system is typically repeated, creating a visual field and a collective pattern of visual interest. When exposed, structure is often emphasized or even exaggerated as a way of realizing its importance as an architectural element. One need only consider any array of exposed beams or columns to understand the power and importance of structurally performative ornament.

Applied Ornament—Performative—Mechanical Much like the aforementioned performative structural ornament, performative mechanical ornament extends from the use of mechanical elements and systems and their exploitation for figural and decoratively ornamental purposes. Buildings have diverse passive and active mechanical systems. How these systems are exposed, manipulated, and articulated can produce a dominant figuration that is responsible for the detailing and expression of ornamentation. This may take the form of a very simple element such as a louver, window shutter, or overhang, but it can also find expression in much more complicated and technical solutions such as operable brise-soleils, photovoltaic panels, or exposed ducting systems.

Applied Ornament—Referential—Organicism The application of ornament with referential organic formalism has a long history and continues to play a significant role in architecture even today. Once architects and sculptors began adding elements to buildings as ornament, the representation and depiction of the natural world formed the primary and initial content. Whether it was the plants, animals, humans, or any other aspect of the organic environment, these subjects have been considered sacred by people for thousands of years. Hence these elements have been immortalized through the use of ornament on the buildings of various cultures. Organic ornament has always been paramount due to the fact that the human body can be included in this topic. The response of built form to the human scale and the anthropomorphosis of the body in static material are essential to the cultural history of man, art, and architecture.

Applied Ornament—Referential—Structural Applied ornament that is referential to structural performance is directly related to performative structural ornament, but is distinctly unique in the authenticity of form. Referential structural ornament represents an aspect of structure. It is not the actual structure itself. This ornament is typically used when an architect is attempting to exploit or celebrate a structure that is covered or concealed, or when there is a reference to a historical practice that set certain architectural rules. While this ornament is selectively used, it is still very common as it has been part of the classical language of architecture for centuries.

Applied Ornament—Historical Applied ornament with historical content is similar to applied religious ornament. This ornament, like religious ornament, is employed primarily as a way of communicating ideas or historical events to the viewer. This ornament, which can be textual or pictorial, is always representational in its attempt to transfer information and meaning. Most secular memorials and monuments, which fall under this category, are adorned with this particular type of ornament.

414	ORNAMENT	MATERIAL	CONSTRUCTION	ELEVATION / DETAIL	ARCHITECTURE
	PRINCIPLE	ORGANIZATION	GEOMETRY	READING	SCALE

Material Construction—Ornament

The mediation of the scale of a building relative to the anthropomorphic dimension requires segmental construction. Weight, size, availability, manufacturing, and transportation constraints each require the use of building components to incrementally assemble buildings. This necessary unitization results in myriad seams, joints, and connections. The understanding of materiality and the associated systems of construction allows for the control and engagement of assembly. This joinery, whether controlled as a design mechanism or simply derivative of the methods of construction, creates a pattern and a visual field of ornamentation. Though functionally derived, the subdivision of the unitization is legible as ornament. In the following projects, the material of construction (both structural and cladding) begins with a utilitarian methodology that evolves into a systematized graphic language of material ornament.

New House, Biddulph Family
Ledbury Park, England 1590
[Tudor, exposed heavy-timber construction with infill, wood]

*The New House, as a representative case study of a much broader style defined by heavy-timber construction, uses mortise and tenon construction of large scale wood timbers infilled with **nogging** of load-bearing capacity (typically masonry, clay, or plaster). The exposed structural lattice becomes developable as an ornamental system, thereby extending the structural functionality of the visual lattice. As representative of a cultural way of building, the density and pattern of the system provide a visual breakdown of the architectural scale and produces an ornamental figuration scribed into the surface.*

416	ORNAMENT	MATERIAL	CONSTRUCTION	ELEVATION / SECTION / DETAIL	ARCHITECTURE
	PRINCIPLE	ORGANIZATION	GEOMETRY	READING	SCALE

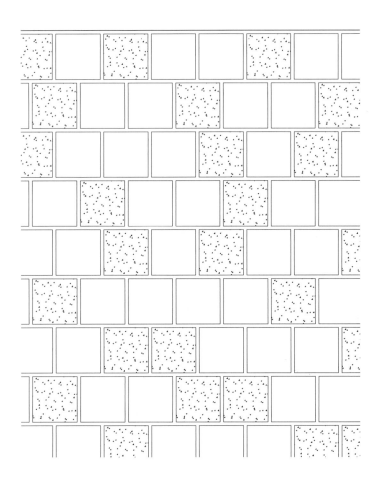

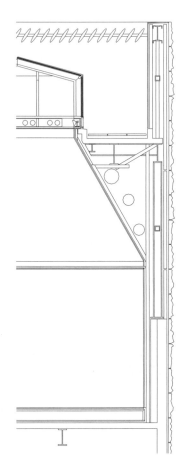

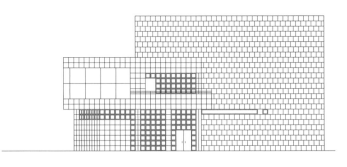

| ORNAMENT | **417** |
| PRINCIPLE | |

The Getty Center, Richard Meier
Los Angeles, California 1997
[Postmodern, museum campus, stone metal and glass]

The Getty Center uses **travertine** *masonry to produce a monumental scale and permanence. The stacked blocks are a stone veneer over a structural steel frame. The back support removes the load-bearing responsibility of the wall, allowing for non-repetitive unitization, variable textures and finishes, interrelated scales of units, and compositional flexibility. The result is a transformation of the veneer surface from a simple, standardized, laminate enclosure to a patterned surface that is locally responsive, aesthetically engaged, and compositionally determined. This material formalism introduces shape, scale, and color as material ornamentation, determining the visual reading and perceptual effect of the resulting buildings.*

| 418 | ORNAMENT
PRINCIPLE | APPLIED
ORGANIZATION | RELIGIOUS
GEOMETRY | ELEVATION / DETAIL
READING | ARCHITECTURE
SCALE |

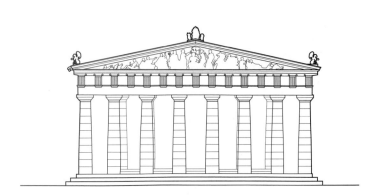
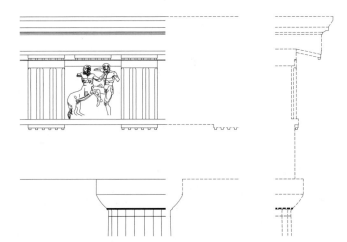

Applied Ornament—Religious

The concept of applied ornament is fundamental to religious buildings. Required to communicate to the mostly illiterate masses, ornament emerged as visual vocabulary in both two- and three-dimensional capacities. Orchestrated to relate the stories and beliefs of a particular religion, the ornament was textual, pictorial, and sculptural. All religious ornament served the same purposes: communication of specific stories and the production of a monumental hierarchy through the complexity of articulation and detail. The ornament gave detail and beauty to the buildings, imbibing them with a sense of meaning and honor that people understood. As an attention-getting device, applied ornament evolved with equal typology and significance as the architecture proper.

ORNAMENT PRINCIPLE | 419

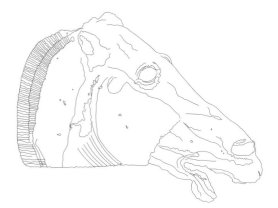

Elgin Marbles, British Museum, Parthenon, Phidias
Athens, Greece 432 BC
[Hellenistic, pictorial ornament, marble]

The Elgin Marbles, which used to adorn the Parthenon in Athens, are a collection of figures and text that occupied the pediments, **metopes**, and friezes of the Classical Greek Temple. The Marbles consist of 17 figures from the pediments. There are 15 metopes panels, which describe the battles between the Lapiths and the Centaurs. The frieze measures 247 feet in length and was originally placed above the interior **architrave** of the Parthenon. These ornaments represent about half of what survives of the ornament from the Parthenon. Their ability to exist as standalone elements and the ease of their removal reveals the fact that they were entirely applied. The significance and value of these marbles is so substantial that the British Museum constructed a dedicated room, the Duveen Gallery, specifically to house them.

420	ORNAMENT	APPLIED	RELIGIOUS	ELEVATION / DETAIL	ARCHITECTURE
	PRINCIPLE	ORGANIZATION	GEOMETRY	READING	SCALE

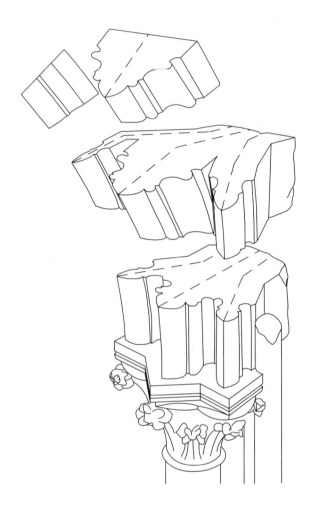
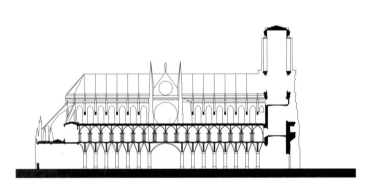
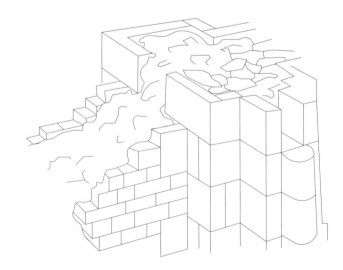

ORNAMENT | PRINCIPLE | 421

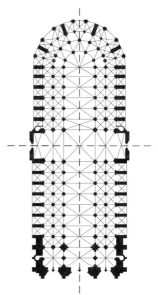

Notre Dame, Maurice de Sully and Viollet le Duc
Paris, France 1240
[Gothic, Cathedral, masonry]

The west facade of Notre Dame Cathedral in Paris contains an impressive collection of applied ornament. Statues of the Virgin and Child occupy the center and are flanked by Adam and Eve. The rose window literally becomes the halo of Mary and Christ. Below this ensemble there is the horizontal frieze, the Gallery of Kings, consisting of sculptures of the descendants of Christ. At the bottom of the façade are the three large portals: The Central Portal (Portal of the Last Judgment), the Portal to Saint Anne, and the Portal to the Virgin. Each of these portals is decorated with a multitude of characters. The purpose of this ornamentation was to communicate to the mostly illiterate users, while adding detail and beauty to the building.

| 422 | ORNAMENT
PRINCIPLE | APPLIED
ORGANIZATION | PERFORMATIVE
GEOMETRY | ELEVATION / DETAIL
READING | ARCHITECTURE
SCALE |

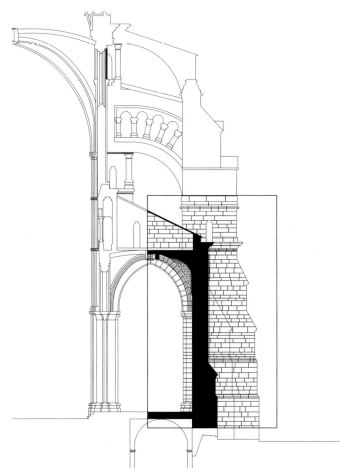

Applied Ornament—Performative—Structural

Applied ornament that is performatively structural refers to externalized systems that produce another system of articulation rooted in structural functional performance. The presence of this system is highly considered and overt, deriving an overlain ornamental system. Beginning with the functional response to a specific structural necessity, the systemization of the necessary physics relative to material form establishes a distinct formal operational rule set. Within this practical parameter, the architect introduces a compositional aspect that ultimately establishes an ornamental effect. The aesthetic arrangement of the structurally performative system establishes the ornament.

ORNAMENT | PRINCIPLE | **423**

Chartres Cathedral
Chartres, France 1260
[Gothic, Latin Cross, stone masonry]

Chartres Cathedral represents the pinnacle of the high Gothic period. The increased height of the nave reached a structural limitation associated with wind loads. As a result, a secondary external structure, the flying buttress that became a hallmark of all Gothic cathedrals, was developed to provide the necessary bracing. With the wall as a structural skeleton, large glazing allowed for the illumination of the spiritual spaces and the extensive use of stained glass through which to relay biblical stories. The rose window was developed at the head of the transept both to resolve the vault of the roof and also to give hierarchy to the tripartite façade. The aggregated field of repetitive elements produces an ornamental exoskeleton that both houses further ornamentation (such as gargoyles) and serves as a repetitive ornamental frame element.

424	ORNAMENT	APPLIED	PERFORMATIVE	ELEVATION / PLAN / DETAIL	ARCHITECTURE
	PRINCIPLE	ORGANIZATION	GEOMETRY	READING	SCALE

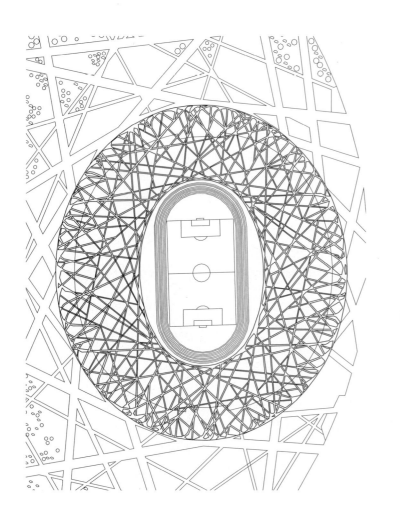
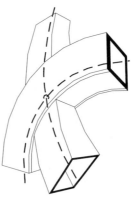
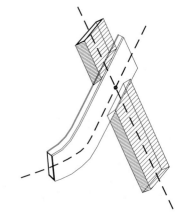
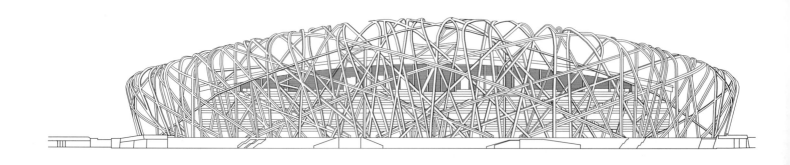

ORNAMENT PRINCIPLE 425

The Bird's Nest Stadium, Herzog and de Meuron
Beijing, China 2008
[Postmodern, irregular lattice frame stadium, concrete and steel]

The Bird's Nest Stadium, built for the 2008 Beijing Olympics, is visually dominated by a series of seemingly random steel members that wrap around the entirety of the stadium, thus giving the complex its name. These steel beams form an exoskeleton that not only acts as an appliqué or ornament, but also performs in a structural manner, helping to support various parts of the perimeter of the stadium. It is essentially an applied aesthetic solution that is governed by compositional sensibilities; however, it substantiates itself further by solving problems of statics and structure.

426	ORNAMENT	APPLIED	PERFORMATIVE	ELEVATION / DETAIL	ARCHITECTURE
	PRINCIPLE	ORGANIZATION	GEOMETRY	READING	SCALE

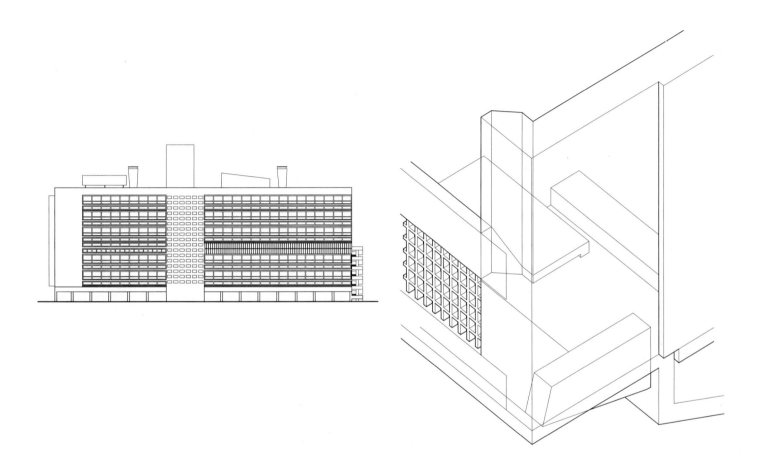

Applied Ornament—Performative—Mechanical

Similar to the methodology of performative structural ornament, performative mechanical ornament operates within a specific functional system to establish an infrastructure that can be compositionally regulated to provide functional ornamentation. Derived from a performative engagement with site and climate, the mechanical systems become articulated for expressionist ornamental purposes. Their formal, organizational, and compositional aspects are exploited for optimal aesthetic figuration.

ORNAMENT	427
PRINCIPLE	

Unite d'Habitation, Le Corbusier
Marseille, France 1952
[Modern, double-loaded corridor, concrete]

The Unite d'Habitation is encrusted with numerous forms of applied ornament, which take on characteristics that emphasize their role beyond mere decoration. Almost every aspect and feature of this building is constructed of cast-in-place concrete, allowing for the various elements to conform and result in a materially holistic form. The brise-soleils, or sun-shading devices, are clearly applied and ornamental; nevertheless, they perform the critical mechanical functions of blocking the sunlight and contributing to the passive cooling strategy of the building. Other mechanical elements, such as the vent-exhaust chimneys on the roof, are also treated with a similar ornamental formula.

428	ORNAMENT	APPLIED	PERFORMATIVE	ELEVATION / DETAIL	ARCHITECTURE
	PRINCIPLE	ORGANIZATION	GEOMETRY	READING	SCALE

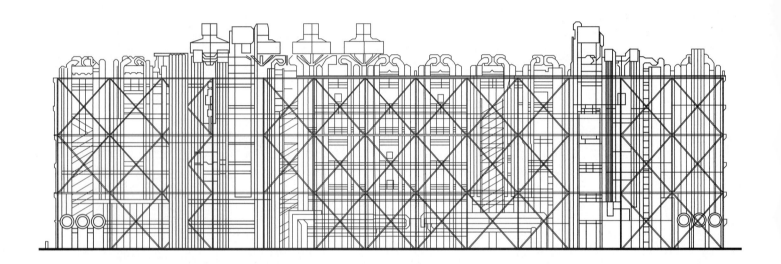

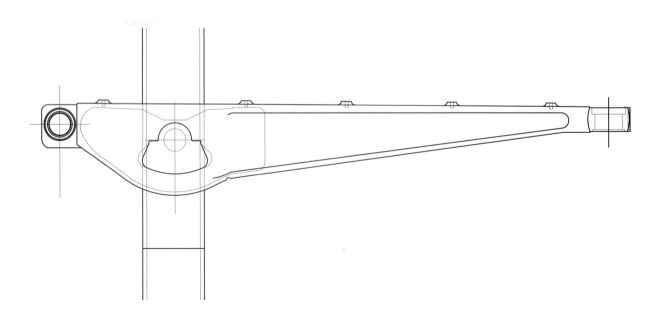

ORNAMENT	
PRINCIPLE	**429**

Centre Pompidou, Renzo Piano and Richard Rogers
Paris, France 1974
[Hi-Tech Modern, free plan museum, steel, glass, concrete]

This museum, which represents the pinnacle of hi-tech modernism, exhibits a novel and graphic type of ornament. Essentially, the building has been turned inside-out, resulting in the mechanical systems, which are typically hidden in a building, now being exhibited on the exterior. Each system is color-coded, using blue for air, green for water, gray for secondary structure, and red for circulation. These mechanical and performative systems have been transformed here into ornament that cannot be removed, lending permanent beauty and balance.

| 430 | ORNAMENT
PRINCIPLE | APPLIED
ORGANIZATION | REFERENTIAL STRUCTURAL
GEOMETRY | ELEVATION / DETAIL
READING | ARCHITECTURE
SCALE |

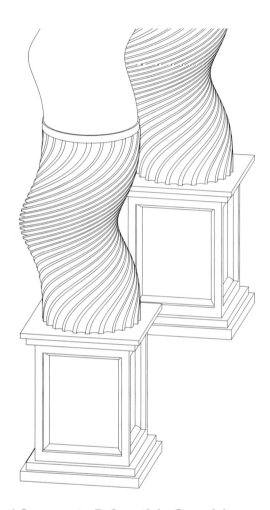 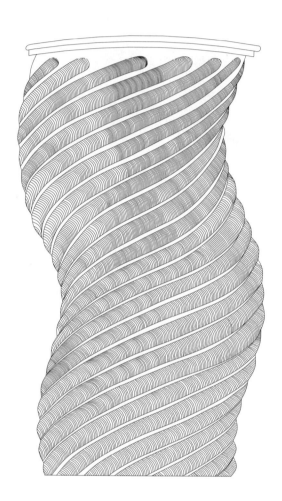

Applied Ornament—Referential—Organicism

Referential organic applied ornamentation is typically in the shape or derivate system of an organic or natural form. Elements such as leaves, flowers, or any other flora or fauna can be included within this type. Organic ornament has been seminal for centuries, linking built form to the natural world and finding wide application in Egyptian, Islamic, Greek, and Roman architecture. Ornament such as the acanthus leaves of the Corinthian capital or the lotus and papyrus leaves of Egyptian columns are clear examples of its application. Organic elements were so valued they were ultimately transformed into stone as a way of communicating their importance and significance.

Baldacchino, Gian Lorenzo Bernini
St. Peter's Basilica, The Vatican 1634
[Baroque, altar baldacchino, bronze]

Bernini's Baldacchino in St. Peter's Basilica is one of the quintessential works of the baroque period. It is a large bronze canopy that hovers above the basilica's high altar. The Baldacchino's large scale allows it to negotiate between the scale of the human and the immense size of Michelangelo's dome rising above. The four spiral columns of the Baldacchino are encrusted with applied organic ornament. Laurel leaves represent poetry, lizards represent rebirth and the search for God, and the bees are a symbol of the Barberini family. Bernini was able to construct a sculpture/building that used organic ornament in such a way that the beauty of the elements, and their meaning, were tantamount to the aesthetics.

432	ORNAMENT	APPLIED	REFERENTIAL STRUCTURAL	ELEVATION /	ARCHITECTURE
	PRINCIPLE	ORGANIZATION	GEOMETRY	READING	SCALE

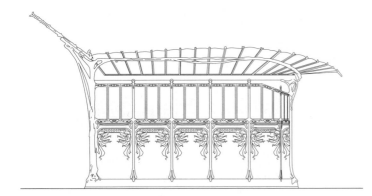
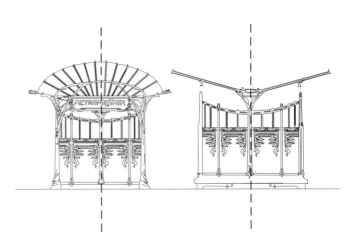

ORNAMENT	
PRINCIPLE	**433**

Paris Metro Entry, Hector Guimard
Paris, France 1905
[Art Nouveau, sculptural gates, iron and glass]

The Metro entries in Paris were executed as a reaction to the then-dominant taste of French neo-classical culture. These gates and entries form a series of organic, almost surrealistic, figures that are in sharp contrast to the rest of the urban fabric. The iron parts of these structures are all cast to resemble organic, plant-like figures. The sweeping curves and bulbous ornament all appear to have grown from the earth, rather than to have been cast in a foundry.

434	ORNAMENT	APPLIED	REFERENTIAL STRUCTURAL	ELEVATION / DETAIL	ARCHITECTURE
	PRINCIPLE	ORGANIZATION	GEOMETRY	READING	SCALE

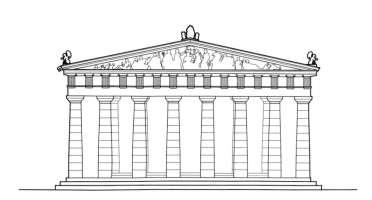
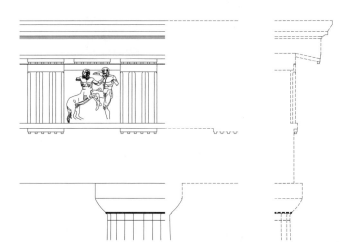

Applied Ornament—Referential—Structural

Some of the most common applied ornament references the structure of the building. This phenomenon began due to the fact that many buildings were constructed of one material and then, over time, were rebuilt using a new material. When building with the new material, there was often an ornamental reference to the original material. There were instances when this change necessitated a change in overall form; however, the design and form principles remained consistent. The disassociation with craft and the re-embodiment of form as ornament established the resulting necessary figures as referential symbols to be redeployed.

ORNAMENT PRINCIPLE | 435

Entablature at the Parthenon, Phidias
Athens, Greece 432 BC
[Hellenistic, structural ornament, marble]

As an example of structural ornament, the entablature at the Parthenon is considered seminal to the argument concerning the transformation of materials of a building over time. It is believed that the Parthenon and, indeed, all Greek temples were originally constructed of wood. The organization of the **triglyphs** *and metopes reflected this wood construction as extensions of roof rafters and ceiling beams. When the temples were rebuilt using stone, this specific detail was kept as a way of referencing the original structure. This feature established the basic form of almost all classical entablatures.*

436	ORNAMENT	APPLIED	REFERENTIAL STRUCTURAL	ELEVATION / DETAIL	ARCHITECTURE
	PRINCIPLE	ORGANIZATION	GEOMETRY	READING	SCALE

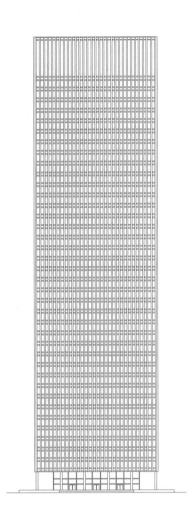
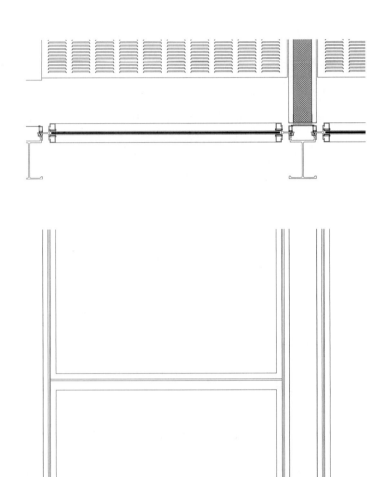

ORNAMENT	
PRINCIPLE	**437**

Seagram Building, Mies van der Rohe
New York, New York 1958
[Modern, office building, concrete and steel]

The Seagram Building contains a singular reference to the idea of structural ornament. The actual structure of the building was steel, but due to code requirements for fireproofing, the steel was encased in concrete and hidden under layers of various materials. Mies decided to replace this concealed structure with an external ornamental I-beam. The reference to the hidden structural material and form on the exterior of each column of the building required a false reapplication. This structural ornament was so powerful that it literally gave the building its aura and its signature; yet it is in seeming conflict with Mies's subscription to an essentialism of form.

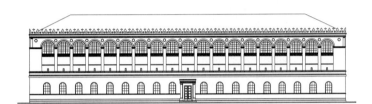
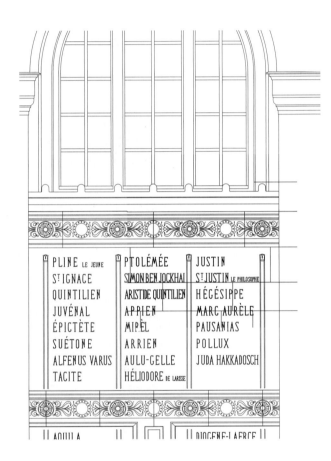

Applied Ornament—Historical

Applied historical ornament is common amongst most cultures and as with most ornament, communication is the primary goal. In this particular circumstance, ornament typically deals with the history or histories of the institution occupying the particular building. This type of ornament is predominantly pictorial, textual, or sculptural. Either literally or metaphorically, this ornament is attempting to relate a story to the user that explains the historical significance of persons or perhaps the institution.

ORNAMENT	
PRINCIPLE	439

Bibliothèque St. Geneviève, Henri Labrouste
Paris, France 1851
[Renaissance Revival, library, masonry and iron]

The Bibliothèque St. Geneviève uses applied historical ornament in the form of the large stone panels that occupy each window niche. The names of 810 illustrious scholars are carved into these panels. This applied ornament is textual, which in itself references the written word and thus the library that contains these words. Interestingly enough, the actual books are stored directly behind in a thickened wall that forms these panels. Labrouste managed to combine history, function, and form into a remarkable ornamental moment.

440	ORNAMENT	APPLIED	HISTORICAL	ELEVATION / DETAIL	ARCHITECTURE
	PRINCIPLE	ORGANIZATION	GEOMETRY	READING	SCALE

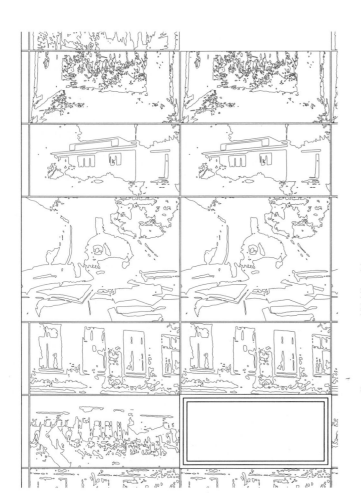
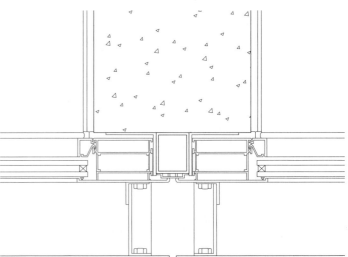

ORNAMENT PRINCIPLE 441

Eberswalde Library, Herzog and de Meuron
Eberswalde, Germany 1999
[Postmodern, library, concrete and glass]

The elevations of the Eberswalde Library display applied historical ornament through a series of images that are printed on the concrete and glass panels. The images, which were collected by artist Thomas Ruff, all take on the quality of newsprint and represent different ideals depending on their location on the façades. At the entrance level, the images are of "man and technology"; through a series of paintings, the second level represents, "love and mortality"; the final level represents "historical events, technology, and knowledge." The simplicity of the building's form allows the complexity of this ornament to be read clearly.

442	PATTERN			
	PRINCIPLE			

Pattern

PATTERN PRINCIPLE **443**

12 - Pattern

Pattern refers to the systems (either visually legible or simply implied) employed to govern the formal organization of relative units. Typically defined through repetitive fields, the variables governing pattern manipulation include shape, material, and color. These variables can reference historical or natural elements, for example, and may occur on varying scales of piece/unit, portion/panel, bay/module, or chunk. The repetition of the units produces a rhythm, the intonation of the form based on the interrelation of parts, allowing the emergence of a legible system. These rhythms govern the rules of pattern itself, which are regulated by the module, frequency, or scale.

Shape Pattern The pattern of shape refers to the individual unit and its geometric form. Encompassing shapes such as the square, triangle, hexagon, and octagon, the consequence of the individuated form on the larger aggregated field is dependent upon the relative shape of unitization. The relationships of part to part and the impact of part on the whole establish geometries that have varied visually aesthetic and functionally performative implications.

Material Pattern Material pattern has two scales of legibility. The first and most dominant across all different material types (wood, masonry, concrete, metal, glass, and plastic) is the intrinsic module. Derived by the synthesis of the fabrication process, the scale of shipping and installation, all relative to the scale of the human body, the material module remains evident in the final object. Mortar joints, panel seams, and construction joints of all types are derivative of the assembly system and remain evident, making the prominence of the material pattern "piece" legible in the final composition. This presence is aesthetically significant in the legibility of architectural form. The second scale is more specific to the material, but is intrinsic to the matter and processes (natural or man made) that facilitate material. The coloration and veining in stone, the texture of concrete, the scale and color of masonry, the grain of wood, are evident in the visual tactility of the material itself. When examined directly, or aggregated as familial arrays, the relativism and interrelation establishes an effectual pattern.

Color Pattern Color pattern refers to the impact of pigment and the systemization of pigmentation to produce visual arrays. The applied pigmentation is an artificial overlay on the natural visual appearance of material and thus requires a specific intention and methodology.

Referential Pattern—Historical Referential pattern refers to the cultural and historical significance of the pattern itself. The creation of pattern through tectonic, material, or cultural rationales has produced a catalog and legacy of its relativism and legibility. The associations of a particular pattern as a historical object provide the referential quality of the pattern itself. These associations reiterate, address, contradict, or simply refer to a historical rationale and cultural significance fueled by the weight of historical precedent. The use of this reference as formal, linguistic play emerged as particularly significant during the postmodern era. Derived from tradition and embodied with association through their repetitive use through history, referential patterns leaned on their legacy of repetition and allusion to achieve their significance and authority.

Pattern—Repetition—Piece / Unit The primal ideal of pattern emerges from the repetition of shape. At its most fundamental level, the piece or unit of repetition is the foundational building block for the system and the intended effect, which are crafted through the varied aggregation of its serial usage. The piece/unit can be materially derived from an anthropomorphic dimension, a system of fabrication, a scale or weight of movement or installation, the relationship to a tool, or a diverse series of other factors. Or the piece/unit may be simply geometrically derived, emphasized through shape, system, or simple aesthetics. The basic figuration of the piece establishes the system of its repetition and the scale and effect of the pattern as a whole. Pattern accomplished through the repetition of a piece, or unit, derives a visual language of repetition through the assembly of the tectonic. Typified through unitized systems such as masonry, the pattern that results from the assembly has a practical construction logic, but one that can be evolved, accelerated, and deployed for ornamental or effectual patterns. The individual pieces of the specific unit create the base elements that build the fundamental pattern module and establish a regimented systemization of the collective unitized pattern.

Pattern—Repetition—Portion / Panel When the piece is scaled to a larger portion of the building or a panel of the surface, the ideal of individuality and variation can occur within a larger field of the portion, allowing for localized responses within a specific framework. This piece is then deployed as a repetitive element that allows the production of a field effect and larger pattern through macro-scaled aggregation. Pattern, through the repetition of a portion or panel, deploys the effect of assembly on a larger scale. Evolving from the intrinsic segmentation necessary for fabrication, transportation, and installation, the portion or panel creates the pattern unit. The repetition and variation of these elements create the visual pattern array.

Pattern—Repetition—Module / Bay Pattern achieved through the repetition of a module or bay is a traditional method of producing architectural rhythm and regularity. This repetition may be horizontal, vertical, or both, as well as either two-dimensional or three-dimensional. Larger than a material module, the bay is used almost exclusively in composition. It is composed through repetition, variation, scale, and seam. The collective aggregation of modules or bays produces a macro-scale pattern to the overall composition.

Pattern—Repetition—Chunk Pattern derived through the repetition of a chunk requires a three-dimensionality to the piece. The exploitation of the depth of the "chunk," along with a typically larger scale, allows for prefabricated units (such as the apartment modules of Habitat 67) to engage fully occupiable, volumetric units to establish a pattern. The multi-dimensionality of the units allows for more complex combinations and assemblies to produce non-linear aggregations.

Pattern—Rhythm—Module Pattern derived from the rhythm of the module utilizes any of the aforementioned scales of elements to develop a repetitive module that relies upon variable frequencies and rhythms to establish a pattern. The use of the module operates independently of the material or construction dimension to establish a compositional basis.

Pattern—Rhythm—Frequency Key to the formal legibility of pattern is the rhythm and ultimately the frequency of its reiteration. The meter, either visual or experiential, establishes the resonance of the field and exposes the legibility of the system. The intonation of the composition established by the increment and frequency of the rhythm regulates the pattern.

Pattern—Rhythm—Scale Pattern and the rhythmic iteration established through the frequency of scale create a grain based on size. The dimension determines the resolution of the effect, whereas the variation itself can establish a pattern through its juxtaposition and contrast.

Pattern—Shape

The pattern of shape is recognized by the use of pure geometries as a device for creating numerous designs and motifs. Shapes such as circles, squares, triangles, hexagons, and octagons have been utilized by diverse cultures throughout time to establish a base module. These shapes often carry associative meanings and significance. Shapes have always formed the basis for much of the pattern employed in architecture; their use continues today. They contain decorative and ornamental associations and are used as attention-getting devices.

PATTERN	
PRINCIPLE	447

Alhambra
Granada, Spain 1390
[Islamic, palace, masonry]

The patterns created at the Alhambra are perhaps more intricate than any that came before or since. These patterns seem to cover every part of this magnificent complex. With meaning embedded in distinct locations, pattern was also deployed as beautiful decoration, allowing the interior walls to feel light and ethereal rather than massive. One of the main patterns used was the Almohad sebka (a grid of rhombuses); this diagonal pattern system is the most prevalent of all the systems used at the palace and, when combined with the arabesques and calligraphy, form one of the most exquisite collections of wall decoration in the world.

448	PATTERN	SHAPE		ELEVATION / DETAIL	ARCHITECTURE
	PRINCIPLE	ORGANIZATION		READING	SCALE

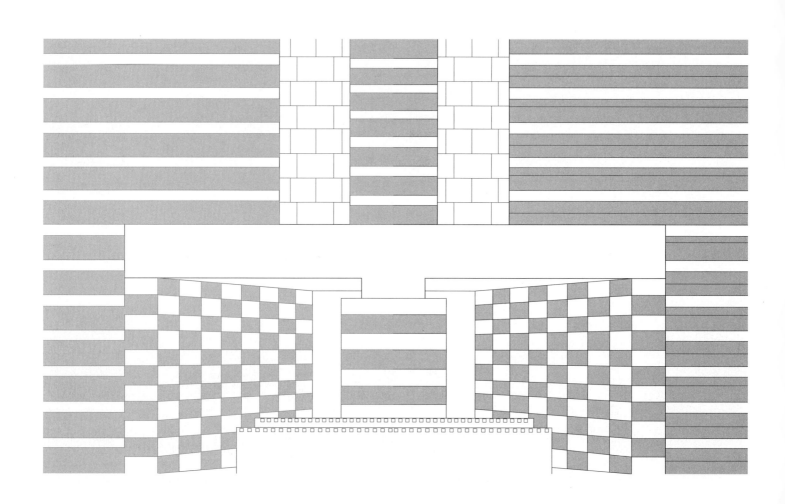

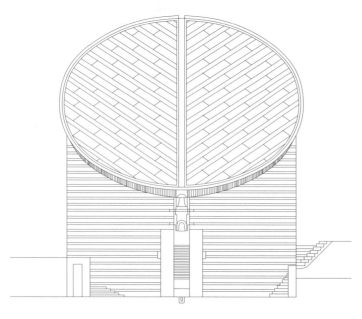

PATTERN | **449**
PRINCIPLE

Church of San Giovanni Battista, Mario Botta
Val Maggia, Mogno, Switzerland 1996
[Postmodern, church, marble and granite facing]

The Church of San Giovanni Battista utilizes a pattern of shapes as a type of ornament or decoration. This formal mechanism is accomplished by alternating native gray granite and white marble throughout the building. The pattern primarily takes the form of simple stripes that dominate both the interior and exterior. This polychrome technique was very common during the Romanesque period and was resurrected during the postmodern era for its graphic boldness and reference. The arch behind the altar offers a moment of more complexity and intricacy to the form of patterning. Each block of granite or marble in the arch becomes smaller, giving the arch a false perspective that is reminiscent of the early perspectival experiments of the Renaissance.

450	PATTERN	SHAPE		ELEVATION / DETAIL	ARCHITECTURE
	PRINCIPLE	ORGANIZATION		READING	SCALE

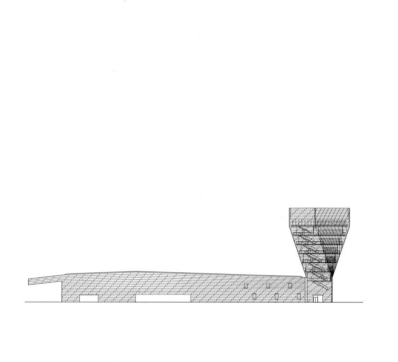
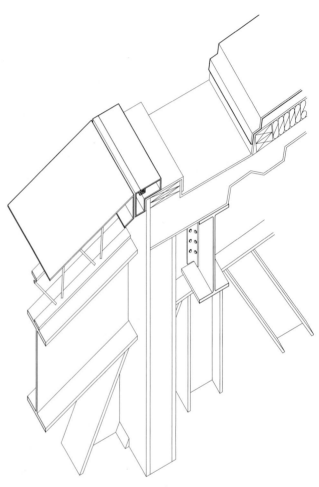
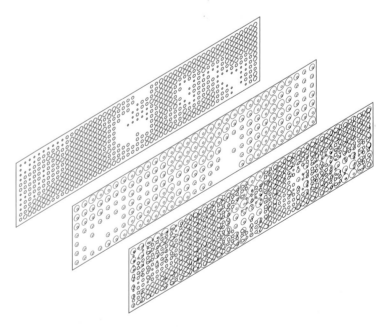

PATTERN
PRINCIPLE | 451

De Young Museum, Herzog and de Meuron
San Francisco, California 2005
[Postmodern, free plan with courtyards, metal and glass]

The De Young Museum uses shapes and patterns as a way of achieving an exterior that is meant to resemble the nearby eucalyptus trees. Through a series of perforations and dimples, the copper exterior skin will, over time, bear a resemblance to the trees in both color and light quality. The entire exterior uses the circle in various sizes and densities as a way of achieving an optical field. The skin is pulled away from the structure, allowing it to produce a moiré quality that changes as one moves around or through this perimeter veil.

| 452 | PATTERN
PRINCIPLE | MATERIAL
ORGANIZATION | | ELEVATION / DETAIL
READING | ARCHITECTURE
SCALE |

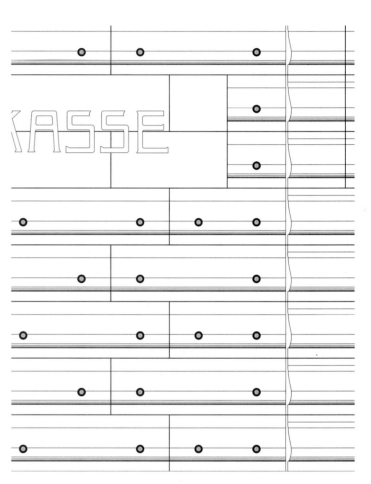

Pattern—Material

Material is an essential generator of pattern. Whether intentionally foregrounded, or simply a residual resultant of the construction process, material and the module of its raw dimension and system of joinery and assembly produce combinational patterns. These patterns are derived by the module of the body (the labor of installation), the module of fabrication (the material limitation and the optimization of the manufacturing process), and the module of transportation (the physical limitations of the forklift, truck, or crane). The combination of these physical limitations establishes an innate segmentation, establishing the need for joinery. The repetition and systemization produced by the assembly defines the material pattern.

Postal Savings Bank, Otto Wagner
Vienna, Austria 1912
[Early Modern, reductive form with technical expression, diverse]

The Vienna Postal Savings Bank illustrates a number of pattern systems by its use of a unique cladding system that is treated in a self-revealing way. Each piece of stone, marble, or metal on the building is attached by a series of aluminum bolts. These bolts, relative to their respective material, are deployed in various patterns across the building. The side elevations are treated in a rather straight forward way, relative to the entrance façade, which is more complex and intricate. The pattern and ornament emerges from the connection system, which is amplified through the graphic nature of the bolts.

454	PATTERN	MATERIAL		ELEVATION / DETAIL	ARCHITECTURE
	PRINCIPLE	ORGANIZATION		READING	SCALE

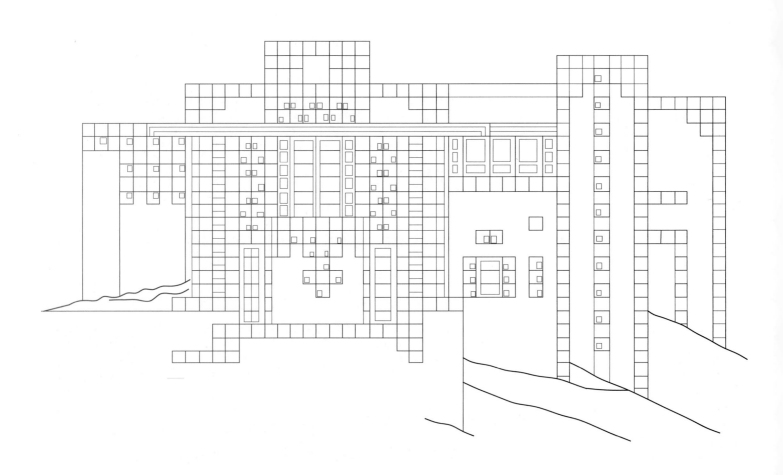

  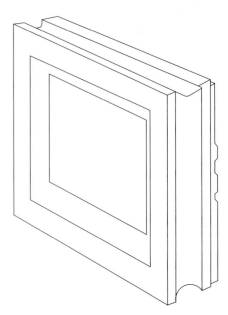

PATTERN	
PRINCIPLE	455

Freeman House, Frank Lloyd Wright
Los Angeles, California 1924
[Modern, textile block house, concrete masonry]

The Freeman House, through its unique construction systems, hybridizes a functional performing tectonic with a highly systematized formal and effectual pattern. The dimension and repetition of the unit is established for functional practicalities, whereas the iterative form of the variable shape produces identity and field effect. Based on a single module of the site-cast concrete, textile block, the variation comes in the amount of detail and openness to surface. The iterative formal module, in combination with a systemization to their position and application, creates a dramatic field effect of material pattern.

456	PATTERN	MATERIAL		ELEVATION / DETAIL	ARCHITECTURE
	PRINCIPLE	ORGANIZATION		READING	SCALE

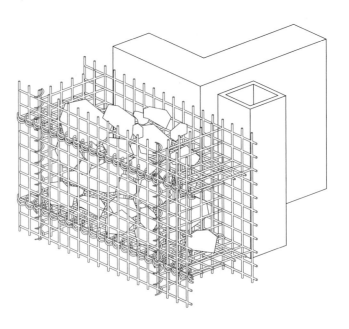

PATTERN 457

Dominus Winery, Herzog and de Meuron
Yountville, California 1998
[Postmodern, rip rap stone bar, glass and stone]

The Dominus Winery utilizes a simple formula to achieve remarkable architectural results. By downplaying the formal aspects of the project, Herzog and de Meuron allow the focus to shift to the articulation of the perimeter skin, which surrounds the long, linear-bar building. This skin is composed of metal cages filled with local rocks of various sizes and then placed on the exterior of a glass wall. This rip rap wall system creates a series of patterns, both in terms of the rocks and the metal cages themselves, but also in the shadows they produce.

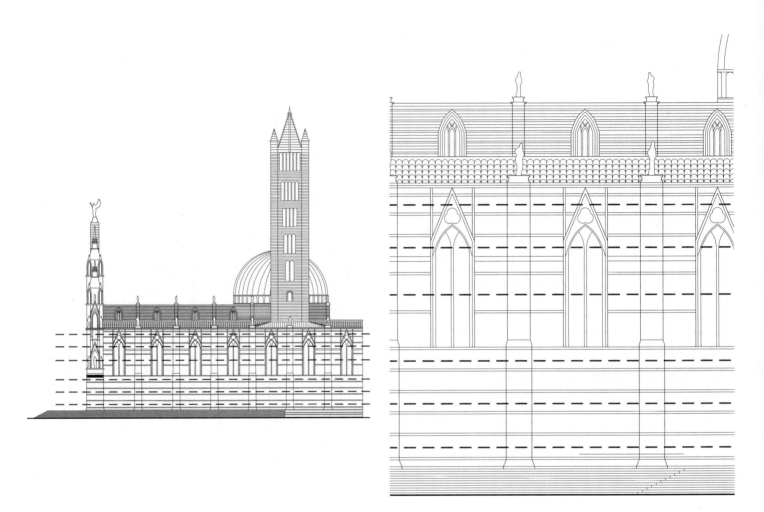

Pattern—Color

Color is one of the most dominant and effective methods of pattern-making in architecture. The patterning systems are as varied as there are colors. These color patterns are used for many things: At times they are used as codes for functional concerns, in other instances they can be used as a way of differentiating between structural or architectural elements. Architects have attempted to use color in pattern as a way of imbibing their buildings with meaning and visual interest. Thus, color may carry specific connotations or merely serve in a decorative sense.

PATTERN 459
PRINCIPLE

Duomo
Siena, Italy 1263
[Medieval, cathedral, black and white marble]

The Duomo in Siena is a visually stunning building that uses a limited color palette for a remarkable effect. The primary color system is focused on the alternating black and white marble stripes that are seen throughout both the interior and exterior. Black and white are the symbolic colors of the city of Siena, representing the mythological black and white horses of the legendary founders of the city, Senius and Aschius. In this particular instance, the color serves both as a visual device as well as an element that carries with it an associative meaning, particularly as related to the building and the city.

460	PATTERN	COLOR		ELEVATION / DETAIL	ARCHITECTURE
	PRINCIPLE	ORGANIZATION		READING	SCALE

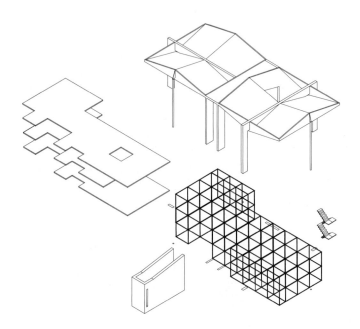

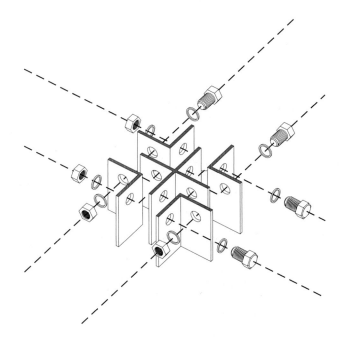

Heidi Weber Pavilion, Le Corbusier
Zurich, Switzerland 1965
[Modern, rectilinear cage under independent faceted roof, metal]

Le Corbusier's Heidi Weber Pavilion in Zurich explores the use of color patterning through the dynamic metal panels of its elevations. The colors of the pavilion are primarily utilized as a bold graphic application of abstract primal fields. Le Corbusier used the traditional primary colors of red and yellow, but also employed green and black. Governed in shape and position by the construction of the building, the panels are infills to the cruciform columns of the building's structural cage. The color provides distinction to the panels, allowing the rectilinear building to be distinguished from the large, faceted, gray-metal canopy that hovers above.

462	PATTERN	COLOR		AXONOMETRIC / DETAIL	ARCHITECTURE
	PRINCIPLE	ORGANIZATION		READING	SCALE

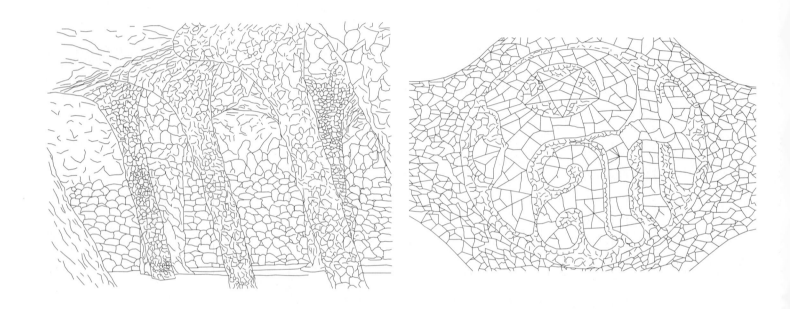

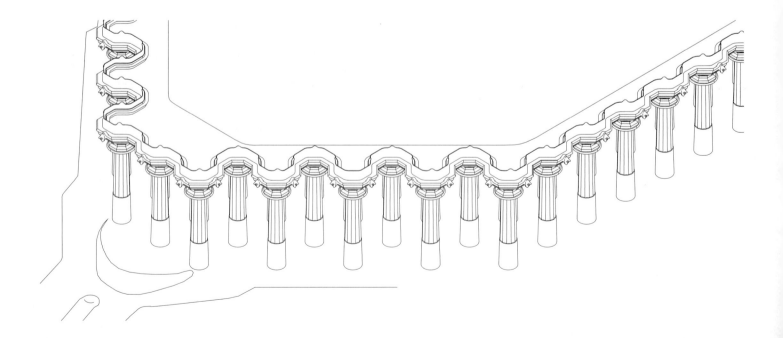

PATTERN 463

Park Guell, Antoni Gaudi
Barcelona, Spain 1914
[Expressionist Modern, structurally expressive park, masonry]

Color is used extensively throughout Gaudi's Park Guell. Gaudi employed ceramic tile as a way of introducing color into this magnificent conglomeration of architecture and landscape. Extending from the Catalonian tradition of ceramics, the color tiles are part of the pure patterns in the architecture and landscape furniture. But they are also used in sculptures of different animals located throughout the park. The colors form so many patterns that it is in fact difficult to focus on one pattern at any time and instead form a kaleidoscope effect that allows the playfulness of Gaudi's design to be better appreciated.

| 464 | PATTERN
_{PRINCIPLE} | REFERENTIAL
_{ORGANIZATION} | | ELEVATION /
DETAIL
_{READING} | ARCHITECTURE
_{SCALE} |

Pattern—Referential—Historical

In an abstract sense, historical referential patterns relate a part or piece of a building's or a culture's history to built form. This can be achieved in a number of ways that range from two-dimensional appliqué to three-dimensional representational ornament. The important aspect of historical referential pattern is the transposition of meanings that it carries with it, contributing to the cultural milieu of any given society.

Majolika House, Otto Wagner
Vienna, Austria 1899
[Art Nouveau, apartment building, masonry with tile facing]

In the Majolika House, Wagner used the technology of ceramic tile as a reference to more traditional three-dimensional ornament. On the main elevation, these ceramic tiles create a referential pattern that represents a type of organic art nouveau decoration. The tiles are repeated on all ten structural elements of the façade, aligning a relationship between ornament and performance. This repetition creates the patterning and removes it from singular ornament or decoration.

466	PATTERN	REFERENTIAL	ELEVATION / DETAIL	ARCHITECTURE
	PRINCIPLE	ORGANIZATION	READING	SCALE

Entablature of Basilica Vicenza, Andrea Palladio
Vicenza, Italy 1614
[Renaissance, loggia, marble]

Palladio's entablature on the Basilica in Vicenza carries with it a number of references. One element that is serially repeated is the cow's skull, which represents death. The other metope contains a circular figure that represents the sun, or rebirth. Hence, on the entablature of the Basilica, both death and life are represented simultaneously as figurative motifs with associative meanings.

468	PATTERN	REFERENTIAL		ELEVATION / DETAIL	ARCHITECTURE
	PRINCIPLE	ORGANIZATION		READING	SCALE

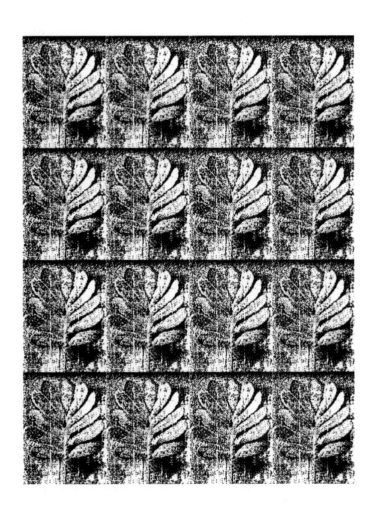

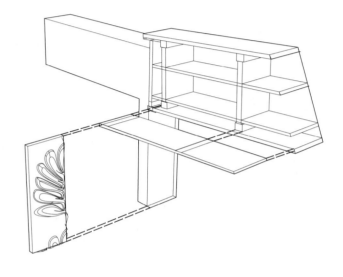

| PATTERN | 469 |
| PRINCIPLE | |

Ricola Storage Building, Herzog and de Meuron
Mulhouse-Brunnstatt, France 1993
[Postmodern, warehouse, polycarbonate panels]

The Ricola Storage Building is graphically founded in the overt study of referential patterns. The building is an extremely simple box that relies on the material qualities of the translucent, polycarbonate panels of which it is constructed. The panels are then patterned, through a silk-screen process, with a repetitive plant motif drawn from a specific photograph. This graphic referential pattern dominates the character of the building and allows the perception of the building to constantly change based on the time of day and nature of light. The serial quality of its appliqué heightens the significance of the image content while denying the presence of any one figure to prioritize the field.

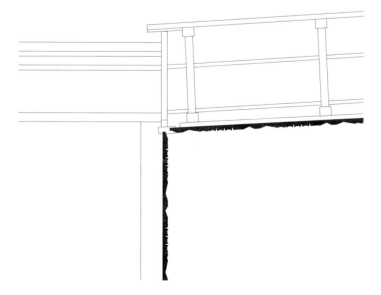

470	PATTERN	REFERENTIAL	PIECE / UNIT	ELEVATION / DETAIL	ARCHITECTURE
	PRINCIPLE	ORGANIZATION	TYPE	READING	SCALE

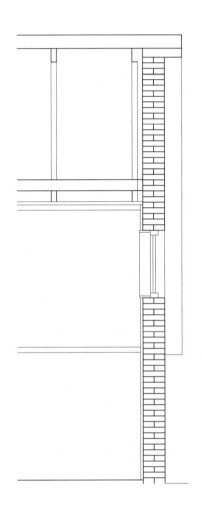

Pattern—Repetition—Piece / Unit

The repetition of pieces or units is fundamental to the creation of pattern. This repetition can occur at multiple scales and can inform the building in a number of ways. At a micro scale, the units can take the form of the actual building materials, as in the module and dimension of bricks or boards used for the construction. At a macro scale, the units can take the form of architectural components such as windows or columns, for instance. This concept is so fundamental that almost any architecture can claim that it is in use. It is important, however, to recognize the difference between an architecture that achieves a pattern by accident and one that does it intentionally.

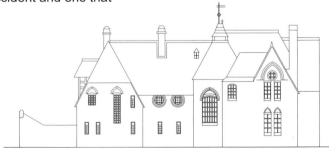

PATTERN	
PRINCIPLE	**471**

The Red House, William Morris and Philip Webb
Bexleyheath, Kent, England 1859
[Arts and Crafts, English romantic house, masonry]

*The repetitive patterns in the Red House are formed by the bricks themselves. The masonry sets the initial pattern that is visible in all portions of the exterior of the house. These repetitive pieces are used in different configurations and levels of articulation to produce varied forms and details. Each aspect of the house's form, no matter the shape or character, is constructed from the same unit, the red brick. This remarkable architectural detailing is apparent upon examination of the brick work around the various windows, which are rendered with both **jack** and Gothic arches.*

472	PATTERN	REPETITION	PIECE / UNIT	ELEVATION / DETAIL	ARCHITECTURE
	PRINCIPLE	ORGANIZATION	TYPE	READING	SCALE

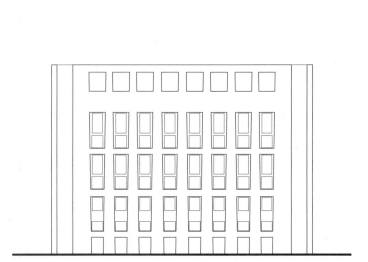

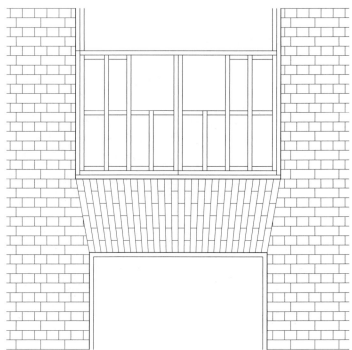

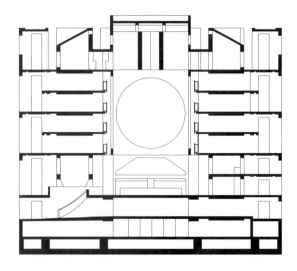

PATTERN **473**

Exeter Library, Louis Kahn
Exeter, New Hampshire 1972
[Late Modern, centralized square, masonry and concrete]

In Exeter Library, Louis Kahn uses masonry to form the outer skin through a repetitive grid façade. Gradated vertically to reduce the structural weight of the load-bearing walls by one brick per level, the aperture size is increased to accommodate the transition and diffuse the mass of the building. The removal is made up through the use of a structural **jack arch**. *The collective tapering-field effect of the repetitive masonry unit dominates the solemn and anonymous nature of the building.*

474	PATTERN	REPETITION	PIECE / UNIT	ELEVATION / DETAIL	ARCHITECTURE
	PRINCIPLE	ORGANIZATION	TYPE	READING	SCALE

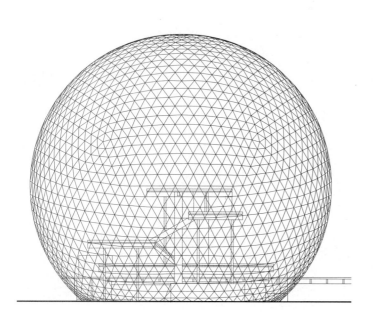
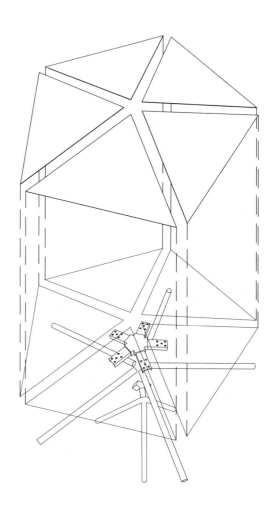

PATTERN	
PRINCIPLE	**475**

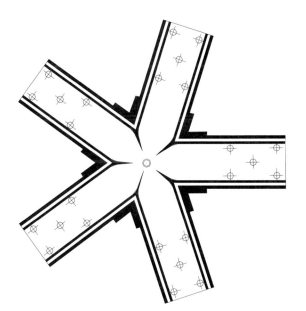

U.S. Pavilion at Expo '67, Buckminster Fuller
Montreal, Canada 1967
[Late Modern, geodesic dome, metal]

The U.S. Pavilion used a clearly repetitive component as its base unit. This unit, which was essentially a series of triangles, combined to form hexagons. The hexagons were joined together to form a large sphere. The complexity of the combination of all shapes formed an overall pattern that challenged the novel formalism and spatial aspects of the dome. This pattern was completely visually engaging and was legible from any vantage point, interior or exterior.

476	PATTERN	REPETITION	PANEL	ELEVATION / DETAIL	ARCHITECTURE
	PRINCIPLE	ORGANIZATION	TYPE	READING	SCALE

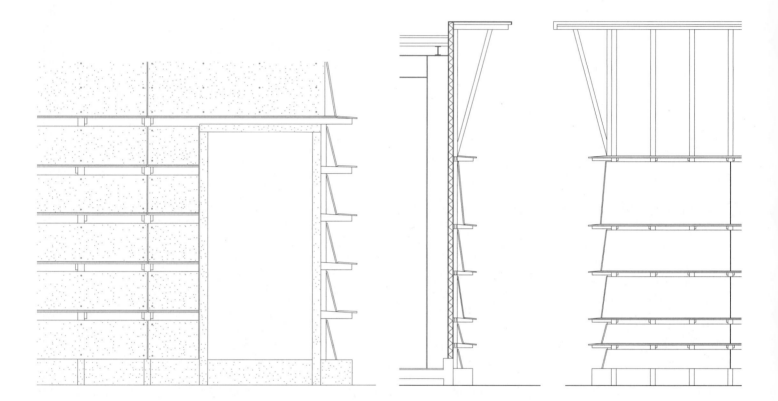

Pattern—Repetition—Portion / Panel

Pattern determined through the repetition of a portion or panel is dependent upon the assembly system. The method of construction determines the aggregation of pieces and creates a larger field. The unit, though larger in size than simple material limitations, establishes a base that through repetition (iterative or otherwise) creates a patterned field. The resulting field creates an aesthetic and perceptual effect that is intrinsically rooted in the tectonic and performance of the assembly.

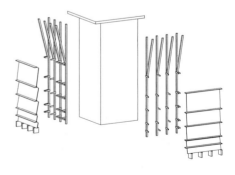

PATTERN	
PRINCIPLE	477

Ricola Storage Building, Herzog and de Meuron
Laufen, Switzerland 1991
[Postmodern, stacked panel storage building, concrete]

At the Ricola Storage Building, Herzog and de Meuron use the tradition of clapboard siding but reinterpret its detailing to produce a widely different performative technique. Using concrete instead of wood, long horizontal planks are openly stacked and individually angled. The result is a layered finning that allows for passive ventilation while creating visual opacity. The clever yet simple design is based in a careful systemization of the tectonic assembly. The resulting field effect, though seemingly familiar, is a uniquely reinterpreted pattern.

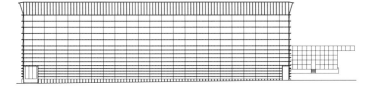

478	PATTERN	REPETITION	PANEL	ELEVATION / DETAIL	ARCHITECTURE
	PRINCIPLE	ORGANIZATION	TYPE	READING	SCALE

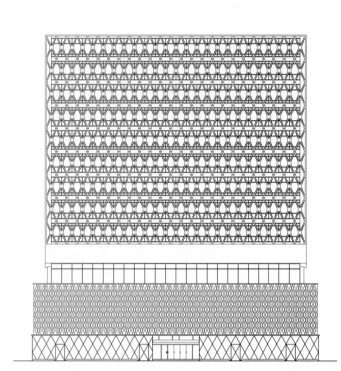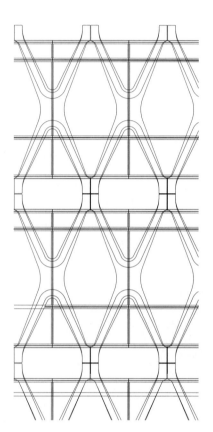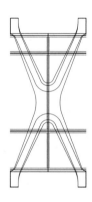

PATTERN PRINCIPLE | 479

American Cement Building, DMJ and Mendenhall
Los Angeles, California 1964
[Late Modern, repetitive precast external support frame, concrete]

The American Cement Building is uniquely structured through its exoskeletal structure. Clad in X-shaped formed precast elements, the pattern of the façade is made through the repetition of the material and the module of fabrication process. The depth and field of the aggregated effect generates an iconographic pattern. The material use is furthered by the formal expression of the repetitive cast of the material process. Using a detailed and repetitive mold, the material form is derivative of the material processes and dominates the collective appearance and perception of the building pattern.

| 480 | PATTERN
PRINCIPLE | **REPETITION**
ORGANIZATION | PANEL
TYPE | ELEVATION /
SECTION / DETAIL
READING | ARCHITECTURE
SCALE |

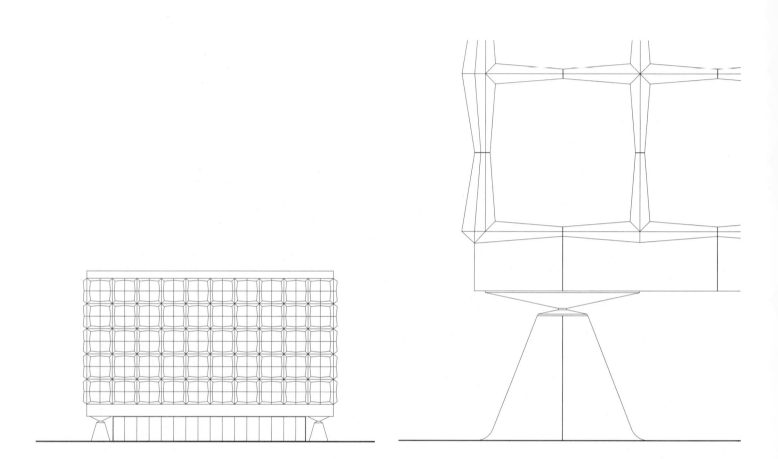

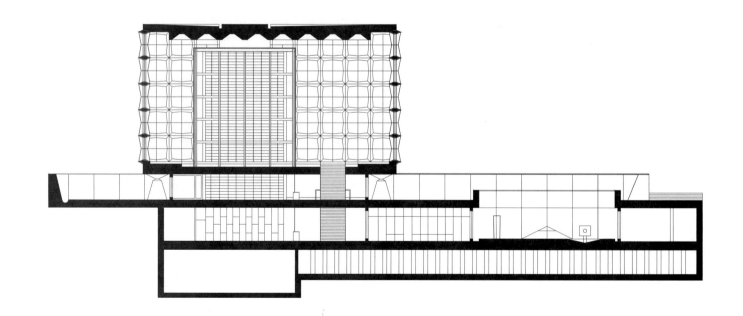

PATTERN	481
PRINCIPLE	

Beinecke Rare Book Library, Gordon Bunshaft, SOM
New Haven, Connecticut 1963
[Modern, centralized glass stack, concrete and masonry]

The Beinecke Rare Book Library is recognizable by its large concrete frame structure, which is infilled with translucent alabaster panels. This architectural system gives the building its language and essential form. This sophisticated cage serves as a covering, which houses the glass and steel pavilion that actually contains the rare books. Each panel is square and identical, giving no hierarchy in any direction. From the exterior, the building appears serene with the pattern of the panels expressing little in the way of function. The interior of the library glows as the sunlight passes through the alabaster panels, filling the room with a golden radiance.

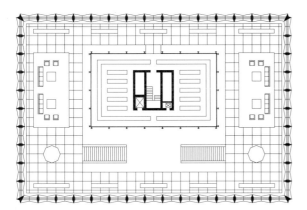

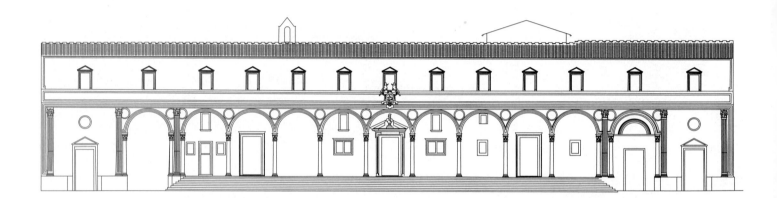

Pattern—Repetition—Module / Bay

Buildings throughout history have been organized with the use of a module or bay system. Such a system is deployed throughout the composition, governing all aspects of the building. A module or a bay can be described as a collective compositional unit from which the design is constituted. Deployed in elevation, section, and three-dimensional space, the module abstractly allows for a geometric organization of elements or, more specifically, the references to repetitive elements that recur as compositional elements. A common module was the Palladian module, which was a set of columns with an appropriate connective entablature. This unit could then be repeated to form entire façades. Architects have often used the idea of this module and its repetition as a way of both organizing a building and giving it aesthetic character.

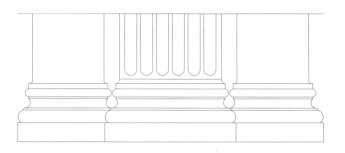

Ospedale degli Innocenti, Filippo Brunelleschi
Florence, Italy 1445
[Renaissance, hospital, masonry]

In the Ospedale degli Innocenti Brunelleschi used a series of nine identical modules across the main façade as a way of establishing order and character to the building. These types of loggias were common to hospitals, but Brunelleschi altered the design by using sail vaults and redesigning the main ornaments of the façade. The modules establish a rhythm and repetition that were so powerful that years later when the other side of the piazza was completed, the architect copied these modules as a sign or respect, using the "principle of the second man," a contextualism through reference.

484	PATTERN	REPETITION	MODULE	ELEVATION / DETAIL	ARCHITECTURE
	PRINCIPLE	ORGANIZATION	TYPE	READING	SCALE

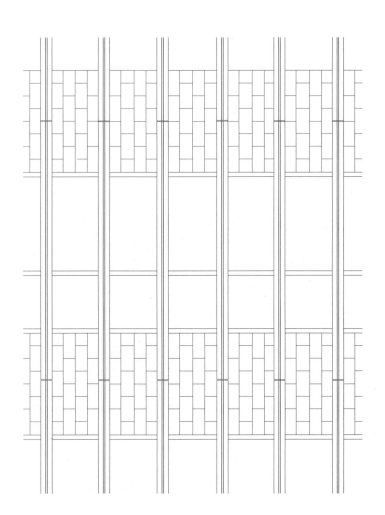
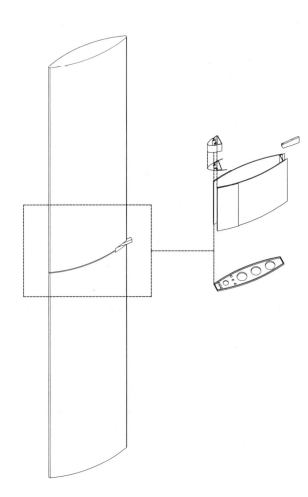

| PATTERN | 485 |
| PRINCIPLE | |

Hall of Records, Richard Neutra
Los Angeles, California 1962
[Modern, office building, steel and glass]

The Hall of Records uses a system of patterns that emerge from different functional aspects of the building. One of the largest and most dominant patterns is formed by a series of vertical louvers that were designed as sun-shading devices. These louvers originally moved with the sun, responsively blocking the sun's rays throughout the day. Other patterns reflect functional aspects of the building as exemplified by the white-tiled circulation tower and stone base with stylized mosaic.

| 486 | PATTERN
PRINCIPLE | REPETITION
ORGANIZATION | MODULE
TYPE | ELEVATION /
AXON / DETAIL
READING | ARCHITECTURE
SCALE |

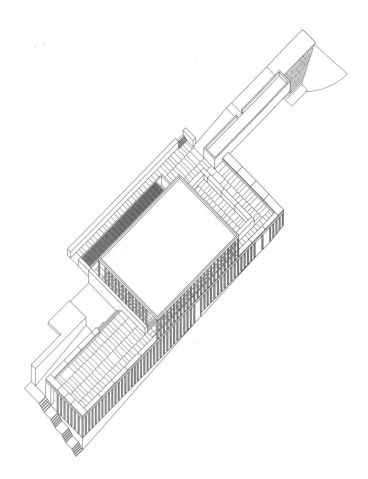
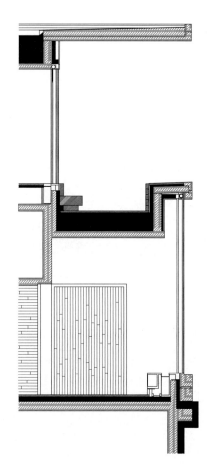
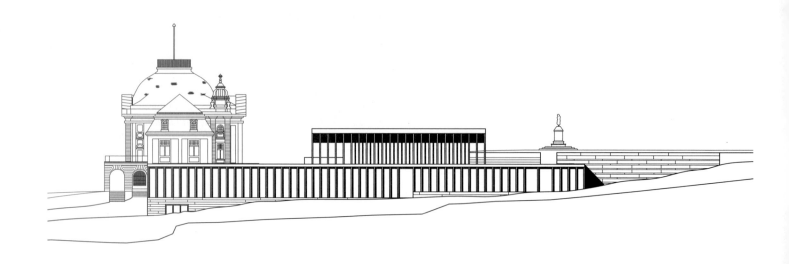

PATTERN	
PRINCIPLE	**487**

Museum of Modern Literature, David Chipperfield
Marbach am Neckar, Germany 2007
[Minimalist, museum, concrete and stone]

The Museum of Modern Literature employs the bay and module in a way similar to that of the Ospedale degli Innocenti. Chipperfield establishes the structural bay and then repeats it to create a pattern that goes beyond mere visual reading, but instead gives a strong spatial character and orderly visual language. The extreme simplicity of the verticality of the columns reads boldly against the dark glass beyond, creating a mirroring effect that extends and intensifies the pattern.

488	PATTERN	REPETITION	CHUNK	ELEVATION / SECTION / DETAIL	ARCHITECTURE
	PRINCIPLE	ORGANIZATION	TYPE	READING	SCALE

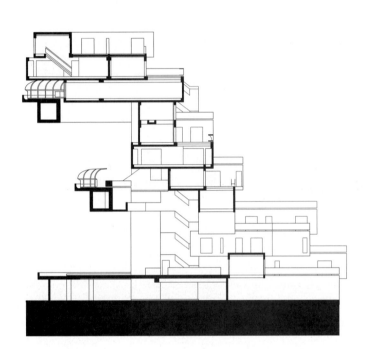 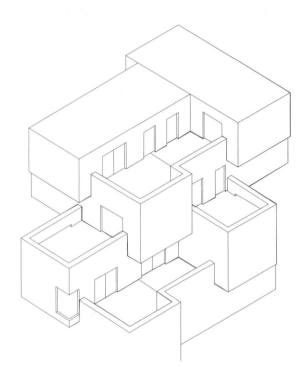

Pattern—Repetition—Chunk

The modern movement brought with it the development of numerous new materials that allowed for different methods of construction and new ways of articulating architectural expression. Conceptually and formally (as a method of fabrication and construction), one of these emerging methods was the use of larger blocks or "chunks" as a visual patterning or spatial device. This occurs when an architect develops a piece, which is then organizationally repeated to establish a system to the building. These elements collectively generate a repetitious pattern that ultimately gives the building its spatial and visual character.

PATTERN	
PRINCIPLE	489

Habitat 67, Moshe Safdie
Montreal, Canada 1967
[Modern, housing, stacked modular units]

Safdie's Habitat 67 is a remarkable housing development that has 354 prefabricated modules, which are stacked and connected by steel cables. These chunks are arranged in such a way that each unit has a balcony, views, and privacy within the tight density of the stacked composition. The resulting, seemingly chaotic, ziggurat form allows for a tremendous amount of localized variety within the complex while allowing a collective continuity. The formal impression of the complex illustrates the visual impact that this design methodology of pattern through repetition can deliver.

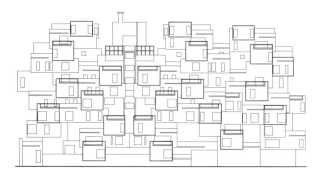

490	PATTERN	REPETITION	CHUNK	ELEVATION / DETAIL	ARCHITECTURE
	PRINCIPLE	ORGANIZATION	TYPE	READING	SCALE

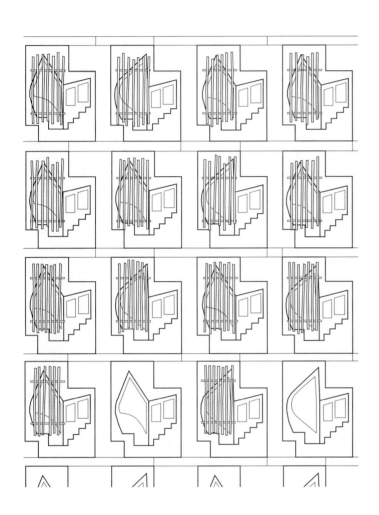
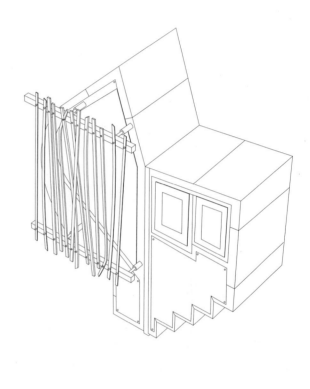
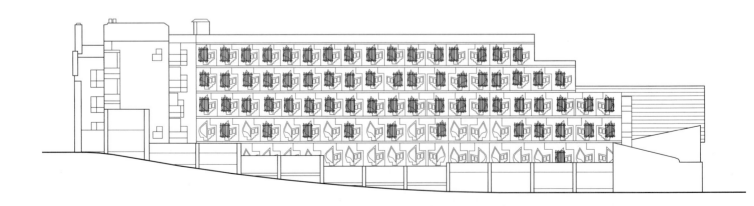

PATTERN | 491

Scottish Parliament Building, Enrique Miralles
Edinburgh, Scotland 2004
[Expressionist Postmodern, Government complex, material]

Miralles used pattern as one of the main design ideas in the Scottish Parliament Building by employing repetitive bays of the widow modules. Bold, graphic, individually articulated, and exquisitely detailed, the resulting pods or chunks serially protrude from the skin of the façade, participating in a graphic ensemble that illustrates the design methodology of pattern. Programmatically, these chunks were window seats that allowed people to inhabit, linger, and view the surrounding landscape. Critics have noted that the expressive nature of the building and its design is a way of illustrating the freedom of thought and debate so cherished by the Scots.

492	PATTERN	REPETITION	CHUNK	ELEVATION / DETAIL	ARCHITECTURE
	PRINCIPLE	ORGANIZATION	TYPE	READING	SCALE

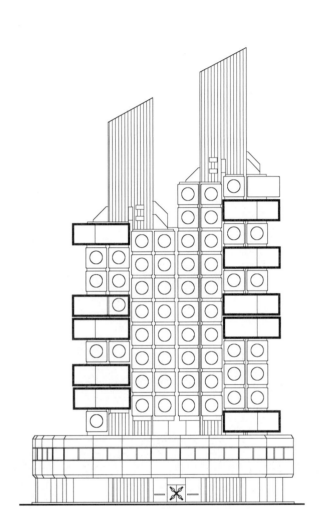
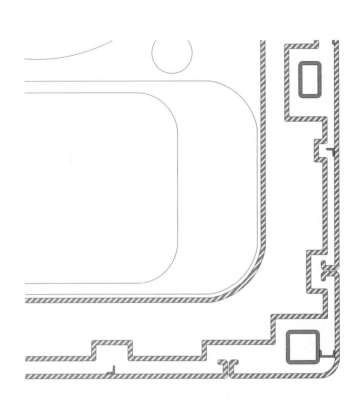

PATTERN	
PRINCIPLE	**493**

Nakagin Capsule Tower, Kisho Kurokawa
Tokyo, Japan 1972
[Metabolist, housing pods, pre-fab concrete with steel boxes]

Kurokawa's Capsule Tower is immediately recognizable for its employment of repetitive pattern of modular chunks or pods as its primary design, or graphic, methodology. Each capsule is constructed of light-weight steel and attached to two concrete towers. They are attached to provide light and privacy to the tenants. The resulting pattern is based purely on functionalism and not composition. This particular building, with its pod capsules and functional appearance, was a catalyst for many other projects that sought to move forward into a futuristic style founded in the serial repetition of mass-produced, cellular units.

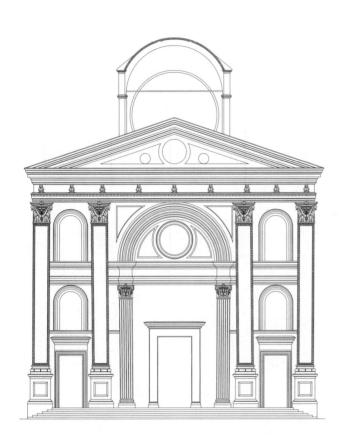

Pattern—Rhythm—Module

The design of a building is often locally articulated through the development of a distinct part or portion of the building. This particular piece or module is then repeated to establish an order and rhythm. The repetition and resulting patterned rhythm then emerge as dominant formal signatures and regulators of the entire building's formal composition and legibility. Nearly all buildings are required to engage the module, its system of repetition, its alteration, and iteration. The emerging patterns can exist throughout the building or merely comprise a certain portion of a plan or façade.

San Andrea, Leon Battista Alberti
Mantua, Italy 1476
[Renaissance, church, masonry with stucco]

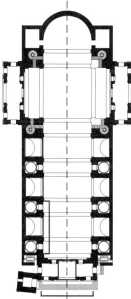

San Andrea in Mantua represents a sophisticated use of the module as a compositional element. Alberti developed the façade through the use of two hybridized pagan symbols: the temple and the triumphal arch. Their combination and layering synthesized to form a single integrated façade. The composition of this façade established the module upon which the rest of the building is based. The chapels on the interior are repetitions of the façade module. The dominance and power of this simple move relates the interior to the exterior in a dramatic way that influences architecture even today.

| 496 | PATTERN
PRINCIPLE | RHYTHM
ORGANIZATION | | ELEVATION /
PLAN / DETAIL
READING | ARCHITECTURE
SCALE |

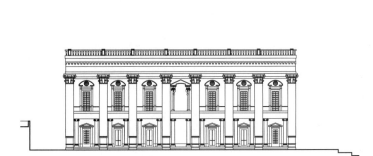

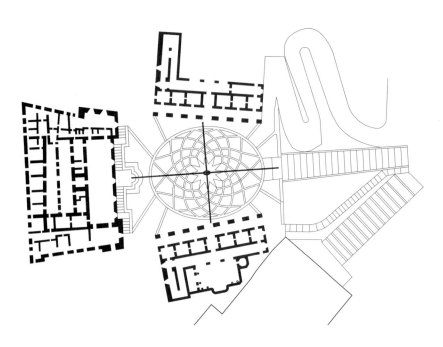

Piazza del Campidoglio, Michelangelo Buonarroti
Rome, Italy 1650
[Renaissance, elliptical piazza, masonry]

The Piazza del Campidoglio, by Michelangelo, remains one of the world's most renowned and dynamic urban spaces. Michelangelo created this unusual space by utilizing the ellipse as the geometric organizer of the space. This geometry was then combined with two monumental façades that flank either side of the piazza. These magnificent façades are composed of a very complicated system of layers that establish a module, which is then repeated across the façades. Linked by a **giant order***, each module (except for the center) is compositionally identical. The upper story of the center module receives more detail and layering as a method of illustrating its hierarchy.*

| 498 | PATTERN
PRINCIPLE | RHYTHM
ORGANIZATION | | PLAN /
AXONOMETRIC
READING | ARCHITECTURE
SCALE |

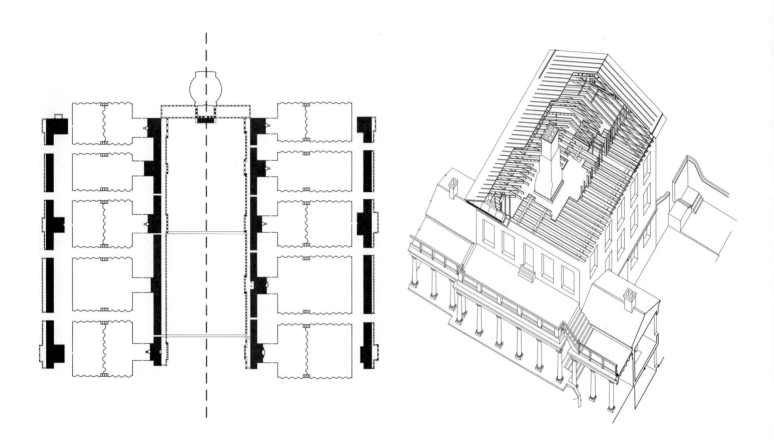

PATTERN	**499**
PRINCIPLE	

The Lawn at UVA, Thomas Jefferson
Charlottesville, Virginia 1817
[Neo-classical, Neo-Palladian central grass area, masonry]

The colonnades of the Lawn are mainly represented by simple brick columns that have been stuccoed and painted white. These columns, which are of the Doric order, make up the module of the lawn. This module marches relentlessly down both sides of the lawn, interrupted only by the elevations of the various pavilions. The distance between the pavilions changes and becomes shorter as you near the Rotunda at the head of the Lawn. This emphasizes the rhythm and perspective of the module and the complex in general, accelerating and expanding as they transition.

500	PATTERN	RHYTHM	MODULE	ELEVATION / DETAIL	ARCHITECTURE
	PRINCIPLE	ORGANIZATION	TYPE	READING	SCALE

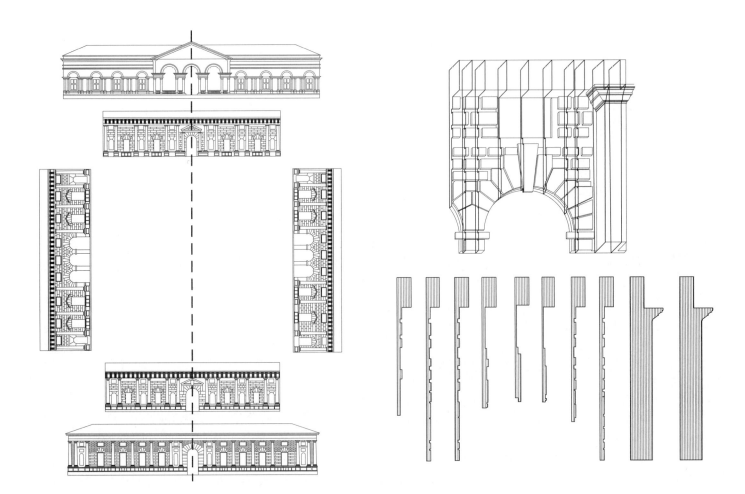

Pattern—Rhythm—Frequency

The use of pattern to form rhythm can be governed through frequency rather than mere repetition. This type of pattern arises out of typical repetition, which is altered to allow a change of frequency and formal movement to create the resultant configuration. This methodology is typically used in the more complex phase of any movement or style, where conventional repetition gives way to more intricate subtleties of pattern based on frequency and rate of recurrence, rather than mere duplication of the module.

Palazzo del Te, Guilio Romano
Mantua, Italy 1534
[Mannerist, palazzo, masonry/stucco]

The Palazzo del Te is an architectural essay on patterns created and governed through the use of rhythm and frequency. The Palazzo del Te is a seminal example of a mannerist palazzo. This reading is evident in the execution of the exterior façades and the elevations of the interior courtyard. In each case, Romano established a perception of a rhythm then changed and altered the frequency, which resulted in a series of complex treatments that only the perceptive and educated eye can detect. Beyond frequency, Romano explored asymmetry and detailing as a way of offsetting what would otherwise be considered a typical palazzo.

502	PATTERN	RHYTHM	MODULE	AXONOMETRIC / DETAIL	ARCHITECTURE
	PRINCIPLE	ORGANIZATION	TYPE	READING	SCALE

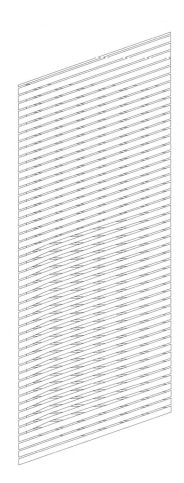
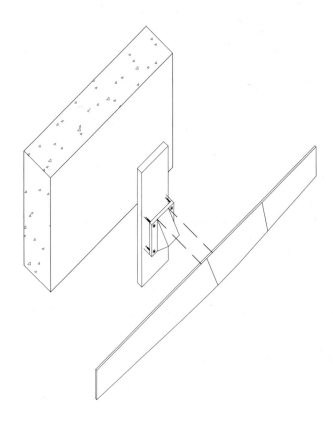
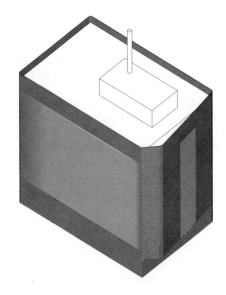

PATTERN	
PRINCIPLE	**503**

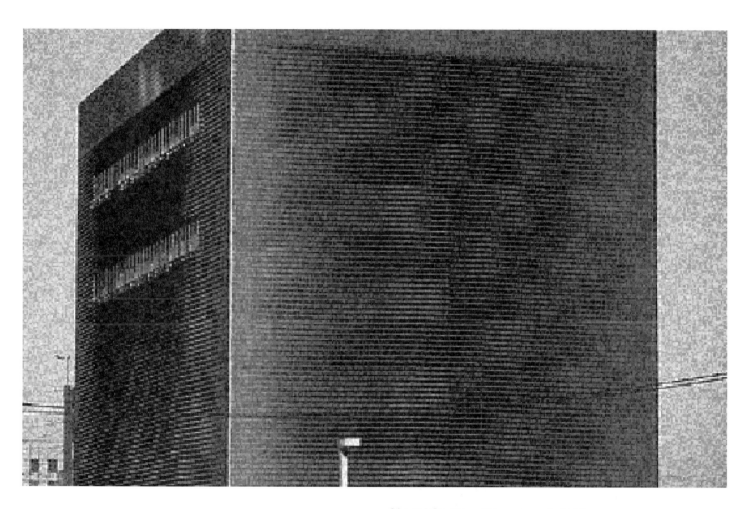

Signal Station, Herzog and de Meuron
Basel, Switzerland 2000
[Postmodern, traffic signal station, copper]

The Signal Station is a simple, slightly distorted, box structure containing the controls of the train signals in Basel. The building's exterior is composed of sheets of copper that are either flat or rotated to create patterns that appear as blurs. The horizontal copper fins are rotated at slightly different angles and frequencies so that visually the pattern is continually moving. Within the copper shell are the electronics responsible for controlling the signals; these too are protected by copper. The entire material makeup of the building references the vibrant quality of the material.

504	PATTERN	RHYTHM	MODULE	PLAN / DETAIL	ARCHITECTURE
	PRINCIPLE	ORGANIZATION	TYPE	READING	SCALE

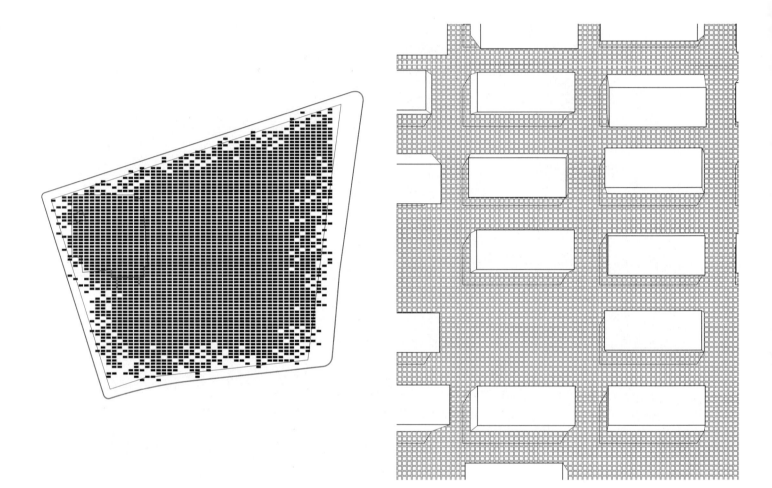

PATTERN	505
PRINCIPLE	

Berlin Holocaust Memorial, Peter Eisenman
Berlin, Germany 2005
[Postmodern, repetitive field, concrete]

The Berlin Holocaust Memorial is a continuous field of concrete monoliths that form a grid across the entire site. Governing the sculptural field is a pattern influenced by rhythm and frequency. Each one of the concrete monoliths is slightly different in height, creating various shadows and imperfections that result in varied frequency and pattern changes. Beyond the localized organization, the grid operates across an unleveled topographic plain, requiring the blocks to respond and creating another rhythm within the gridiron.

506	PATTERN	RHYTHM	SCALE	ELEVATION / AXON / DETAIL	ARCHITECTURE
	PRINCIPLE	ORGANIZATION	TYPE	READING	SCALE

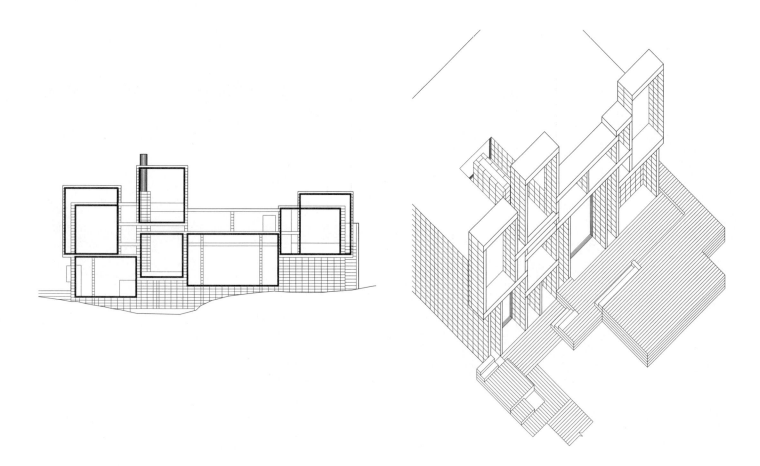

Pattern—Rhythm—Scale

Pattern created through variation in rhythm and scale allows for a singular figure to be iterated with one aspect of controlled variation. The transitions of scale and position establish the rhythm and meter of their relative relationships. The effect is often more powerful through the limitation of rules within a controlled variable system.

| PATTERN PRINCIPLE | 507 |

Milam House, Paul Rudolph
Jacksonville, Florida 1961
[Late Modern, extruded concrete brise-soleil, concrete]

The Milam House develops its identity through the exaggerated formulation of its brise-soleil. As a thickened wall plane, the cellular nature of the sunscreen extends as a field. Varied, iterative rectangles with diverse scales allow for the establishment of a repetitive field that extends beyond the edge of the house itself. As a thickened frame, the depth provides screening from the harsh Florida sun while creating funneling sub-frames that produce a visual mosaic of the beach view. The pattern of varied scale provides the unified formal composition and the experiential effect of the architecture as a whole.

508	PATTERN	RHYTHM	SCALE	ELEVATION / DETAIL	ARCHITECTURE
	PRINCIPLE	ORGANIZATION	TYPE	READING	SCALE

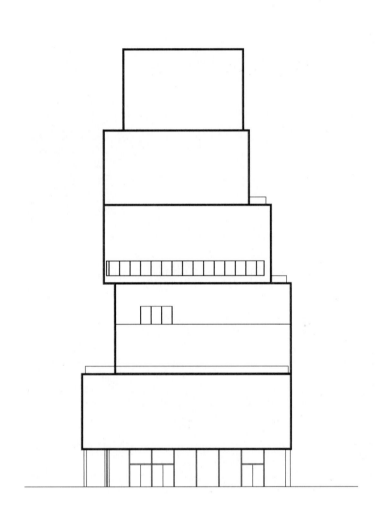
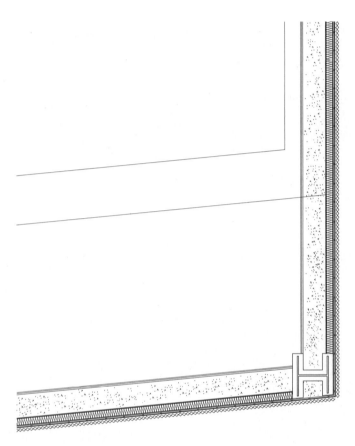

PATTERN 509

New Museum of Contemporary Art, SANAA
New York, New York 2007
[Postmodern, shifting stacked gallery boxes, metal]

The New Museum of Contemporary Art, by SANAA, deploys a singular simplicity to its formal agenda. As a series of shifting, stacked boxes, its solemn vagueness, lacking in detail, adds to its juxtaposing and enigmatic presence. The repetition of the figuration, simple iterative pattern of the individual massing, and varied scale define the building.

510	PATTERN	RHYTHM	SCALE	ELEVATION / DETAIL	ARCHITECTURE
	PRINCIPLE	ORGANIZATION	TYPE	READING	SCALE

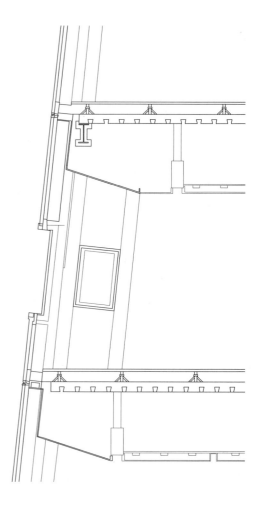
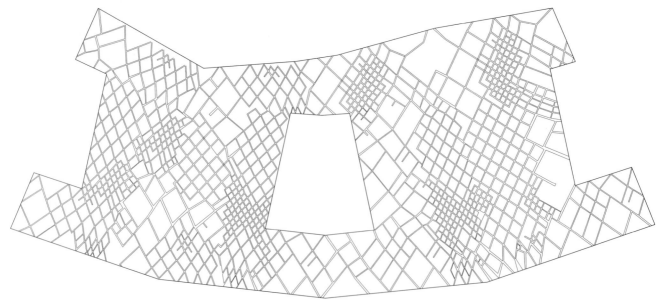

| PATTERN PRINCIPLE | 511 |

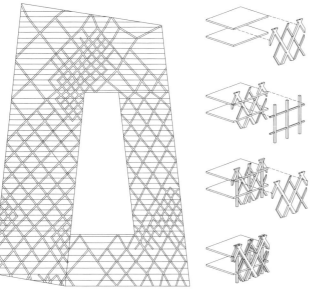

CCTV Building, Rem Koolhaas, OMA
Beijing, China 2009
[Postmodern, linked tower, metal and glass]

The CCTV Tower employs an external, diagonal, structural grid that creates a variably scaled pattern. Responding to the variable structural forces of the irregular building form, the skin pattern branches and densifies as necessary. Derived by a combination of overall formal decision with the functional practicalities of physics, the pattern overtly expresses the intrinsic natural forces. This pattern is regulated by the simplicity of scale as the response variable. It creates a surface pattern that is functionally ornamental.

512 PERCEPTION

PRINCIPLE

13

Perception

13 - Perception

Perception is simultaneously the most significant factor in the understanding of architecture, but equally one of the most elusive and difficult to govern. It occurs through the delicate balance of the physical and the mental, residing somewhere between the biological senses of our bodies and the cognitive processes that engage, relate, process, and philosophize. It engages emotions, yet frames them relative to the physical world and our relationship with it. Founded in the five senses of sight, smell, hearing, touch, and taste, the "information" provided aggregates into much more through the physical relationship and experience of space. The following factors are the primary categories relative to perception: light, color, focal point (via vision), perspectives (both real and false), material, sound, memory and environment.

Light Light is the basis of architecture as its projection allows for the legibility of forms. The use of light for effect, color, visual illumination, or environmental control regulates perception. The consideration and design of these factors (both natural and artificial) for primal effect are essential to the perception and experience of space.

Color The use and perception of color are an architectural inevitability. Through the natural, tonal coloration of a material palette or the applied orchestration of pigment through paint, stain, or other processes, the use of color as an atmospheric effect or a sign system allows for design and systemization to affect perception.

Perspectives—Constructed Focal Point The visual perception of forms in space and the perspectival limitations of our optical abilities establish the cone of vision as an essential perceptory tool. The focal point as an intrinsically hierarchical position establishes a significance and relativism to perception.

Perspectives—False Building on the optical effects of how we typically perceive spaces and forms, false perspectives provide an opportunity to generate by design a projective reality. Extending the visual lines to produce a shallow space, the constructed artificiality creates an extension of actual space and furthers the choreography of an orchestrated perception.

Material Material perception refers to the spiritual effect derived from a material use. The weight, mass, effect, presence, and relationship to the space create material perception. As a quiet but essential factor in the experience of the space, the material selection and its application are paramount to the perception of architecture.

Sound The acoustical perception of space is an integral sensorial factor. The audio experience of space is one that is varied and more typically event-based, yet remains framed by the architecture, orchestrated by the built environment, and facilitated by the design intent and effect.

Memory / History Historical perception, or the associative relationships of either personal memory or broader cultural references, relies upon memory to establish a framework for one's experience. The relativism of such perception is rarely universal; rather it is more personal in parameter.

Environment Environmental perception refers to the associative relationship of experience to place. Relying upon the surrounding context (either natural or man-made), the milieu in which a building occurs affects its formal, spatial, and cultural reading.

Light

The perception of light, which is essential to the legibility of form, can use illumination as a compositionally objectified form-maker or a more ambiance-based experiential effect. As a substantial effectual element, the perception can occur as a representational or more abstract composition. The emphasis on either a visual object or perceptual ambiance allows light to be an active or passive element in the perception of space.

La Sainte-Chapelle, Pierre de Montereau
Paris, France 1248
[Gothic, sanctuary with ornamental glass, masonry and glass]

La Sainte-Chapelle is a private chapel that maintains an intimate scale while projecting a powerful presence. Defined by its exquisite and large stained-glass windows, the quality of light and color defines the space. The delicate scale of the colored glass and the massive field of the window surface engulf the viewer. The glass wall inverts the presence of the physical mass with the ephemeral ambiance of the light. The visual presence of the light tells a physically and pictorially descriptive story while simultaneously establishing an ambient, illuminated field. The result is an enlightened and spiritual space that begets its religious ceremony.

PERCEPTION
PRINCIPLE 517

Notre Dame du Haut, Le Corbusier
Ronchamp, France 1955
[Expressionist Modern, free plan church, masonry and concrete]

The effectual and formal choreographies of Notre Dame du Haut are produced and defined through the orchestrated use of light. The curvilinear roof levitates above a delicate slit of light, hovering ephemerally in transformed weightlessness. Dominating the space and producing the primary light effect is the thickened entry wall that was constructed using the stones of the original church, which had previously occupied the site. The mass of the wall is optimized through the diversely scaled funnels that transform the tradition of stained glass into a projective field. The compositional and atmospheric effects establish the perceptive aura of the space and amplify the formal and spiritual effect.

St. Ignatius, Steven Holl
Seattle, Washington 1997
[Postmodern, variably shaped light funnels, concrete]

At St. Ignatius, Holl deploys a clearly Corbusien sensibility in terms of form and light effect. He developed the building's concept as a series of top-lit light funnels bundled in a singular rectangular perimeter. The diverse figures establish a sequential series of spaces that are choreographed through their identity, which is defined by their light. Amorphic roof figures bend and twist to respond to and harness the light, creating their form and individuation through their local effect and defining the building as a whole through the dramatic figuration of the roofscape.

Color

The choreography and experience of color have a dramatic effect on the perception of space and form. Rooted in conceptual theory or simply emotively deployed, the use and effect of color to orchestrate the reading and legibility of an architectural composition is a powerful perceptory influence and design element.

Casa Gilardi, Luis Barragan
Mexico City, Mexico 1977
[Late Modern, simplified form with colored walls, masonry]

Casa Gilardi extends the cultural use of color to the Mexican vernacular. Barragan's bold, personal use of color enables him to translate the power of pigmentation into a mechanism of modernist spatial and effectual vocabulary. Bold fields defining primitive planes, surfaces, and forms establish the dominant presence of an architecture of color. The space, as an extended reading of the use of color, becomes reflective of the surface as a doppelganger to the light. The color, through light, brings ambient complexity to the simple forms.

PERCEPTION	
PRINCIPLE	**519**

Loyola Law School, Frank Gehry
Los Angeles, California 1991
[Postmodern, collection of disparately formed buildings, diverse]

The Loyola Law School combines the discrete use of color with the individuation of form and material vocabulary to define the collection of separate pavilions. As a dispersed campus of buildings gathered in a dense urban site, each building exhibits a specifically personal form. The individuation of these forms is furthered by their construction method and material application and ultimately articulated by their color. As an associative sign system, the use of color creates individual identity and juxtaposes the collection of pieces while simultaneously unifying them. Not generative of a spatial effect, the two-dimensional sign-like nature makes the effect postmodern in application and graphic concept.

Laban Dance Centre, Herzog and de Meuron
London, England 2003
[Postmodern, simplified form with colored skin, plastic]

The Laban Dance Centre, clad in a polycarbonate skin, uses color as a surface effect. Varying color tones make a shifting play of a transitioning surface. The simplicity of the form and the tautness of the skin are activated by the material application of effectual color. Giving diversity and life to the shallow surface of the skin, the color allows for the static to become dynamic and the permanent to become variable. Color, as a perceptory effect dependent upon position, light conditions, and time, allows for a layered reading to the building. This in turn allows for a compositional and atmospheric re-presentation of the building as an object. The effect of color can dissolve and change with its environment, creating a performance to the perception of the building itself.

Perspectives—Constructed Focal Point

The perceptual use of perspective with a constructed focal point was most common and widely used during the Renaissance. The concept of creating a space around a constructed perspective with one focal point was so ubiquitous during the Renaissance that the technique and the movement came to be considered one and the same. This form of representation was initially explored by painters, who constructed views of ideal cities. Formed and composed through this methodology, constructed perspectives were used as a way of designing space. Architects adopted these methods and techniques to lay out actual buildings and urban spaces. The result was a hybridized drawn and constructed reality governed by the order of the geometric system of construction.

Piazza Pio II, Bernardo Rossellino
Pienza, Italy 1462
[Renaissance, piazza, masonry]

The Piazza Pio II is located at the highest point in the city of Pienza. This space was commissioned by Pope Pio II to demonstrate the urban design methodology of the early Renaissance. The space is the first attempt at constructing the ideal city popularized in paintings of the early Renaissance. The piazza is a trapezoidal space with the cathedral at the larger end. The two sides are occupied by the Palazzo Borgia and the Palazzo Piccolomini; while the smaller end is occupied by the Palazzo Comunale. The Piazza Pio II represents one of the first attempts at using perception with a single focal point, in this case, the cathedral.

PERCEPTION	521
PRINCIPLE	

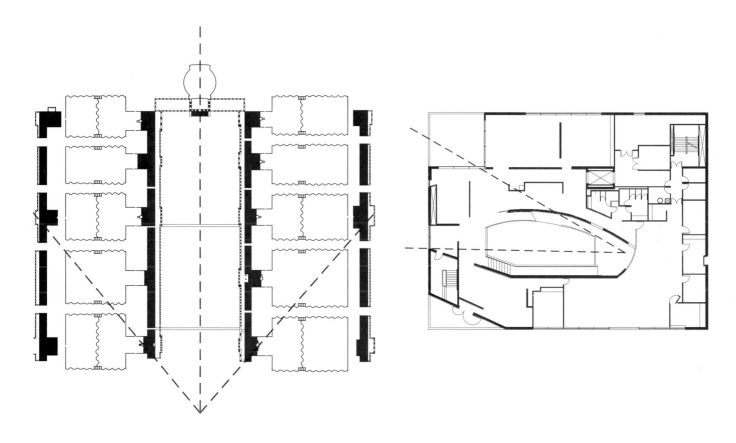

UVA Lawn, Thomas Jefferson
Charlottesville, Virginia 1817
[Neoclassical, university, masonry]

The Lawn at the University of Virginia is a seminal example of a constructed perspective with a single focal point. The focal point of the Lawn is the Rotunda, which is located at the head of the space. From any position on the Lawn, there is little doubt that this building is the main focus. This is accomplished primarily through the use of the columns, which repetitively march up each side of the lawn. This focus is also achieved by the articulation of ten pavilions and their subsequent placement, which emphasize, by decreasing the space between them, the perspective toward the rotunda.

Bellevue Arts Museum, Steven Holl
Bellevue, Washington 2000
[Postmodern, museum; concrete, aluminum, and glass]

The Bellevue Arts Museum forms a series of three, linear, slightly curved galleries that use light as a primary focus point. These galleries stack next to each other, forming an exterior that is expressive of the interior configuration. The promenade through the galleries is governed by an orchestrated sequence whereby the user follows the galleries upward, moving back and forth, continually drawn by both curvature and light. The space is a result of the constructed vantage of Holl's signature, perspectival watercolor-design process. The culmination of this sequence occurs in a light court situated on the roof.

Perspectives—False

During the Renaissance, architects were fascinated by the idea of the false perspective. False perspective is the construction of a perceptual recession of field through the compositional manipulation of focal point. Like a stage design, it offers the opportunity to emphasize and construct an aspect of a project through the optical effect of perspective. Perspective and the ability to mathematically construct it were new ideas during the Renaissance and architects were searching for ways to utilize this new knowledge. Typically this perspective was used as a way to extend a view or to make an element appear farther away than it actually was. At times it was used to emphasize an object or room. At other times it was used purely as a pleasure principle to amuse or trick the viewer. In any scenario, the construction of a false sense of space produced a constructed optical effect that blurred the boundary of reality through perception.

Santa Maria presso San Satiro, Donato Bramante
Milan, Italy 1482
[Renaissance, church, masonry]

Santa Maria presso San Satiro replaced a ninth century place of worship. The constricted site on which the church sits would not accommodate the typical Latin Cross plan; instead Bramante employed a truncated T-shaped plan. To maintain the continuity of the nave, he used perspective as a way of optically extending the space. The result was one of the first examples of architectural trompe l'oeil. This false perspective makes it appear as if the nave continues for another three bays. It is a marvelous and novel use of the false perspective to restore a spatial typology.

PERCEPTION 523

Teatro Olympico, Andrea Palladio
Vicenza, Italy 1584
[Renaissance, theater, masonry and cut stone]

The Teatro Olympico is one of the last surviving examples of Renaissance theater design. Palladio based the design on his studies of ancient theaters. He created a large, wooden proscenium that acts as a middle ground, the stage proper as the foreground, and the city behind the screen acting as the background. This background city is formed by a series of streets that are false perspectives, appearing to continue off into the distance. The overall effect of this remarkable sequence is a truly singular and spectacular architectural moment orchestrated with careful control of visual perception.

Scala Regia, Gian Lorenzo Bernini
St. Peter's Basilica, Vatican 1666
[Baroque, staircase, masonry and plaster]

The Scala Regia is a staircase that joins St. Peter's Basilica to the Vatican Palace, providing part of the formal entry into the Vatican. Bernini used a series of baroque elements as a way of creating dramatic spatial effects while fitting the staircase in a narrow slot of space. The staircase begins at Bernini's colossal equestrian statue of Constantine and terminates at the Sistine Chapel, thereby linking two of the most important spaces of the complex. As a large, barrel-vaulted hallway, the stair quickly ascends and seems to become smaller through the use of a false, forced perspective. While making the distance between the two spaces seem longer, the false perspective heightens their importance.

Material

The perception of material is so significant that it is explored in much more detail in Chapter 10. The physical, tactile, and performative qualities of materials impregnate space and form with perceptory presence. The relationship between tectonics and the compositional intentions of the forms and spaces produced is rooted in the physical manifestation of thought through material. The perception of material is the perception of architecture itself.

Barcelona Pavilion, Mies van der Rohe
Barcelona, Spain 1929
[Modern, free plan, steel, glass and masonry]

The hyper-articulated material palette of the Barcelona Pavilion introduces a richness and regality to the formally simplified and minimally detailed components. The celebrated marble walls are book-matched, a process through which the stone is cut in parallel slabs and positioned as pages of a book. The result is a line of symmetry at the seam. Spatially, the marble is the only material with an illustrated grain, scale, and detail. The dominance of this mirroring presents a vertical reading of the entire building as a sectionally symmetrical and equal space. This bold, formal organization is reminiscent of the order and systemization of monumental classical architecture; but, it is adapted as highly modern through the reorientation of the axis.

PERCEPTION	525
PRINCIPLE	

Villa Mairea, Alvar Aalto
Noormarkku, Finland 1939
[Modern, L-shaped house, diverse]

Villa Mairea uses an artful deployment of material and clever juxtaposition to produce its composition. Collaging wood, masonry, glass, metal, plaster, and even rope, each material is individually expressed and collectively related to produce a dynamic choreography of space, form, and experience through material perception.

Thermal Baths at Vals, Peter Zumthor
Vals, Switzerland 1996
[Postmodern, field of pools and spaces, concrete and masonry]

The Thermal Baths at Vals is premised on the physical experience of water. Through changes in temperature, varied scales of the different pools, and contrasting light conditions, each space creates an atmospheric chamber that allows for an essential relationship between body, material, and space. The focus on the effectual quality of the space is enhanced through the use of indigenous stone. Deployed in thin, elongated, horizontal ribbons, the natural stone creates a pattern, rhythm, and scale that couples with the dappled reflection and wetness of the water. The collective composition creates an engulfing sensorial effect. Through its simplicity and clarity of deployment, the material, function, and experience collaborate to create a powerful perceptorial environment.

Sound

The perception of sound is rooted in acoustics, but extends to augment and collaborate with the experience of architecture. By marrying primal effectual relationships with space and light, sound has the potential to govern the perceptorial condition and fundamentally engage architecture. The auditory sense creates scale, reflects function, responds to materiality, and evokes memory as vibrantly present as the more traditional use of sight. The following represent both passive and active examples of the use of sound in the production of architecture.

The Kimbell Art Museum, Louis Kahn
Fort Worth, Texas 1972
[Modern, parallel galleries, concrete]

At the Kimbell Art Museum, Kahn used sound to augment and choreograph the entry sequence. As a grove of trees provides shade, the rustling of leaves begins to muffle the sounds of the city. The crunching of the gravel ground plane further dissipates the noise of the city. The sound of shallow, elongated fountains of falling water erases the stresses of daily life. Upon reaching the entrance of the museum, the visitor is prepared to engage with the building and the visual experience it offers.

PERCEPTION	
PRINCIPLE	**527**

Fort Worth Water Gardens, Philip Johnson
Fort Worth, Texas 1974
[Postmodern, terraced urban water park, concrete and water]

The Fort Worth Water Gardens create an oasis of an urban park. The variable surface of terraced layers of irregularly formed concrete is activated by water. Cascading down the surface, the water aerosolizes and engulfs the space, creating an atmospheric and acoustical harbor from the hot Texas environment and the density of the city. A series of pedestal steps allow the visitor to float down in, through, and across the waterfalls. This physical orchestration and direct association with sound govern the feel and order of the space and material.

Kalkriese Archaeological Park, Gigon and Guyer
Osnabruck, Germany 2002
[Postmodern, pavilions dedicated to the senses, metal]

The Kalkriese Archaeological Museum consists of a series of pavilions, each dedicated to one of the five senses. Designed to engage visitors with their surroundings, the pavilions highlight aspects of the landscape. The pavilion predicated on sound uses the vocabulary of orthogonal, corten-steel buildings, modified to formally and functionally express the auditory sense. A massive, organically formed earphone rotates, allowing the visitor to focus on a specific moment of the surrounding perimeter and experience the parkscape through sound. The functional experience and the exaggerated formal expression of the acoustical properties combine into a passive and active participant in the perception of sound.

Memory / History

Memory serves as one of the most powerful architectural devices. A building or structure can mark an event by direct association, formally manifesting a connection to a place, time, and event. History, as a compelling architectural stratagem, can be resurrected through the careful use of form, materiality, and concept. Critical factors in the use of perception as an architectural device, memory and history trigger emotions untapped by other design strategies.

Museum of Roman Art, Rafael Moneo
Merida, Spain 1985
[Postmodern, museum, masonry]

Moneo's Museum of Roman Art uses architecture to convey a sense of history through form and materiality. The building presents the concept of an excavation, with a series of parallel brick walls that occupy the site and provide the structure for the main exhibits. These walls recall the brick construction techniques of the Romans and thus are self-referential in their application. The brick materiality and the large arches provide a language immediately recognized as a historical reference to the Roman art being displayed.

Oklahoma City Memorial, Hans and Torrey Butzer
Oklahoma City, Oklahoma 2000
[Postmodern, memorial, various]

The Oklahoma City National Memorial and Museum, which commemorates the terrorist attack on the Arthur P. Murrah Federal Building in 1995, recalls memory and history through an episodic sequence of architectural moments. Two gates of time symbolically mark the minute before and after the blast; a reflecting pool mirrors the visitor's face, subsequently touched by the memory of this event; finally, 168 glass and bronze chairs represent the victims of the blast. The arrangement of the chairs records each victim's place at the moment of the explosion. All of these architectural moves honor the memory of the event and the victims as a way of orchestrating a meaningful and poignant response.

Berlin Holocaust Memorial, Peter Eisenman
Berlin, Germany 2005
[Postmodern, repetitive field, concrete]

The Berlin Holocaust Memorial commemorates the Jewish victims of the Holocaust perpetrated by the Nazis during World War II. Peter Eisenman designed a field of 2,711 concrete slabs, or stelae, that are arranged in a grid pattern covering the site. The individual slabs, which march across the changing contours of the site, vary in height. This memorial is designed to produce an uneasy feeling and convey the idea of an ordered system that has gone astray or has been corrupted. The solemn scale and scope convey the atrocity of the Holocaust to the viewer.

Environment

The environment can play a crucial role in how an architectural project is perceived. Architects have often emphasized natural elements as a catalyst for their work and as a focal point of their projects. Though this can be achieved in a number of ways, the most common is through the orchestrated use of vision. An architect might use a series of views as a way of focusing attention on specific moments, highlighting the natural beauty of a vista. The architecture itself is controlled through the use of framing. Similarly, a designer may employ natural elements such as water, wind, and materiality to govern form and experience.

Salk Institute, Louis Kahn
La Jolla, California 1965
[Late Modern, symmetrical laboratories, concrete and wood]

Kahn utilized the compelling view of the Pacific Ocean as the primary natural device on which the design of the Salk Institute was based. The project was then divided into two symmetrical portions that flank this commanding view out to the western horizon. The perception of this environment is immediately established upon entering the central void. It pervades the project, controlling the user experience. The surroundings and their incredible power are established through the formal hierarchy of the central axis. The windows of the office towers that edge the central space are individually angled toward the ocean as a way of recognizing, framing, and deferring to the strength of this natural environment.

PERCEPTION **531**
PRINCIPLE

Curtain Wall House, Shigeru Ban
Tokyo, Japan 1995
[Postmodern, house, various]

The Curtain Wall House uses the elements of light, wind, and view to allow the natural environment to affect perception. This small house is located on a narrow street in Tokyo; the second floor cantilevers out to the curb. To retain the Japanese tradition of openness within the space of a house, Ban used only a fabric curtain to separate the interiors from the immediacy of the city. Light filters through this sheer wall, the wind moves the fabric, and, when the curtain is open, the views of the city are welcomed into the house. The natural environment orchestrates the perception of the architecture.

Bibliotheque Nationale de France, Dominique Perrault
Paris, France 1997
[Postmodern, library with central courtyard, glass and wood]

The Bibliotheque Nationale de France uses natural landscaping in an unusual way, creating an objectified primeval forest at its heart. The project is composed of four separate towers that rise out of a large, wooden plinth. In the center of this massive base is a sunken courtyard garden encircled by reading rooms. This natural garden is inaccessible, secluded, and quiet in contrast to the density of the Parisian urban fabric. It is the sharp contrast between this serene garden and the frenetic pace of the city outside that creates a heightened appreciation of this objectified environmental perception.

532	SEQUENCE			
	PRINCIPLE			

14

SEQUENCE	
PRINCIPLE	**533**

Sequence

14 - Sequence: Circulation

The use of a sequence in a building's or city's circulation is one of the fundamental principles in architecture and urban design. In the most basic terms, sequence / circulation can be thought of as a journey through any building or city. The connection of perception through the chronological experience of movement through space defines the architectural promenade or sequence. A sequence of events, stages, and perceptions evolve into the experience of the building. The architect orchestrates these various experiences to create a reading of the overall design. A holistic approach to sequence is needed to appreciate how a series of spaces add up to a composition that is greater than the sum of the individual parts. Some sequences are based on typologies, histories, patterns, or programs. Or they can be based on more sensorial effects of texture, light, scale, and material, for example.

Horizontal Sequence—Axial Horizontal axial sequence exists as a dominant type at both architectural and urban scales, innately establishing a dramatic hierarchy to the system. This type of organization follows an axis that transports the viewer through a series of spaces (indoor or outdoor) along a linear path that culminates in an honorific space. These systems typically occur under religious or political circumstances and require operation at the total architectural or urban scale of intervention in order to accomplish such a singular and forceful organization. On the urban scale this design intervention is typically accompanied by the destruction of large areas of a city in order to render the proper axial sequence.

Ceremonial Sequence A ceremonial sequence is typically designed around a religious ritual or a very specific set of performance-based experiential ideas that dictate the sequence. The ceremonial spaces used in any religion tend to form over time and grow to represent certain rituals, the spatial needs of which are founded in associative meanings of each religion. Due to this association with programmatic typology, churches and temples often share similar, common sequences. Of course, ceremonial sequence be found in other building types; however, the common link is the prioritization of the ceremony to govern experience.

Experiential Sequence Experiential sequence emerges from a series of perceptual spaces that lead from one to the next, collectively creating a dynamic experience. The programmatic type that this sequence corresponds to can encompass almost any function. The most critical concepts in this sequence are that the experience has meaning and that the sequence works to govern, orchestrate, and determine the collective perception of the architecture.

Programmatic Sequence A sequence based on program is typically laid out in such a way that the functional requirements of the building are clear and the chronology of experiencing this program establishes the sequence. Programmatic sequence can be applied to almost any building function; however, they are typically dependent upon functions that include rituals, chronologies, and ceremonies. The most graphic examples emerge when an architect bases the entire parti of the primary concept on how the program of the building is arrayed. This arrangement goes beyond mere functionalism, adding up to a program-based idea that is not necessarily a form-driven idea, but rather a sequence-driven concept.

Vertical Sequence Vertical sequence engages the Z-axis to operate sectionally in the vertical dimension and ascension through space. Requiring the collaborative engagement of other sequential devices in use for the horizontal staging, vertical sequence engages heavily the concept of moving up or down as a means to an end. This concept encompasses the natural sequence, which can engage sectionally through landscape, as well as the entirely constructed efficiencies of a skyscraper. This vertical dimension adds another layer of meaning to projects and typically results in a rich and complex system.

Horizontal Sequence—Axial

Axial horizontal sequence is the most common and ubiquitous type of formal circulation. It innately produces a hierarchy and dominance of procession and organization. Its geometric power establishes a series of spaces, or sequential episodes, that lead from one to the next along a horizontal axis, culminating in an honorific space. Occurring at both the urban and architectural scale, the scope and scale of such axial sequences are typically part of a large political, religious, or social plan being put in place by governmental or ecclesiastical forces.

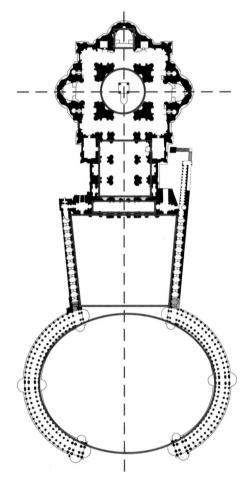

St. Peter's Basilica, Various
Rome, Italy 1547
[Renaissance, dome and Greek Cross plan, masonry]

The sequence begins on the Ponte Sisto that leads to the front of the Castel Sant Angelo and continues along the long axis created by Piacentini. This path arrives at the Piazza Pio XII, which hovers between the end of the Via della Conciliazione and the majestic space of the Piazza San Pietro, formed by the two large semi-circular arms of Bernini's colossal columns. Moving first into the oval Piazza San Pietro and then the forecourt between the Basilica and the piazza, the sequence finally enters the Basilica and ultimately the narthex, which is a long room running at right angles to the Basilica proper. At one end of the narthex sits Bernini's equestrian statue of Constantine, marking the beginning of the Scala Regia. Passing through the narthex, the sequence arrives in the nave of the Basilica, which then leads to the Baldacchino and the papal altar at the crossing under Michelangelo's dome. The ultimate climax of the sequence is Bernini's Cathedral Petri, located at the final wall of the transept.

	SEQUENCE
	537
	PRINCIPLE

Paris, Georges-Eugène Haussmann
Paris, France 1870
[Second Empire, connecting boulevards]

Haussmann considered the city of Paris a solid block from which he could excavate a series of boulevards that would connect the city in a new, more ordered, and honorific way and also allow him to destroy the more dilapidated areas of the city. Haussmann understood that he needed to take advantage of buildings and monuments that were already in place. He carefully designed boulevards and streets that strategically connected many of these spaces, creating not one but many horizontal axial sequences through the fabric of Paris. The collective result was a network of large avenues, a series of rings of boulevards, large north-south and east-west crossings, squares at each of the crossroads, acknowledgement of the monuments and the railway stations, and the incorporation of green spaces. Haussmann was able to weave all of these factors into a series of episodes that define modern Paris.

EUR, Piacentini, Piccanato, Rossi and Vietti
Rome, Italy 1936
[Italian Rationalist, axial new town, diverse]

EUR, which stands for Esposizione Universale Roma, is a manifestation of the rationalist architectural principles favored by Mussolini and his architects. This new city brought Italian rationalism to urban form. An ordered symmetrical system that is embodied by wide roads, and dominated by a grid, the EUR used a strong north-south axis interrupted by a number of smaller east-west axes. The white-marble buildings imparted a beauty and austerity; however, the unfinished project was abandoned in 1942, not revisited again until the 1950s. The EUR in its planning and architectural content contains some of the best architectural efforts of the Italian rationalist movement.

Ceremonial Sequence

Ceremonial sequence is inextricably tied to a ritualized ceremony or historical event. The architecture is fully defined and dependent upon the activity and tradition it supports. These sequences tend to be linear, imparting a form to the chronology of the ceremony. The architecture is developed to impart an engagement and understanding of the ceremony and its cultural references. Typically there are defined rooms or spaces that carry meaning. These spaces, aggregated along a path, allow for the experience of the ritual to coincide with the experience of space. The movement through the architecture provides an essential reenacting of a story or sequence that is critical to the culture that it serves.

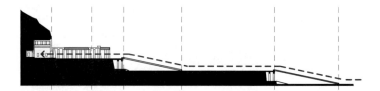

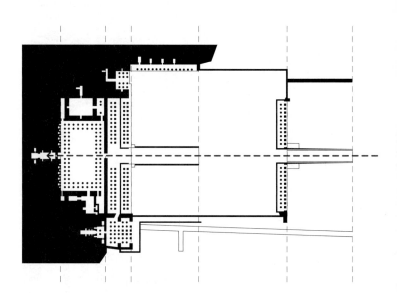

Temple of Queen Hatshepsut, Senenmut
Deir El-Bahri, Egypt 1400 BCE
[Egyptian, funerary temple, masonry]

The Temple of Queen Hatshepsut is considered to be the closest the Egyptians came to classical architecture. This temple is dominated by lengthy, open colonnades that correspond to various layers that were traversed by long ramps surrounded by lush gardens. The first space of the complex is the central court, which allowed access to the Punt Colonnade and the Birth Colonnade. Within this vast temple is a sequence of chambers containing representations of the Queen's life and accomplishments, not least of which is the construction of the temple itself. The sequence ends at the upper court, which offers access to various altars, sacrificial halls, and the innermost sanctuary. The temple, which is perfectly aligned with the east-west axis, is a seminal example of ceremonial sequence.

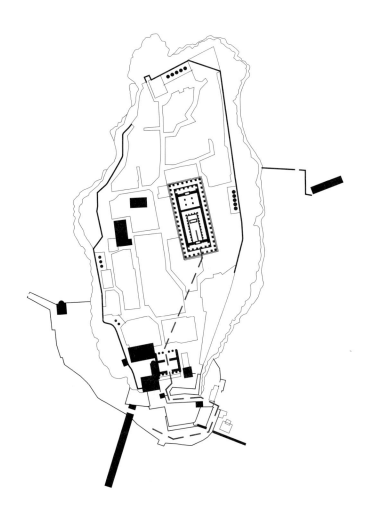 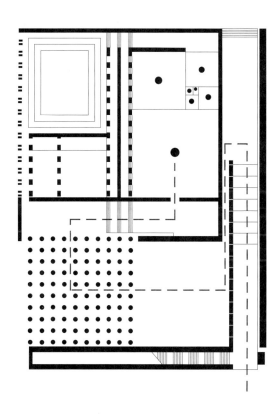

The Acropolis
Athens, Greece 432 BCE
[Classical, circular rotunda temple, masonry]

The Acropolis is situated on a hilltop above Athens, Greece; it is a sacred part of the city containing the most important structures and temples. Set atop a flat-topped rock that rises 490 feet above sea level, the Acropolis comprises three main buildings: the Propylaea, the Erechtheum, and the Parthenon. For centuries this was the site of very specific religious ceremonies dedicated to the Goddess Athena. The citizens of Athens would follow a procession up the stairs to the Propylaea and then negotiate the various angle changes towards the Parthenon. Although the ceremonial sequence was linear and axial, it was not straight; it was instead a series of vectors that culminated in the cella of the Parthenon.

Danteum, Giuseppe Terragni
Rome, Italy 1942
[Italian Rationalist, grid hall of glass columns]

The Danteum was an architectural expression of Dante's "Divine Comedy." Terragni referred to the spaces and paths of the poem to construct the design of the building. The sequence moved through the rooms of the inferno, purgatory, and heaven. This unbuilt, theoretical project serves as a tremendous example of ceremonial sequence due to the adherence to a well-known and highly descriptive text that also served as a catalyst for its design.

| 540 | SEQUENCE
PRINCIPLE | CIRCULATION
ORGANIZATION | EXPERIENTIAL
TYPE | PLAN
READING | ARCHITECTURE
SCALE |

Experiential Sequence

Experiential sequence is defined by the overt manipulation of the user through a series of spaces or rooms to produce a perceptual effect. Unlike a ceremonial sequence, experiential sequence emerges out of more emotive or conceptual means. Whether it is orchestrated through an abstract construct or evolved from a base perceptual method, experience, through movement and time, creates a collective engagement of diverse spaces founded in memory. Across an orchestrated chronology of experiences, the perception and cognition of body to place, gleaned from movement through space and form, creates a collective composition.

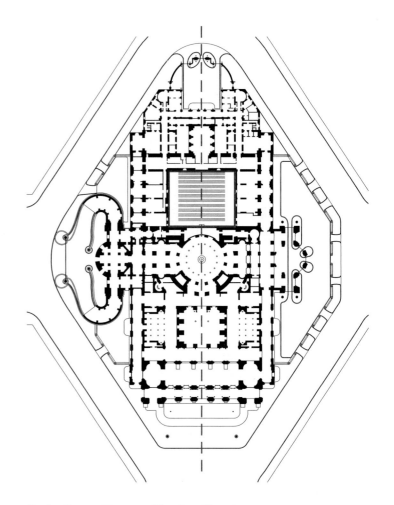

Paris Opera House, Charles Garnier
Paris, France 1874
[Neo-Baroque, horseshoe-shaped indoor theater, masonry]

This extravagant complex was designed to engage the civic fabric of the city of Paris. The experiential sequence begins at the entry where the visitor is greeted by a large colonnade that stretches across the main façade, negotiating between the frantic pace of the city and the fantasy world of the theater. The first lobby of the theater mirrors the exterior colonnade. From the lobby, the visitor continues to smaller antechambers. Continuing up through the stairways and hallways, the visitor finally arrives in the theater itself, prepared for the show. Garnier arranged a series of spaces that act very much like scenes from an opera performance, where the visitor may view and be viewed. These sequences occur on all levels of the building and certainly identify this building as a seminal example of experiential sequence.

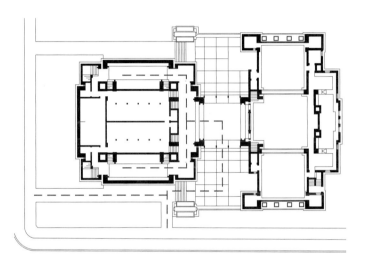
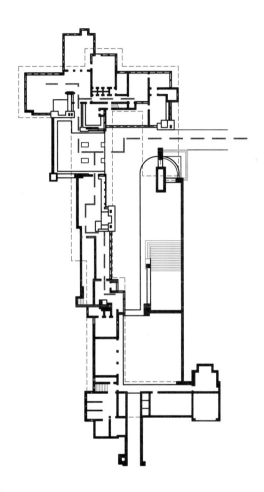

Unity Temple, Frank Lloyd Wright
Oak Park, Illinois 1906
[Early Modern, church and parish house, site-cast concrete]

At the Unity Temple, Wright used a series of complex experiential sequences to allow the visitor to uniquely perceive the temple. The entrance is a glass colonnade that separates the sanctuary from the parish house. Moving toward the temple, the visitor is forced to circumnavigate the sacred space through a lower cloister, which provides selective views into the central space. Wright skillfully heightens the visitor's expectations until finally allowing entrance into the temple and ascent into the church proper.

Taliesin East, Frank Lloyd Wright
Spring Green, Wisconsin 1909
[Modern, house, stone and wood]

Taliesin East draws on its natural environment as a focal point in its experiential sequence. Wright designed the house specifically to offer views of the surrounding landscape at every turn. This constant referencing of the landscape, in combination with the form and materials used to respond to the site and program, gives discrete local experiences that are emblematic of the complete experience of the project as a whole. Wright employed exaggerated horizontals, deep-set overhangs, anchoring hearths, and dynamic planning to draw the viewer through and along the sequence, while engaging the building, space, and site as a fluid whole.

Programmatic Sequence

Programmatic sequence is common as it is predicated on the idea that each space is preceded or followed by a necessary functional or programmatic counterpart. These programmatic diagrams can come from pure functionalism, issues of typology, or applied pattern. The premise of the functional organization and practical requirements of sequence based in program has been fundamental to the practice of architecture. The simple act of establishing a chronology of experience of two or more programs that are adjacent to one another engages an associative relationship that requires new ways and methods of mediation.

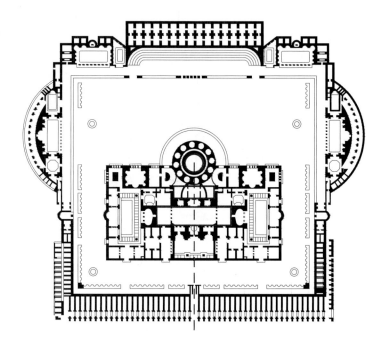

Baths of Caracalla
Rome, Italy 216
[Roman, baths, masonry]

The Baths of Caracalla were a massive complex that spread over many acres and comprised a number of functions and programs dominated by the Frigidarium, Tipidarium, and Caldarium. Architecturally and programmatically, these spaces represented the very essence of the baths themselves. Each user followed a specific sequence through these three spaces. The act of bathing in three different temperatures of water was believed to be critical for health and the Romans practiced it regularly. The axial strength of the architectural solution allowed for a linear sequence with programmatic nodes that tolerated diverse methods of occupation and allowed users to move between the programs at will.

SEQUENCE
PRINCIPLE **543**

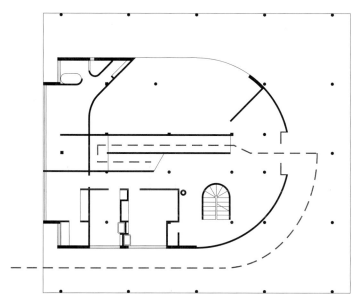 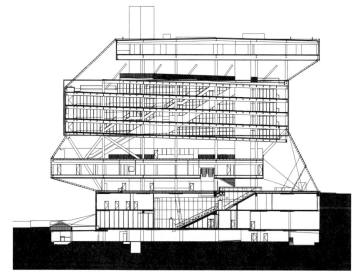

Villa Savoye, Le Corbusier
Poissy, France 1929
[Modern, free plan, concrete and masonry]

The Villa Savoye, when considered in terms of programmatic sequence, immediately illustrates the efficiency of the ramp and its ability to separate the program into diverse levels. The entrance level is used by the servants and the automobiles, though it engages the lobby and begins the ascent up and through the villa. The main level is almost entirely for the use of the family with the exception of the kitchen placed in the corner. As Le Corbusier allowed the lobby to be an intruder of sorts on the ground floor, here he does the same with the kitchen on the main floor. The ramp continues to the roof, arriving at the roof garden, a place of serenity and relaxation. Corbusier has skillfully used the vertical ramp as a way of connecting and distinguishing the three programmatic floors of this singular villa.

Seattle Public Library, Rem Koolhaas OMA
Seattle, Washington 2004
[Postmodern, continuous free plan, concrete and glass]

The Seattle Public Library has four distinct programmed areas (each containing a number of floors) that correspond to the four overt pieces in the shifting form of the building. The four large parts are shaped to reflect their program and then unified by a continuous steel and glass skin. These parts are organized using a free plan, which ultimately makes the reading of the various functions of the forms more complex and intricate. The stacks are located in a four-story, continuous spiral ramp that blurs the distinction between the floors, allowing various programs to exist in one space. The programmatic sequence allows for a single path to thread through each of the divergent functions of the building along a single, formal, circulation route.

Vertical Sequence

Vertical sequence employs a series of spaces or levels organized in section as a way of distinguishing important moments in the progression. This formal condition can occur in a natural setting where an architect is using typography to transition and orchestrate the sectional experience. Or this sequence can be fully constructed and more empirically derived through the staking and layering of architectural spaces. Verticality adds a new Z-dimension to the previously discussed sequences, producing an inherent complexity and difficulty in the physical and visual transitions through space. Due to the intrinsic physical difficulty attached to climbing, a clarity of design intention that matches the responsibility of the exertion is required. For centuries, shallow vertical sequences were employed in church typologies to instill hierarchy and visual dominance. Later, architects began to design and build intense vertical ascensions for both efficiency and effect to attach importance to form (as exemplified by the skyscraper).

Villa Lante, Jacopo Barozzi da Vignola
Viterbo, Italy 1568
[Mannerist, house and garden, masonry]

The pleasure gardens of the Villa Lante use a vertical sequence as a way of separating the various sections and levels while using gravity as a utilitarian water distribution system and a metaphor of time. Each garden carries a metaphor or narrative. The beginning of the sequence is the Quadrato, which is a square **parterre** *garden with the large Fountain of the Moors, representing the ocean and the lakes, at its center. Farther up the sequence is the Fountain of the Lamps, which represents culture. On the third level is the large, stone dining table, which represents agriculture. Above this is the fountain of the River Gods, representing the rivers of the world. The next level holds the unusual water chain or stair, which represents the stream. At the uppermost are the Fountain of the Dolphins and ultimately the Fountain of the Deluge, which represents the origin of the earth. Vignola has crafted a unique vertical sequence that allows pleasure and metaphor to mingle simultaneously.*

SEQUENCE	545
PRINCIPLE	

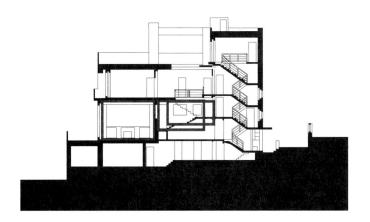

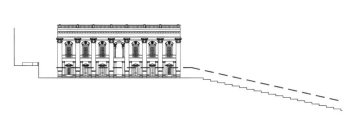

Piazza del Campidoglio, Michelangelo Buonarroti
Rome, Italy 1650
[Renaissance, elliptical piazza, masonry]

The Piazza del Campidoglio and its elongated approach stair create a dynamic and ceremonial sequence through the urban fabric of Rome. Michelangelo played with the perspectival vision relative to the sectional ascent, creating a constant dialogue of the viewer's relationship with the centralized equestrian statue. The slow ascent to the piazza allows the statue to incrementally emerge, anthropomorphize, and finally arrive at the dominance of the mounted scale. The perspectival play relative to the vertical sight line and the viewer's body position creates an expansive relationship to the ovaloid piazza. Entering on the elongated end, the visual vantage collapses the figure, making a circular figure of the piazza and moving the perceived depth of field from the actual position of the statuary. The movement of the bracketing buildings, the ground pattern, and the centralized statuary produce a ceremonial experience along a cinematic pathway through the vertical sequence.

Villa Müller, Adolf Loos
Prague, Czechoslovakia 1930
[Modern, Raumplan house, masonry]

The Villa Muller develops a compelling vertical sequence directly related to the organizational and experiential ideas of the Raumplan. The user moves from the exterior porch into the entry foyer and then proceeds up a set of stairs to a landing adjacent to the main living room. From this point, the sequence splits three ways: straight out into the living room, to the right to a flight of stairs leading to the dining room, and to the ladies sitting room to the left. These vertical sequences are based on the hierarchical organization of the Raumplan, which provides the dining room and the ladies sitting room with visual connections to the living room.

| 546 | SEQUENCE
PRINCIPLE | CIRCULATION
ORGANIZATION | VERTICAL
TYPE | PLAN SECTION
READING | ARCHITECTURE
SCALE |

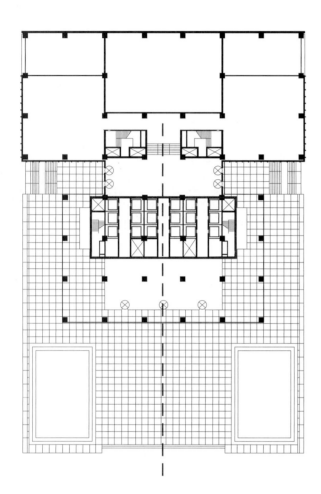

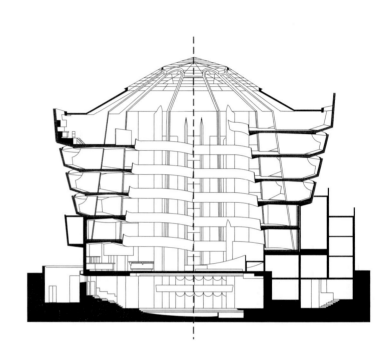

Seagram Building, Mies van der Rohe
New York City, New York 1958
[Modern, free plan with central service core, steel and glass]

The Seagram Building is the modernist high rise that came to typify the corporate office building through its conception of uniform universal space. Its form and detail has been mimicked and iterated in nearly every city of the globe. Celebrating the industrialized forms innate in the material along with the open plan and centralized office core make the spaces highly subdividable. The exterior articulation of the structure and the elimination of the variation between base, middle, and top, make the elevation a uniform field that perceptually extends infinitely. The vertical sequence runs through a central core, which creates jump-cut transitions from floor to floor and disembodies, yet individuates and equally prioritizes each level.

The Guggenheim Museum, Frank Lloyd Wright
New York City, New York 1959
[Modern, ramping spiral gallery around atrium, concrete]

Wright's Guggenheim Museum contains a unique and formal vertical sequence. Upon entering the museum the patron immediately takes the elevator to the uppermost gallery. From this vantage point, the patron has a complete view of the central atrium. Then begins the descent through the museum, which traverses the continuous spiral ramp that forms the shape of the museum and expresses its concept. The art is often displayed on or along the ramp; however, the ramp also pauses to lead to more traditional, peripheral gallery spaces. This is a tremendous and dynamic sequence that certainly remains one of the more seminal of the modern movement.

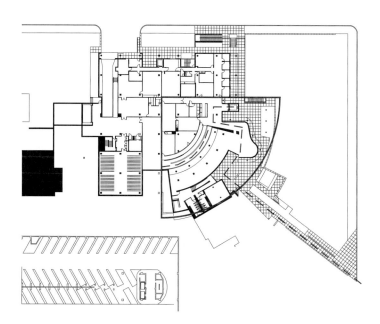
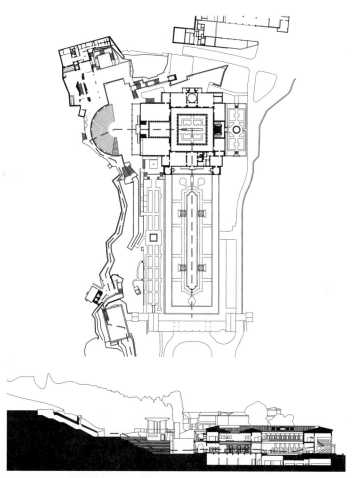

High Museum of Art, Richard Meier
Atlanta, Georgia 1983
[Late Modern, L-shaped galleries with circular atrium, metal]

The High Museum utilizes a vertical sequence as a way of organizing the galleries and the museum proper. The complex is formed by an L-shaped building that contains the main galleries, and support spaces. These elements are consequently joined by an ascending ramp that carries the patron from floor to floor. Each ascent of the switchback ramp allows the user to refocus and visit the galleries on that particular floor. This repetitive but compelling system orchestrates a vertical sequence that visually and spatially allows the user to comprehend and appreciate his or her relative position in the museum. The vertical sequence organizes the galleries and museum into a clear, concise, and centered organization that acknowledges the power of the art and encourages its pleasure.

The Getty Villa, Machado Silvetti
Malibu, California 2006
[Postmodern, museum addition and renovation, masonry]

The Getty Villa addition and restoration project by Machado Silvetti engages a ceremonial vertical circulation sequence. It begins with the parking garage that is nestled into the Malibu hillside, moves through a series of nodal outdoor rooms, and ultimately arrives at an amphitheater built into the natural topography. This landscape descends to the existing replica of a Roman villa, weaving in the historical sequence of the Roman house and garden. As articulated individual spaces defined by pure forms and controlled positions, relative to the dramatic topography and to one another, the nodes create moments in a complex sectional ascension up the hillside. Moving through lavishly articulated earthen rooms, which are banded in diverse materials and treatments to simulate the geologic strata and textured to reference archaeological excavations of Roman antiquity, the visual hierarchy relative to the ground plane is extended into the materials and details. The position along the sequence is governed by a constant and legible relativism of place and perception.

548   MEANING

                PRINCIPLE

Meaning

15 - Meaning: Theory / Themes / Manifestos

Architecture can be subdivided into a series of movements or themes. Below are the subdivisions of conceptual thought typified by discrete thinkers and manifestos that this text has attempted to explain and illustrate. Each movement, though seemingly isolated, is a truly interrelated response and evolution, emerging out of its cultural, technological, and philosophical context. Each movement comprises a distinct knowledge base that contributes to the overall meaning and lineage of architecture.

Classicism Classicism may represent the original architecture. It is a number of formal elements and design principles that were regularized and recognized as given conditions. Column orders, pediments and friezes, for instance, were all utilized in similar and referential ways. They were not employed at the whim of each individual architect; rather, they established a collective language. Vitruvius, a Roman writer and architect, published the established tenets of classicism in the first book about architecture—*De Architectura*.

Romanesque Romanesque architecture was a hybridization of classical architecture mixed with elements from the Byzantine. It adopted many of the Roman construction techniques (and associated forms); however, it lacked much of the ornament and sophistication that was typical of classical architecture. Romanesque architecture was fundamentally an architecture of mass, utilizing the wall rather than the column as its main structural form.

Gothic Gothic architecture is best known for its extreme verticality and accentuated expression of structure. Gothic architecture was typically used in sacred buildings and carried with it aspects of religion not seen in later styles. The accentuation of height and light was coupled with an increased expression of diverse and prevalent ornamentation.

Renaissance The Renaissance proved to be one of the most influential of all architectural periods. It is marked by a number of critical factors, including the separation of labor, an intense interest in the past (including the architecture of the ancients), a compelling understanding of perspective and geometry (typified by a purity of symmetry and proportion), and an increased desire to stretch the boundaries of structure as an architectural system. The architects Brunelleschi and Alberti were the early primary practitioners of the Renaissance, setting a new standard of architecture. These tenets were documented in Alberti's numerous texts, which focused on both architecture and painting. This was furthered in texts by Serlio and Palladio who advanced and streamlined the thinking.

Mannerism Mannerism was essentially a modified continuation of the Renaissance. Mannerists took a language that had very specific rules and then manipulated those rules to produce new compositional conditions. As it was dependent upon a vocabulary that could still be understood, it was not a complete alteration, merely an iterative one that produced new levels of experimentation, dynamism, and even folly. Mannerism produced some of the more fantastic moments of architecture, despite its limited popularity.

Baroque The baroque era and style formed a natural continuation of the mannerist tendencies begun during the late Renaissance. This style pushed architecture to its limits, connecting architecture and the arts by synthesizing space, structure, geometry, and ornament. Introducing an element of theater, ideas of lighting and perspective evolved to new levels through experimentation. The use of curvilinear lines and figures, particularly the oval, allowed for an accelerated complexity of geometry that had a profound effect on form and figure. Ornamentation claimed a new importance that it had not enjoyed since the Gothic. Collectively, the baroque style was an architecture of expression, dynamism, and theater.

Neo-Gothic Neo-Gothic style occurred as a renewed interest in medieval forms grew. This formal interest, combined with a reaction against neoclassicism, spawned an architecture that coupled political and religious consequences. The neo-Gothic became associated with and representative of Christian values and the monarchy. Unlike Gothic, the neo-Gothic expanded to represent both secular and sacred buildings. Like the Gothic, it was infused with detail, ornament, and verticality. The English architectural theoretician John Ruskin popularized this style through his seminal texts, whereas visionary designers like Viollet le Duc synthesized the style with technology and new materiality to advance its structural underpinnings and capabilities.

Neoclassicism Neoclassicism, which began in the mid-1800s, was developed as a reaction to the excesses of the baroque and rococo styles. It was an attempt to return to the splendor and glory of Rome, Greece, and the High Renaissance. Neoclassicism relied on symmetry and simplicity, extolling the virtues of the iconic. It was interested only in establishing a continuity with the past; yet, it was remarkably adaptable to culture and geography and thus can be found throughout the world.

Art Nouveau Art nouveau developed as a reaction against the many revivalist styles that developed during the nineteenth century. Architects sought to create a new style, one that represented a new language and used nature as its inspiration. It was also an architecture that expanded the scope of design to synthetically include furniture, fabric, fixtures, and even clothes. Despite its popularity, art nouveau was relatively short-lived. Its extreme use of ornament, craft, and fantasy continues to inspire architects and artists.

Arts and Crafts The arts and crafts movement began as a reaction to the Great Exhibition of 1851. This exhibition was renowned for displaying very ornate objects that were machine-made thereby eliminating their craft and material character. The arts and crafts movement believed in simplicity and materiality. The arts and crafts movement opposed superfluous ornament and any form of mass production or industrialization. The movement preferred traditional and vernacular forms of architecture, believing that there was an essence to all things, and that this quality was critical.

Industrialism Industrialism refers to an architecture that was essentially the product of mass production. Components of the building were no longer particular to only one building, but were instead, part of many buildings. It created a market for mass consumerism that continues today. The architecture of industrialism marked one of the first movements toward modernism in its ideology and aesthetics.

Modernism Modernism arose as a reaction against the replication of historic styles in architecture. This movement developed with an honesty of structure and materials, embracing the new principles of abstraction and industrial production. Dominated by distinct figures, each operated under a new aesthetic sensibility. Le Corbusier used painting, in addition to his five points, as a way to further his architecture and formalism. Walter Gropius founded the Bauhaus, extolling a fundamental relationship between geometry and functionalism. Adolf Loos relied on simplicity and materiality. Mies extolled universal space and material sensuality out of minimal forms. All modernists approached their architecture in distinctly unique ways. Modernism was a complete break on nearly all levels with what had gone before. It was an architecture that called upon materials, structure, construction, fine art, industrial techniques, and theory to shape many of its primary tenets.

Rationalism Rationalism was practiced primarily in Italy during the early twentieth century. It was a movement that valued not only the new ideas of modernism, but also the historic classical ideas of a movement known as Novecento Italiano. It was in this middle ground, that the rationalists found an architecture they believed solved many problems of the modern condition, while at the same time remaining true to their classical Italian heritage.

Brutalism Brutalism was a term applied to architecture that was noted for massive, geometric forms constructed of cast-in-place concrete. This type of architecture celebrated its construction and tectonic techniques, often exposing its circulation and mechanical systems. Brutalism was known for its use and exploitation of rough and exposed concrete. There was a sense of honesty in the construction that was revealed and celebrated. The rawness of its articulation resulted in criticism for its lack of warmth and response to the human scale and needs.

Hi-Tech Hi-tech refers to an architectural movement wherein the technical components (mechanical and structural systems) were highlighted and integrated as a visible and essential language of the architecture. Each technical element was articulated and exposed so that the form and language of the building revolved around this overtly technological expression. Hi-tech was extremely expressionistic and proved to be extremely expensive, making its architecture not easily reproduced. It was a movement that was truly integral to the building and thus was able to rebuff any purely stylistic accusations.

Postmodernism Postmodernism developed out of a reaction to the blandness and abstraction of the late modern movement. Most noted for its reinvestigation of history, postmodernism brought back elements of ornamentation and playfulness that had been systematically reduced throughout the modern movement. This ornamentation was often applied strictly as a appliqué with little or no thought to the consequence beyond graphic iconography ultimately giving it a poor reputation and heralding a new strictly, style concentrated period of architecture.

Structuralism Structuralism developed as a reaction against the functionalism and abstraction of high modernism. The structuralist believed stylistically in modernism; however they believed that architecture could be approached from a more humane sensibility. Structuralism allowed for the introduction of intangibles, looking to new methodologies to solve problems giving significance to user groups and how they would interact with the architecture and urban context.

Post-Structuralism Post-structuralism was a brief, but powerful architectural movement marked by an understanding that all theories of architecture can be questioned. The post-structuralists considered all aspects of architecture as relative, not as truths. The intellectuals at the foundation of this movement were extremely critical of each other, which resulted in an inability to form a precise and viable core to the architectural philosophy. The work is marked by geometric abstraction, dense theoretical diatribes, and a loss of the value of materiality and construction in order to privilege form.

Deconstructivism Deconstructivism, like postmodernism before it, became the style of choice during the late twentieth century. As a style, it favored non-rectilinear forms and attempted to question certain absolutes that had traditionally been a part of architectural thought. Issues such as envelope, structure, and function were all examined and challenged. Ultimately, the movement degenerated into a pure stylistic exercise that, like postmodernism, lost its initial essentialism.

Regional Modernism Regional modernism examined the local context and culture of an area and then utilized those principles in combination with the general concepts of modernism. This approach allows for continuity of both modernism as a language and also the culture, climate, and place in which the building is built. This movement has proven to be extremely rich and provocative in that the architecture remains modern, but is vehemently against an international style and instead exploits local tradition and form as its primary catalyst.

Globalism Globalism as an architectural style marks a movement that reflects the global economy that is dominated by western thought. This has manifested itself into the concept of the "star architect." These architects and their projects have become commodities and are seen primarily as brand names, which are exploited by the clients for recognition. The architecture of globalism is typically a type of formalism that exists in the individuated style of the author regardless of program or place. This includes a relatively small and elite group of architects that now practices architecture on a world stage and creates a signature that has become powerful through its iconography.

Classicism

Classicism, as an architectural style, is ubiquitous throughout the world. Classicism has its origins in Greece, but was subsequently resurrected numerous times throughout history, most notably in Rome. The classicism of Greece was based on the Doric, Ionic, and Corinthian orders along with other regularized architectural elements, geometric and proportional systems, and their collective systemization of employment. This original classicism was pure and honest in its construction. The buildings were typically constructed of solid marble and generally took the form of temples dedicated to the pagan gods. The architectural system of proportions of the temples was worked out over centuries and has been in use since. Ideas of order, symmetry, hierarchy, proportion, and even perception were fully developed and considered in classicism. The Romans developed classicism to a different level, expanding the language, building typology, and construction techniques. The Roman architectural theoretician Vitruvius wrote what is considered to be the first architectural treatise, *De Architectura*, laying out the rules and concepts for architecture and classicism, which were at the time inseparable.

MEANING	555
PRINCIPLE	

Parthenon, Phidias
Athens, Greece 432 BCE
[Classical, pagan temple, masonry]

The Parthenon is often considered the quintessential building in classical architecture, perhaps even all of Western architecture. The Parthenon established the ideal model used for thousands of years; it continues even today to have a large influence on architecture. As the largest building of the Acropolis, the Parthenon had a commanding presence over the city of Athens through its scale and elevated position, coupled with the perceived perfection of the building, of which lead to its fame and influence. The appropriation by the British of much of its ornament, and the ensuing London exhibition of the Elgin Marbles, added to the building's Western importance and significance. The consequence of this architecture can be seen by the thousands of imitations of temple fronts that have been built throughout the world since its completion.

Romanesque

Romanesque architecture began in the sixth century and continued throughout the Middle Ages. The Romanesque style dominated Europe for approximately 600 years, making it one of the most ubiquitous and longest lasting of all the styles. It can be found in every country in central Europe, making it one of the first pan-European architectural styles after the classicism of the Roman Empire. Essentially, Romanesque architecture was developed through the hybridization of Roman buildings with Byzantine buildings. Romanesque architecture typically employed many of the same construction techniques that the Roman's used; however the Romanesque buildings lacked the refinement of materiality and ornament that were common in Roman architecture. The buildings were typically constructed by guilds with master builders, but were never associated with one single architect. Whereas the Roman or classical style utilized the column as a major structural element, the Romanesque relied primarily on the wall as its main structural element. Characteristics of Romanesque architecture included: massive walls, arches, vaults, towers, simplicity, and symmetry. Churches represent the majority of buildings constructed in the Romanesque style. The system of proportions that was common in classical architecture was not as critical in the Romanesque. The Romanesque builders were also known to reuse elements salvaged from other buildings and styles, creating a number of architectural hybrids.

San Miniato al Monte
Florence, Italy 12th century
[Romanesque, rectangular basilica, masonry and wood]

San Miniato al Monte, above Florence, is considered one of the most influential examples of Romanesque architecture. It was constructed rather late within the Romanesque style and is best known for the green and white marble, polychrome articulation that dominates the entire complex. The plan is a straightforward rectangle with two rows of piers dividing the space into a nave and two side aisles. There are smaller chapels branching off on the sides, forming a series of interesting outer buildings. This type of plan was common early in the Romanesque period rather than the Latin Cross, which ultimately would emerge as more popular. The striking polychrome of the façade continues into the interior of the church, creating a visual ornamentation that affected Florentine architecture for centuries. Even today, San Miniato is considered the original Church of Florence and as such will always be seen as one of the most important in the world.

Gothic

The Gothic style flourished after the Romanesque. It was noted for its verticality, which was expressed through its structure, ornament, and space. It was the main architecture of Europe from the twelfth century until the sixteenth century. There was a moral aspect to Gothic architecture that was expressive of its connection with Christianity. The buildings were typically designed to be as tall as possible, stretching the structural constraints of architecture. Gothic architecture was more refined than the Romanesque, allowing the buildings to be far more expressive of the skeletal structure. This formal expression was required by the increased height and large openings for intricate stained glass windows, which communicated through their pictorial content to the mostly illiterate public. Gothic architecture is noted for its unique features, which include the pointed arch, ribbed vault, spires, and flying buttress. Its material, which was typically stone rather than brick, was an identifying characteristic. All these elements added up to a soaring vertical architecture that reached closer to the heavens than anything previously built, thereby expressing the architecture's connection to God. The plan of many Gothic cathedrals took on the shape of the cruciform, again using Christian forms as a way of expressing the glory of God.

Chartres Cathedral
Chartres, France 1260
[Gothic, Latin Cross, stone masonry]

Chartres Cathedral stands as one of the best preserved high Gothic cathedrals in the world. It contains all the elements that are so common in Gothic Cathedrals including the rich use of stained glass, the soaring vaults, and the flying buttresses. Like all Gothic Cathedrals it was constructed over hundreds of years and is the not the work of one architect, but of many guilds and master builders. It illustrates a rich and complex use of Gothic ornament and tracery, resulting in an intricate and visually exciting stone and glass tapestry. The west façade is asymmetrical and exemplifies one of the first uses of the tripartite façade as well as the proportional ideas of the root-2 square. Although perhaps not quite as famous as Notre Dame in Paris (due largely to location), Chartres remains the quintessential high Gothic cathedral, and the zenith of the Gothic architectural movement.

Renaissance

The Renaissance was arguably the most influential of all the styles, due in part to the fact that it affected so many different aspects of architecture, art, and urbanism. Subjects such as science, geography, and astronomy were developing at a tremendous pace during the Renaissance. It was during the Renaissance that architects began re-examining the architecture of the ancients, thus marking the first time that history and the concept of the building as a teaching model became critical. Architects such as Alberti and Palladio spent years in Rome measuring and documenting the ruins and buildings of ancient Rome, learning from its forms, spaces, geometries, and construction techniques. The emergence of the extensive use of proportions allowed buildings to be liberated from merely responding to the shape of the available plot (as was often the case during the Middle Ages), but were instead, subscribing to the pure geometries of perfect squares or golden sections. The elevations of the Renaissance were constructed using established rules of typology and proportions. There was a culmination of the arts during the Renaissance, which allowed direct connections between architecture, sculpture, and painting to emerge and flourish. A fascination with the laws of optics through perspective allowed for a mathematical bridging of representation with form and space through perception. Brunelleschi, often considered the first architect of the Renaissance, established the idea of the Renaissance man, a person who was able to operate in numerous fields and was able through sheer will power and invention to solve problems that were previously considered unsolvable. A sense of order and refinement was prevalent throughout the Renaissance. Architecture was no longer a series of subjective and personal opinions, but a body of thought and knowledge that informed every aspect of culture and the profession.

MEANING	561
PRINCIPLE	

Palazzo Medici Riccardi, Michelozzo di Batolomeo
Florence, Italy 1460
[Renaissance, rectangular courtyard palazzo, masonry]

The Palazzo Medici Riccardi is regarded as one of the archetypal examples of early Renaissance palace design. As a model it established many of the rules that governed the design and construction of the other later Renaissance palazzi. It is constructed of stone and has three different levels, each of which is articulated differently through the use of rustication. The base or bottom level has heavy rustication whereas the upper floor is very smooth. On the exterior, each of the three levels is divided by a stringcourse, and the entire composition is crowned with a classical cornice. The exterior of most of the Florentine palazzi are a combination of late medieval and early Renaissance architecture. The interior courtyard emerges as a perfect square with the rooms of the palace surrounding and relating to this central and primary element. These courtyards are direct descendants of the Roman courtyards that were so common in the typical Roman house. The overall consideration of order, history, and proportions separated this palazzo from earlier iterations.

Mannerism

Mannerism only existed for a short time; however its impact and importance cannot be overlooked. Mannerism represents a stage of architectural development that immediately followed the Renaissance. As such, its relationship to the Renaissance is so close that to the untrained eye its differences might prove indiscernible. The mannerists were pushing the boundaries of what architecture was and what it could achieve. Mannerism dealt primarily with understanding the rules of the Renaissance and then manipulating them to form an architecture that was more free, more dynamic, more complex, and even more playful. While the classical language of the Renaissance remained critical to the mannerist, it was their prerogative to experiment with its language and to alter preconceptions. These manipulations occurred through formal machinations such as layering, distortion of perspective and scale, the use of asymmetry, and the search for an overall compositional and perceptual dynamism. The result of all these elements was an incredible mixture of classicism with a hint of the freedom and ingenuity that was yet to come in the baroque. Many historians have found the mannerist period more interesting due to its subtle relationship to the Renaissance, which was one fully intact yet one where the architectural inventions were subtle, requiring a higher sense of understanding. A mannerist building can be seen as a continuation of the classical language, whereas the later movements served as complete breaks. This is exemplified by the ability of a layperson to clearly tell the difference between a baroque and classical building, whereas this is far less evident with a mannerist building.

Palazzo del Te, Giulio Romano
Mantua, Italy 1534
[Mannerist, rectangular courtyard palazzo, masonry]

The Palazzo del Te is generally considered to be the first building to cavalierly examine and challenge architecture. Previously architecture had always been considered a serious business, usually extolling the grace of God or the hierarchy and status of royalty. Giulio Romano, through this suburban palace, was able to approach architecture in a completely new way and with a sense of humor. Romano essentially designed a building that to an untrained eye would appear as a traditional classical palace. However one learned and trained in architecture could witness that many of the laws and rules of the classical language were questioned, broken, or ignored. In reality these moves were often very subtle, resulting in dropped keystones, strange gaps between stones, broken pediments, slipped symmetries, cartoonish rustication, or inconsistencies in ordering systems or proportions. These manipulations were enough to make the building extremely popular; though it was considered something of an anomaly.

Baroque

The baroque represents the epitome of dynamic architecture. Representing a natural continuation that began with mannerism (defined by a manipulation of the classical language to form a new style of architecture), the baroque fully engaged the geometric and formal evolution of the natural complexity and development of the lineage of the arts. More than any other style, the baroque came to be associated with the concept of all the various art forms coming together. Architecture, sculpture, and painting combined to form one synthetic sensibility. Unlike mannerism, which was to remain relatively close to the architecture of the Renaissance, the baroque looked to establish a language that truly stretched that boundary. Curvilinear elements that had previously been the exception now became the rule. This complexity of geometry and dynamism of movement became the defining principle, evident everywhere in the buildings of the baroque period. The order of the architecture became more complex, resulting in buildings that seemed to dance and swell. The desire for this differentiating geometry privileged the use of the oval as an architectural element. A fascination with this shape occurred as a response to an influence from astronomy. During the Renaissance, astronomers were discovering that planets traveled along an elliptical path and not a circular one as previously speculated. This discovery led to a fascination with the complexity of shape that was then converted into all manner of architectural convention. Interestingly enough, symmetry was never truly questioned during the baroque. The buildings, despite their organic forms, remained symmetrical.

San Carlo alle Quatro Fontane, Francesco Borromini
Rome, Italy 1641
[Baroque, oval, masonry]

Despite its small size, the Church of San Carlo alle Quatro Fontane articulates many of the principles that are synonymous with baroque architecture. The plan of the church is an articulated series of overlapping ovals that undulate and create an overall sense of dynamism and unrest through the use of geometric plasticity. The space of the church can be described as an oval influenced by a cruciform. Smaller ovals help create two side chapels as well as a small narthex and an altar. These parts are combined to form one space rather than the typical separation common during the Renaissance. This sense of movement and plasticity is evident throughout the building but is particularly evident in the façade, which heaves and swells along a curve that negotiates between the street and the interior of the church. Though the façade has almost no relationship to the interior, it creates a more theatrical setting in which these layered elements hint at the space behind it.

Neo-Gothic

Neo-Gothic architecture began in the mid-1700s in England and continued to be used well into the nineteenth century around the world. It occurred due to the intense revival of interest in medieval forms and was subsequently a reaction to neoclassicism. Besides the revival of Gothic forms, neo-Gothic also carried with it political and moral values. Neoclassicism was seen as a celebration of liberalism and republicanism, whereas neo-Gothic represented monarchism and conservatism. Neo-Gothic came to stand for proper Christian values, and hence any organization that utilized it would be associated with those particular values. Unlike Gothic architecture, which was almost exclusively the realm of religious buildings, neo-Gothic was used in many different building types, both religious and secular. Neo-Gothic architecture tended to be more consistent in that a building was typically the responsibility of one architect, rather the product of the ancient guild system, which was the system that produced Gothic buildings. In general, however, the forms for both were similar, relying heavily on excessive detail and verticality.

Houses of Parliament, Barry and Pugin
London, England 1867
[Neo-Gothic, multi-courtyard mat building, masonry]

Charles Barry designed the Houses of Parliament in London in what was termed the perpendicular style, which was a subset of neo-Gothic. Barry, who was actually more recognized as a neoclassicist, relied on help from the renowned Gothic expert Augustus Pugin, who served as an advisor on the details and for the interior furnishings. Barry compositionally crafted an immense symmetrical complex that revolves around numerous courtyards and primarily the House of Commons and the House of Lords. Each House forms the central piece of each side. The neo-Gothic style worked well as a symbol of the monarchy and conservatism that was so popular in Great Britain at the time. It also served as an opposition style to France and the United States, who had both adopted the neoclassical style as their national architecture. The Houses of Parliament are an interesting example of classical sensibilities merged with Gothic detailing. The overall sense of order combined with the intricacy and quality of the details perfectly represents the government and people of Great Britain.

Neoclassicism

No style was as popular or as influential as neoclassicism. Beginning in the mid-eighteenth century, it survived and dominated until the end of the nineteenth century. Neoclassicism originated as a response to baroque and rococo excesses. These styles were seen increasingly as shallow and overindulgent. Neoclassicism was seen as a method of returning to the architecture that was associated with the glory of Greece, Rome, and the High Renaissance. Rococo epitomized decoration, movement, excess, and the beginnings of asymmetry, whereas neoclassicism extolled the virtues of simplicity and symmetry. Neoclassicism became the first true international style and was used around the world. It was a heroic style that faded in the twentieth century as modernism gained dominance. Like many of the other movements and styles, neoclassicism had parallel movements in the allied arts, such as music and painting. Politically it signified a sense of liberalism and republicanism, ultimately becoming the preferred styles of both France and the United States governments. This style continually relied on the past as a way to move forward. It was not interested in the idea of the new, but instead believed in a body of work that had become iconic and unquestioned. Neoclassicism had the ability to adapt to almost any culture or function. It was and perhaps remains the most important and ubiquitous of all the styles.

La Madeleine, Pierre-Alexandre Vignon
Paris, France 1842
[Neoclassical, colonnaded rectilinear church, masonry]

La Madeleine, originally designed as a monument to Napoleon's Army and later consecrated as a church, was heavily influenced by the Maison Carre in Nimes, one of the best-preserved Roman buildings in the world. Interestingly enough, Vignon did not believe that the relationship between the exterior and the interior was as critical as true classical structure. The interior of La Madeleine is comprised of three domed spaces that are aligned toward the altar. From the exterior, the pedimented façade gives no indication of this layout, nor does the Parthenon-like colonnade that surrounds the entire structure. It was critical that the simplicity of the exterior not be compromised by the complexities of the interior. The exterior participated in the city of Paris as a monument; thus it had further responsibilities as a piece of architecture and urban planning. This illustrates how flexible neoclassicism could become when faced with difficult planning and architectural issues.

Art Nouveau

Art nouveau as a style lasted only from 1890 until 1905; however during these years, it proved extremely popular and influential. Art nouveau developed as a reaction to the many revivalist styles so common during the nineteenth century. Instead of imitation, art nouveau looked to other inspirations as a way to create an entirely new language. A large part of the new inspiration came from the forms of nature. The curved and sinuous lines, the bulbous forms, and the delicate details of the natural world all found their place in this new and expressionistic style. The plans of art nouveau buildings began to challenge the norms established by classicism; however typically they typically remained firmly within the classical canon. The inventiveness and character of the art nouveau was seen in the extensive use of ornamentation, detailing, and the concept of a complete design. Each object within a project necessitated the attention of the architect. Van de Velde brought this practice to an extreme level, going so far as to actually design the dresses and shoes for the women who resided in his designs. The fixtures and furniture were a large part of this equation, ultimately supplying the movement with a sense of the fantastic that lasted well beyond the brief history of the architectural movement. Art nouveau served as a middle ground between the revivalist styles and modernism. It was a critical step in convincing architects that they could in fact invent entirely new forms that were to herald the arrival of the modern movement.

MEANING | PRINCIPLE | 571

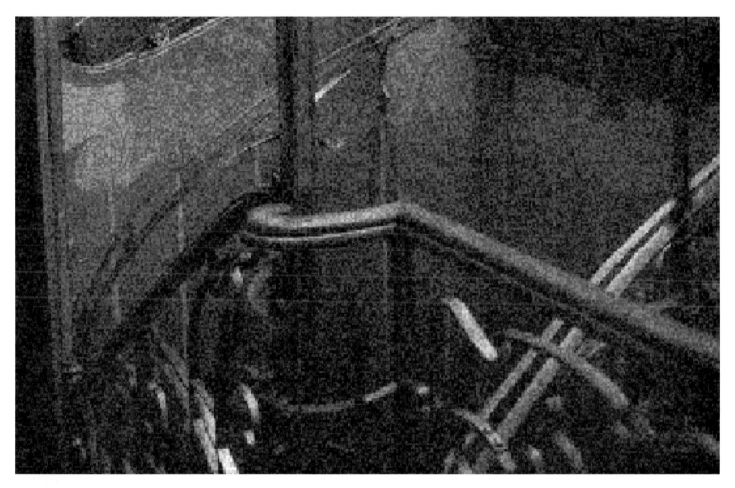

Horta House, Victor Horta
Brussels, Belgium 1898
[Art Nouveau, organic vegetal forms, masonry and iron]

The Horta House serves as a stunning reminder of the power and ingenuity of the art nouveau movement. This house, which was both a residence and an office, was essentially two townhouses that were carefully joined. The combination of these two elements speaks volumes about the Horta's mindset and his ability to conceive of this hybridization of functions while allowing the two entities to remain separate both spatially and formally. This individuation is clearly seen on the main façade, helping to break any relationship to symmetry. The two sides are joined only on the ground and first floors. The truly remarkable parts of the buildings emerge in the ornament and details that are evident in the façade iron work and the interior main staircase. These elements proved one of the catalysts of modernism, where an architect no longer relied on the past for inspiration, but instead invented a new language based on observations of nature and society.

Arts and Crafts

The arts and crafts movement flourished for approximately 50 years, between 1860 and 1910. It was begun in England by the writer William Morris and was influenced by the writings of architectural theorist John Ruskin. The beginning of this movement came about in part as a reaction against the Great Exhibition of 1851 in London, which advertised many items that were extremely ornate, artificial, and made no recognition of the quality of materials. The arts and crafts movement believed in the use of simple forms and traditional craftsmanship and was generally against any form of mass production or industrialization. They favored traditional or romantic styles of decoration and vernacular forms in architecture. This movement attempted to do away with any superfluous decoration and tried instead to concentrate on the essence of things. Often the details of construction were left exposed and natural materials were used and exposed as a way of communicating the truth or honesty of both materials and construction. They did not believe that ornament itself was bad, only that ornament had to be appropriate and should never become more important than the object itself. Like many artistic movements, arts and crafts influenced many aspects of life, including painting, sculpture, illustration, photography, domestic design and the decorative arts, including furniture and woodwork, stained glass, weaving, jewelry and ceramics. Beyond these disciplines, the arts and crafts movement had a large influence on social philosophy and politics.

Gamble House, Greene and Greene
Pasadena, California 1909
[Arts and Crafts, asymmetrical central hall house, wood]

The Gamble House is one of the finest examples of arts and crafts. Constructed entirely by hand in approximately one year by only a dozen workers, it ushered in an entirely new language (that of the southern California bungalow), heralding the arts and crafts movement. The house itself is a large wooden structure that has a particularly casual feeling in that it is has lower ceilings, rich sensual materials, an informal floor plan loosely based on the central hall house, and, due to the permissive weather, a remarkable relationship between the interior and exterior. The majority of the furniture and fixtures was also designed by Greene and Greene, illustrating and extending the influence that this particular style achieved. Informal asymmetries are favored over classical symmetries, resulting in a plan that attempts to retain clear separations of functions but without the strict axes. The house is equally famous for its use of exterior space through numerous sleeping porches and various patios. Collectively, the house stands as a testament to craftsmanship and detail.

| MEANING | INDUSTRIALISM | | ARCHITECTURE |
| PRINCIPLE | MOVEMENT | | SCALE |

Industrialism

Industrialism refers to an era when mechanization allowed mass production. The advent of technology permitted the repetitive production of products and materials, allowing for regulation of production, an increase in quantity of available objects, and a decrease in the cost. The impact on architecture was an expansion of materials beyond traditional craftsmanship and repetitive machine production of components. The transition in production brought about an associative transition in social norms, moving from trade and farming to manufacturing and industry. This movement introduced new programmatic building types in response to the new technology, methods of production, and knowledge. The emergence of a middle class created a new consumer culture defined by a worker that could now purchase the products that the assembly line produced. The material production allowed the emergence of large-scale production and standardization, which introduced predefined building products and transitioned the architecture from a hand-made, craft-based object to an assembly based, machined product in both form and tectonic.

| MEANING | 575 |
| PRINCIPLE | |

Fagus Works, Walter Gropius
Alfeld an der Leine, Germany 1913
[Industrialism, functionalist factory, steel, masonry and glass]

The Fagus Works typifies industrialism through both the function and the form. As a factory, it programmatically houses the industrialized processes of mass production that the architectural movement was responding to. Its formal expression was a clear rejection of craft, material traditions and ornament for functionalism, streamlined forms, and revealed, modern materialism of steel, brick, masonry, and glass. The industrialized aesthetic illustrates a rational formalism with an expressive honesty to the materials, personified through their repetitive unitization, resulting in a taught surface. The result is a mechanized product in spirit, appearance, and material.

Modernism

Modernism emerged as a response to conservative realism and the traditional historical forms that relied upon overt historical references. Employing abstraction, new materials, and the associated forms and structural capabilities of the new technologies, machine-like forms with open plans emerged to embrace and personify the new industrialized formalism. New technologies rendered old building styles obsolete, rejecting traditions and deferring to an emergence of form and function as dominant principles. Following the machine aesthetic, modernism rejected ornament for materials and geometry. The significance of materiality as ornament fully emerged and solidified during modernism.

Villa Savoye, Le Corbusier
Poissy, France 1929
[Modern, free plan, concrete and masonry]

The Villa Savoye exhibits the tenets of modernism not only through the utilization of Le Corbusier's five points, but also by using a different sensibility and language that was previously undetermined. This new movement relied on various aspects, including ideas of modern art and abstraction, new materials and technologies, and a general concept of the machine aesthetic. The Villa Savoye serves as a primary example of these concepts in terms of its planning, structure, and aesthetics. The building floats on a series of columns, illustrating new structural techniques. It includes the automobile in its planning. Its overall form and language are abstract and embody the new machine aesthetic. Historic ideas of weight, mass, and symmetry are transplanted with concepts of pure simple forms that seemingly float and dematerialize, allowing nature to penetrate into interiors that are markedly similar to the exterior.

Rationalism

The term rationalism refers to an architecture that is primarily based on reason and logic, rather than illusion or fantasy. During the early twentieth century, architects such as August Perret were experimenting with new ideas and materials, attempting to approach architecture in a much more structured method, resulting in buildings that were straightforward and honest. In Italy, the rationalist movement was thriving through the works and writings of the Gruppo Sette, a number of young Italian modernists who described their position as a middle ground between the classicism of the Novecento Italiano movement and the industrialized architecture of futurism. The rationalists understood the power of history and were not afraid to use it in their work and yet they remained fully versed in the tenets of modernism as practiced by Le Corbusier and other Northern Europeans. This particular movement, perhaps more than any other, had the ability to examine both classicism and modernism and combine the paramount virtues of each into an architecture that was engaged in the present, looked to the future, but also relied on the past. As a movement, it lasted a relatively short time, but produced remarkable results.

Casa del Fascio, Giuseppe Terragni
Como, Italy 1936
[Italian Rationalist, urban object, masonry]

The Casa del Fascio represents the most complete and compelling result of the rationalist movement. The simplicity of its cubic form and the rationalism of its exposed structural frame place it firmly within the theories of the modern movement; however, its relationship to earlier Roman houses and Renaissance palaces (formed primarily through the plan) mark it as a building sensitive to its locale and traditions. The building employs the use of proportional systems, geometry, materiality, and even constructed views as a way of engaging the site and the context. The building is often referred to as a masterpiece of propagandistic architecture in that it was seen as an elegant set piece for Fascist rallies during the 1930s. The building forms a perfect combination of the northern modernity of Le Corbusier and the more historic Novecento style preferred in the south. The language is strictly modern; however, the sensibility seems more historic and less reactionary.

Brutalism

Brutalism is a movement defined by its dominant, rough, blocky, and angular geometries and its typification of exposed and process-textured, cast-in-place concrete. The expressive forms emerged from a desired honesty in the building's construction and articulation. Revealed and celebrated building functions, in conjunction with celebrated tectonic articulation of the simple, durable material and its construction are hallmarks of the movement. This same bluntness of form, massiveness of building, opacity of figure, and roughness of materiality extended the modernist sensibility with a brutal honesty of simplified and overt expression.

MEANING	581
PRINCIPLE	

Whitney Museum, Marcel Breuer
New York, New York 1966
[Brutalism, terracing massing of galleries, concrete]

The Whitney Museum of American Art typifies brutalism with its massive presence. Made of cast-in-place concrete (covered in granite), its large, opaque figure is cloaked in detail-less anonymity, broken only by the figuration of one funneled window. Defined by massing that steps out above an open sunken sculpture court, the starkness of the form and the rawness of the exposed concrete make it a distinctly iconographic yet highly controversial building. The nature of the gallery, which requires opaque vertical surfaces, naturally lends itself to blank form, whereas the stacked and terracing nature of the form reveals the sectionally layered levels as a product of the density of the site, creating a vertical sequence and strata.

Hi-Tech

Hi-tech architecture expressively incorporates technology into detail and form. It returns to modernist principles for general building form and organization, but extends the expression and articulation of the technology. Typified by the revelation of structure, elaborate articulation of building systems, and a tendency to fetishize the componential nature of construction, high tech is clearly associated with the machine-like relationship of building performance. Though the origins of the design lie in functional practicalities and performative responses, the expression extends well beyond necessity to establish an overt identity. The articulation of the piece, the assembly, and the collective system establish a componential sensibility to the form. The overt expression of function through a machine tectonic results in a distinct iconography of formal legibility.

Centre Pompidou, Renzo Piano and Richard Rogers
Paris, France 1974
[Hi-Tech Modern, free plan museum, steel, glass, concrete]

There are few buildings that embody an architectural movement more precisely than the Centre Pompidou in its representation of hi-tech. The tenets of this architectural style are expressed throughout this museum. The overall expression and language of the building is articulated through the exaggerated formalism of the exposure of the various mechanical organizations of the architecture. Each of these systems is color coded and displayed as a methodology and ultimately a language and style. Ideas of hyper-functionality and mechanization, which reflect the architectural culture of the time, are revealed in the extremely complex and neo-machine aesthetic that pervades the entire complex. Concepts that organized historical or conventional architecture are completely dissolved in the Centre Pompidou, giving way to new ideas of space, structure, and systems and novel ways of displaying these various designs through highly articulated forms.

Postmodernism

Postmodernism emerged as a response to the reductivism of modernism. Rebelling against the removal of historical reference and the inhumanity of the abstract composition, postmodernism returned to an overt historical reference to aid in its formal composition and ornament. Whimsically drawing upon historical references with a collagist compositional sensibility, the eclecticism of formal reference with new materials and construction systems often left the forms compositionally awkward. The return to history allowed and required a deep understanding of the visual, stylistic, and referential meaning of history, which kept legibility at arm's length for the layman. The esoteric nature of forms devoid of historical significance left the movement with non-functional ornamentation, which alienated the user and undermined the intent of the movement.

Portland Building, Michael Graves
Portland, Oregon 1980
[Postmodern, block mass with decorative façades, various]

The Portland Building, as one of the icons of postmodernism, adopts a basic cubic form and clads it in abstracted historical references. Changes in material, historically referential patterns, and three-dimensionally figured application define the building's façade. The surface nature of the postmodernist application of architectural identity leaves the design as a two-dimensional application that fails to impact spatially. The translation of architecture into graphic appliqué advanced the movement dramatically as it fit well with a market demand for iconographic, corporate recognizability and was easily integrated into conventional building forms and construction practices. Ultimately, the application of esoteric architectural references left the style artificial and bankrupt.

Structuralism

Structuralism began as a reaction to the CIAM conferences that took place from the 1920s until the late 1950s. The structuralists could no longer support many of the tenets of the CIAM manifesto; they decided to move away from what they saw as the functionalist doctrine of the modernists and create a movement that was more sympathetic toward the users. Structuralism dealt with the idea of signs and systems and how these combined to develop an architecture that was both modern and also allowed for certain intangibles, which were seemingly frowned on by the functionalists. The structuralists did not believe that architecture could be configured like a science and that logic could not dictate how a problem might be solved. The first structuralists were members of Team 10, a splinter group from the CIAM conferences, formed in the late 1950s. These architects later split due to differences. The new brutalist movement was formed by Alison and Peter Smithson and the structural movement was formed by Dutch architect Aldo Van Eyck and others. The structuralists believed in a more human approach to urban planning and architecture. They felt that Le Corbusier and the modernists had gone too far in their desire for functionalism and logic. They felt that the concepts of a sense of place and user participation were critical if architects were going to be successful in designing for the masses.

Sonsbeek Pavilion, Aldo Van Eyck
Arnhem, Netherlands 1966
[Structuralism, parallel-walled pavilion, masonry]

The Sonsbeek Pavilion, a temporary structure, extolled many of the principles of the structuralist movement. Recognized by its plan, the composition appeared extremely ordered while remaining playful at the same time. The use of order and pattern as a design methodology infused with a sense of freedom was one of the hallmarks of structuralism. The Sonsbeek Pavilion achieved this by taking on the overall shape of a circle in plan with a modified rectangle placed in the center. The circle was a platform on which the rectangular pavilion was situated. The pavilion took the form of a series of parallel walls that were then manipulated to form hierarchical spaces where the sculpture and other functions took place. It was a remarkable building in that it combined certain aspects of modernism with a new, more casual design mechanism. Prior to structuralism, modernism was being cast as a purely functional architecture. The pavilion was constructed of concrete block, which accentuated the curves of the walls at hierarchical moments.

Post-Structuralism

Post-structuralism in architecture represents a movement that understood and valued structuralism, but profoundly disagreed with many of its tenets. Post-structuralists were never really able to form a cohesive group because not only did they believe in critical thought, but also in criticizing others. Aside from their beliefs in structuralism and their advancement beyond the movement, there was little to hold them together. They shared the understanding that there was no right or wrong, but everything was relative. This anti-humanist view informed much of the work, resulting in an interesting, albeit brief, period of production. To the post-structuralists, everything should be challenged. Ideas that had been part of architecture since the beginning were now being questioned. Post-structuralism was intrinsically suited for philosophical conjecture and discussion, not for the practice of architecture, which required a reliance on absolutes innate to physicality.

MEANING PRINCIPLE | 589

House III, Peter Eisenman
Lakeville, Connecticut 1971
[Post-structural, rotated geometries, wood and stucco]

House III was an exercise in geometries and how the collision of two separate geometries might form a third condition that existed between the two original conditions. One of the original geometries was a square that was subsequently divided into three separate bays, which collided with a rectangle that contained its own systems and hierarchies. This parti forced Eisenman to negotiate the collisions and the results. The house was instrumental in challenging typical modern ideas of function and structure. It was purposefully complex and broke free of constraints placed on architecture by modernism, functionalism, and even the simplicity of structuralism.

Deconstructivism

Deconstructivism (a subset of post-modernism) was the final stage of architecture prior to the introduction of the computer. Deconstructivism is characterized by elements of fragmentation and distortion. This architecture is consumed by the use of non-rectilinear shapes and the constant questioning of typical architectural elements such as structure, enclosure, function, and history. History has little meaning for the deconstructivist; instead, all must be invented, often from very little. The deconstructivists, like the structuralists before them, did continue the tradition of looking to other fields as a way of furthering architecture and thereby achieving new forms and ideas. Deconstructivism began with the 1988 exhibition of Deconstructivist Architecture at the New York Museum of Modern Art, organized by Philip Johnson and Mark Wigley. It soon became the style of every school and seemed to dominate the architectural landscape. It is not an absolute architecture like most movements that came before it; instead it constantly changes with minimal theoretical underpinnings, making it difficult to critique in that its only rule is that there are no rules.

MEANING | PRINCIPLE | **591**

Parc de la Villette, Bernard Tschumi
Paris, France 1987
[Deconstructivist, iterative nodal grid field, metal]

This project is especially interesting because it was designed by Bernard Tschumi, who, up until this point in his career, was more of a theoretician than a designer. The project is a gridded series of bright red pavilions, which contrast and engulf the surrounding context. The grid informed the project so that the pavilions could be different and yet form one complex. Each pavilion pays tribute on some level to the Russian constructivists of the 1920s, while at the same time accelerating a deconstructivist agenda. The overall order and pattern of the red cubes seems to present both structuralist and deconstructivist tenets. Tschumi was continuing from what he felt was architecture's last zenith, excusing some of the experiments of the 1970s and early 1980s. The project was instrumental in its engagement with the landscape and its incredibly large scale. Its sense of ordered chaos will always position it as one of the essential deconstructivist projects.

Regional Modernism

Regional modernism combines modernist principles with local contextualism, climatic and cultural responses, and vernacular forms. Using locally available resources and traditions to address regional needs, circumstances, and traditions, regional modernism takes the global principles of modernism and adapts them to reflect the environmental, cultural, and historical contexts in which the architecture is built. This hybridization allows for the generic formalism of the modernist movement to locally adapt, creating waves of geographically based subcategories. Global cultural diversity allows for localized responses laden with local traditions and specific climatic and material responses. Within the United States, there are diverse, emergent, regional zones in which each has adapted and integrated their specific localities. These hybridizations of vernacular historical principles and modernist sensibilities establish distinct families of regional modernism tailored to location, yet integrally joined with broader trends. The mixture allows for a local sensitivity while obtaining a synthetic, updated ethic of contemporary culture and technology.

Marika-Alderton House, Glenn Murcutt
Northern Territory, Australia 1994
[Regional Modernism, vernacular formed house, wood and metal]

The Marika-Alderton House, even in its very small footprint, uses an evolutionary spatial planning. The simplicity of form derives from a vernacular understanding of culture and climate. Derived from a synthesis of practicality with responsiveness to the specifics of place, the form bridges the principles of modernism with the local necessity of the region. The public spaces are organized in a single free-plan room that allows for diverse configurations and interconnections of its disparate programs. The private zone, in opposition to the fluid public space, is highly regimented, deploying a single-loaded corridor to connect and access the clearly bounded and specifically segmented bedrooms and bathroom. The combination of the two allows for their juxtaposition: public vs. private, day vs. night, fluid vs. fixed, and open vs. closed. This simple program of a single-family house responds to its site, creating a sophisticated response to place and building.

Globalism

Globalism is a movement that extends the influence and practice of individual architects across a global stage. Ushered by the emergence of a single global economy with a dominant Western vision of architecture, the architect with celebrity status emerged. Individual practitioners with signature styles make brand-name buildings, which serve as commodities. Their identity alone validates the architecture, ensuring publicity, attention, and significance regardless of quality, appropriateness, or design. This idea of a practice that emerges out of a personal style and formalism and then reoccurs regardless of geography, culture, and place was heralded by the success of Frank Gehry's Guggenheim Bilbao. This concept of celebrity now extends to a collection of architects such as Zaha Hadid, Rem Koolhaas, and Norman Foster. The result is a predictable product and the further advancement of select individuals and anomalous buildings over shared principles and cultural ethics. The premise of this movement forces and rewards extremity and differentiation regardless of reason, rationale, or effect. The result is a contemporary architectural discourse that focuses on a small, eclectic, and ultimately esoteric group of architects. The identity-based building emphasizes brand over place, space, experience, and all the fundamental principles of architecture.

CCTV Building, Rem Koolhaas, OMA
Beijing, China 2009
[Postmodern, linked tower, metal and glass]

The CCTV has a dynamic tower form that climbs, bridges, and folds back down. An external, diagonal, structural grid creates a variably scaled pattern that emerge out of the variable structural forces of the irregular building form. Built in China by the internationally renowned Dutch architect Rem Koolhaas, the CCTV typifies in image, intent, and execution the globalist principles of the practice, iconography, identity, and brand as architectural trademarks.

anthropomorphic scale. The use of the scale of the human body relative to an architectural component or composition.

architecture parlante. The concept of buildings that explain their own function or identity. Literally translated as "speaking architecture."

architrave. The lintel or beam that rests on the capitals of classical columns. *See also* **lintel**.

axis mundi. A formal hierarchy through which there is a symbolic connection where travel and correspondence can be made between higher and lower realms (heaven and earth). This is typically accomplished through a strong vertical figure and/or and unoccupiable axis, which is reserved for a spiritual, not physical, occupation.

bay. A unit of form in architecture. This unit can be defined in myriad ways based upon the compositional or material technique. Traditionally it is defined as the repetitive zone between a series of columns, pilasters, or posts.

belvedere. Italian for "beautiful view." An architectural structure built in an elevated position (in the upper part of a building or as a separate structure) to provide lighting and ventilation and to command a fine view. Typically roofed but open on one or more sides, it often assumes the form of a loggia, or open gallery. *See also* **loggia**.

béton brut. French term referring to concrete as it appears when the formwork is removed, revealing a surface reflects the form joints, wood grain, and fasteners around which it was poured. *See also* **formwork**.

brise-soleil. Refers to a variety of permanent sun-shading techniques. Typically, they are based in a horizontal projection extending from the sun-side facade of a building to prevent areas of glass from being exposed to direct light and overheating.

caldarium. In a Roman bath complex, a room with a hot plunge bath. Typically heated by an underfloor heating system (a hypocaust) this was the hottest room in the regular sequence of bathing rooms. After the caldarium, bathers would progress back through the tepidarium to the frigidarium. *See also* **frigidarium** and **tepidarium**.

cella. Either the principal enclosed chamber of a classical temple or the entire central structure of a classical temple.

cladding. A material used as an external wrapping; typically used as a durable and often weather-proof layer.

columbaria. A place for the respectful and usually public storage of *cinerary urns* (urns holding a deceased's cremated remains).

crenellation. A pattern along the top of a wall, most often in the form of multiple, regular, rectangular extensions between which arrows or other weaponry may be shot (especially as used in medieval European architecture).

cross-section. A section made by a plane cutting transversely, typically perpendicular to the longest axis.

cupola. A light structure on a dome or roof, serving as a belfry, lantern, or belvedere. *See also* **belevedere**.

datum. Any system that maintains regular continuity used as a reference in measuring or experiencing more divergent forms.

dendriform. Having the shape or form of a tree.

Doric peripheral temple. A temple type that consists of a rectangular floor plan with a series of low steps on every side and a colonnade of Doric columns extending around the periphery of the entire structure.

drum. A cylindrical or faceted construction supporting a dome.

earthworks. A military construction formed chiefly of earth for protection against enemy fire and assault; used in both offensive and defensive operations.

ETFE. An abbreviation for *ethylene tetrafluoroethylene*. A fluorine-based plastic with high corrosion resistance and strength over a wide temperature range, making it a prominent emerging building material.

façade. Any face of a building finished accordingly. Typically these are front facing and engage a public passageway or space.

fenestration. A reference to openings in a structure that includes the design and/or collective position of these openings within a building's enclosure. Openings include windows, doors, louvers, and vents, for example.

field. An expanse of anything, typically consisting of unit, arrayed through regularized increments of repetition.

focal cone. The visual region displayed by the eye or a drawing that relates to a person's normal vision peripheral vision removed. It is the area of sight or the angle of sight. Also known as the "cone of vision."

formwork. The temporary or permanent molds into which concrete or similar liquid materials are poured.

frigidarium. A chamber containing a cold pool in a Roman bath, entered after the caldarium and the tepidarium, which were used to open the pores of the skin. The cold water would close the pores. *See also* **caldarium** and **tepidarium**.

genius loci. The spirit of a place.

giant order. An order whose columns or pilasters span two (or more) stories.

grain. The self-similar direction of a series of elements that create a collective pattern.

grid. A network of horizontal and perpendicular lines, uniformly spaced.

hypostyle hall. A roof that is supported by columns. Typically it uses a large field of oversized and densely spaced columns that, through their repetition, produce a drama and hierarchy.

iris. A thin, circular structure in the eye, responsible for controlling the diameter and size of the pupils and thus the amount of light reaching the retina. Here the term refers to an architecturalization of this natural system in a mechanized aperture.

jack arch. A structural element in masonry construction that provides support at openings in the masonry. Also known as a "flat arch" or a "straight arch," the jack arch is not semicircular in form, but instead is flat in profile and used similarly to a lintel. *See also* **lintel**.

lintel. A structural, horizontal block that spans the space or opening between two vertical supports. An element of post and lintel construction.

loggia. A space within the body of a building open to the exterior on one side, serving as an open-air room or as an entrance porch. It is often associated with a gallery or arcade open to the air on at least one side.

martyrium. A place where the relics of a martyr are kept.

mat building. A composition defined by the organization and propagation of spatial and programmatic modules on a relatively flat plane. The organization and connection of the modules are rule-based and the propagation within the site is inevitably affected by the rules of organization. The resulting compositional aggregation creates a cohesive horizontal and multidirectional singularity despite the modular parsing.

metopes. Rectangular architectural elements that fill the space between two triglyphs in a Doric frieze, which is a decorative band above the architrave of a building of the Doric order. *See also* **architrave**.

module. A separable component, or unit, for assembly into compositions of differing size, complexity, or function.

monocoque. A construction technique that supports structural load by using an object's external skin.

mullion. A vertical element that forms a division between units of a window, door, or screen and can be used for functional or decorative subdivisions.

multivalent field. A field that extends in two or more directions.

nogging. The infill used between timber framing.

oculus. A circular opening, especially one at the apex of a dome.

opisthodomos. A small room in the cella of a classical temple, typically used for a treasury. Also called a *posticum, See also* **cella**.

palimpsest. Something reused or altered but still bearing visible traces of its earlier form. A superimposition of multiple layers that allows for traces of both the new and the old.

parterre. A formal garden built on a level surface consisting of geometrically formal planting beds, edged in stone or tightly clipped hedging, and crosscut by gravel paths typically arranged in symmetrical patterns.

piano nobile. The principal story or main floor of a large building such as a palace or villa.

pilaster. An engaged or partially embedded column directly attached to the wall surface.

piloti. A column that elevates a building, detaching it from the ground plane and allowing the removal of the structural responsibility of the wall.

plinth. The base or platform upon which a column, pedestal, statue, monument, or structure rests.

poche. French for "pocket." A representational technique, typically used in plans or maps, in which materials such as walls, columns, or urban fabric are completely filled in (usually in black), visually providing a dominance to the figure ground of space and object.

porte-cochere. A porch or portico-like structure at an entrance to a building through which a horse and carriage or motor vehicle can pass in order for the occupants to enter under cover, protected from the weather.

program. The functional use of a specific space.

proportion. (1) A comparative relation of quantitative attributes such as size, measurement, or quantity, for example. (2) A ratio.

quoin. The cornerstone of masonry walls. They may be either structural or decorative and are often used to imply strength and firmness to the outlining figure of a building.

radial geometry. Forms or geometries that use organizational structures that radiate from a central point.

rip rap. Rock or other material used to protect shorelines, streambeds, bridge abutments, pilings, and other shoreline structures against scour, water, or ice erosion. This material system can and has been translated to architectural tectonic and material applications.

sallyport. A secure, controlled entryway, as employed at a fortification or a prison. This term evolved, however, to include large ceremonial openings in buildings that create a threshold to a complex or adjacent space.

spandrel. In a building with more than one floor, the space between the top of the window in one story and the sill of the window in the story above. The term is typically employed when there is a sculpted panel or other decorative element in this space, or when the space between the windows is filled with opaque or translucent glass, in this case called *spandrel glass*.

stringcourse. A horizontal band or course, as of stone, projecting beyond or flush with the face of a building, often molded and sometimes richly carved.

tectonic. Of or pertaining to building or construction and the techniques of fabrication and assembly that determine architecture.

tepidarium. The warm (tepidus) bathroom of the Roman baths, typically heated by an underfloor heating system (a hypocaust). The specialty of a tepidarium was the pleasant feeling of constant, radiant heat, which directly affected the human body through the walls and floor.

travertine. A form of limestone with a fibrous or concentric appearance and existing in white, tan, and cream-colored varieties.

triglyph. The three angular channels (two perfect and one divided) of the Doric frieze. The rectangular, recessed spaces between the triglyphs on a Doric frieze are called metopes. *See also* **metopes**.

voussoir. A wedge-shaped stone or brick that is used with others to construct an arch or vault.

wythe. A continuous vertical section of masonry one unit in thickness.

INDEX

Numbers
21st Century Museum of Contemporary Art, 245
885 Third Avenue, Lipstick Building, 143

A
Aalto, Alvar, 24, 177, 249, 401, 524
Aalto Summer House, 401
Abe, Atelier Hitoshi, 369
Abraham, 373
The Acropolis, 539
Adirondack House, 287
Adler and Sullivan, 236
AEG Turbine Factory, 242
Air Force Chapel, 318
Alberti, Leon Battista, 14, 79, 179, 325, 334, 495
Alcatraz Prison, 211
Alhambra, 447
Altes Museum, 85, 108, 193, 335
American Cement Building, 479
Ando, Tadao, 383
Andrews and LeBlanc, 294
Apollodorus of Damascus, 218
Applied ornamental
    historical, 413, 438–441
    performative—mechanical, 413, 426–429
    performative—structural, 413, 422–425
    referential—organicism, 413, 430–433
    referential—structural, 413, 434–437
    religious, 413, 418–421
Architrave, 419
Art Nouveau, 552, 570–571
Arts and Crafts, 552, 572–573
Asilo Sant'Elia, 207, 321, 323
Asplund, Gunnar, 139, 184, 314
Assemblages, 92–105
Asymmetry, 331, 338–339
AT&T Building, 102–103
Athens, Greece, 35
Atlantic Center for the Arts, 53
Atticus, Herodes, 213
Aurelian Wall, 275
The Avenue de Champs-Élysées, 26
Axial formalism, 233
Axial hierarchy
    architecture, 354–355
    defined, 349
    urban, 356–357
Axis mundi, 134
Azuma House, 91

B
Baeza, Alberto Campo, 127
Baker House, 24
Baldacchino, 431
Baltimore, Maryland, 35
Ban, Shigeru, 319, 531
Barcelona Block, Spain, 129, 157
Barcelona Pavilion, 45, 302, 342, 525
Bari, Italy, 37
Baroque, 551, 564–565
Barragan, Luis, 518
Barrio Troglodyte, Guadix, Spain, 267
Barry and Pugin, 567
Basilica at Paestum, 110
Basilica Giulia, 76
Basilica of Old St. Peter, 76
The Basilica, 340
Baths of Caracalla, 116, 542
Bauhaus, 207
Bavinger House, 57, 163
Behrens, Peter, 242
Beinecke Rare Book Library, 204, 351, 481
Bellevue Arts Museum, 521
Bentham, Jeremy, 11, 211, 367
Bergstrom, George, 155
Berlin Holocaust Memorial, 505, 529
Berlin Philharmonic Hall, 33
Bernini, Gian Lorenzo, 144, 147, 174, 261, 366, 431, 523
Bibliotheque Nationale de France, 531
Bibliotheque St. Genevieve, 111, 202, 439
Biddulph Family, 415
Bilateral symmetry
    architectural elevation, 334–335
    architectural plan, 332–333
    defined, 331
    urban, 336–337
The Bird's Nest Stadium, 143, 303, 425
Blobwall, 409
Bo Bardi House, 60
Bo Bardi, Lina, 60
Bordeaux House, 61, 261
Borden, Gail Peter, 247
Borden Partnership, 369
Borromini, Francesco, 80, 165, 173, 565
Boston Public Library, 203
Botta, Mario, 449
Boullée, Étienne-Louis, 234, 312, 316
Box in box, 59
Bramante, Donato, 10, 74, 134, 180, 522
Breuer, Marcel, 581
British Museum, 194
Brother Claus Field House, 83, 279
Brown, Venturi Scott, 239
Bruges, Belgium, 132
Brunelleschi, Filippo, 13, 40, 172, 320, 324, 344, 362, 483
Brutalism, 552, 580–581
Buckhead Library, 205
Bullfinch, Charles, 372
Bunker, 269
Bunshaft, Gordon, 137, 204, 351, 481
Buonarroti, Michelangelo, 11, 15, 75, 333, 344, 497, 545
Burbage, James, 214
Burgee, John, 223
Burnham, Daniel, 148
Burnham and Root, 219, 273, 392
Butzer, Hans and Torrey, 528

C
Caltrans District 7 Headquarters, 405
Campo, Sienna, 161
Campus typology, 169, 226–229
Canberra, Australia, 161
Candela, Felix, 394
Capsule House K, 48

Carré d'Art, 239
Carson Pirie Scott Building, 220
Casa da Musica, 217
Casa del Fascio, 107, 129, 221, 579
Casa Gaspar, 127
Casa Gilardi, 518
Casa Mila, 341
Case Study House No. 22, 189
Case Study Houses, 91
Castel del Monte, 158
Castle of Canossa, 269
Castle of Lastours, 260
Castle Sant Angelo, 271
Cathedral of Our Lady of the Angels, 83, 177
CCTV Building, 511, 595
Ceausescu, 363
Central, 7
Centralized systems, 6–17. *See also* Organizational systems
   central, 7
   central with axis, 7
   centrifugal, 6
   centripital, 7
   circle, 7, 10–11
   defined, 4
   diamond, 6
   Greek Cross, 7, 14–15
   Latin Cross, 7, 12–13
   radial, 6, 16–17
   radial geometry, 6
   square, 6, 8–9
   types of, 6–7
Centre Pompidou, 197, 240, 371, 429, 583
Centrifugal, 6
Centripital, 7
Ceremonial sequence, 535, 538–539
The Chancellery, 355
Chandigarh, 100–101, 109, 351
The Chapel of Thanksgiving, 163
Charreau, Pierre, 60, 277, 386
Chartres Cathedral, 77, 171, 273, 423, 559
Chengdu Building, 369
Chinese Courtyard House, 88
Chippendale furniture, 102–103
Chipperfield, David, 487
Chunk, 445, 488–493
Church
   facade, 72
   lineage, 77
Church of San Giovanni Battista, 449
Church of San Lorenzo, 344
Church on the Water, 383
Ciel Rouge Creation, 265
Circle, 7, 10–11
   defined, 125
   as figure / form, 299, 312–313
   as figure / form—architecture, 134–135
   as figure / form—urban, 136–137
   as space / void, 300, 316
   as void—architecture, 138–139
   as void—urban, 140–141
Circle-to-drum, 300, 314–315
The Circus, 141

Clark and Menefee, 343
Classicism, 551, 554–555
Cleveland, Ohio, 133
Cologne Cathedral, 78
Color
   hierarchy, 349, 370–371
   pattern, 444, 458–463
   perception, 515, 518–519
Comparatives, 106–119
Computational complexities, 301, 326–327
Concrete
   detail, 402–403
   form—geometry / system / pattern / ornament, 382–383
   structure, 394–395
Context
   approach to, 285
   cultural, 285, 294–295
   defined, 284
   historical, 285, 290–291
   material, 285, 292–293
   natural, 285, 286–287
   types of, 285
   urban, 285, 288–289
Context formalism, 232, 238–239
Contino, Antonio, 210
Cortona, Pietro, 358
Courtyards, 89
Cram, Ralph Adams, 228
Crenellation, 223
Crowe, Frank, 279
Crystal Cathedral, 406
Crystal Palace, 404
Cultural context, 285, 294–295
Cultural precedent, 67
Curtain Wall House, 531

D
da Sangallo the Younger, Antonio, 106, 130, 181, 306
da Vignola, Giacomo Barozzi, 72, 86, 87, 164, 172, 183, 250, 544
Danteum, 39, 323, 539
The de Menil Collection, 199, 289
de Montereau, Pierre, 516
de Sully, Maurice, 421
de Toledo, Juan Bautista, 128
de Vauban, Marquis, 157
De Young Museum, 63, 385, 451
Deconstructivism, 553, 590–591
Dendriform, 221
Design 3, 383
Detail
   concrete, 402–403
   glass, 406–407
   masonry, 400–401
   material, 376
   metal, 404–405
   plastic, 408–409
   wood, 398–399
D.G. Bank, 373
di Batolomeo, Michelozzo, 179, 561
di Cambio, Arnolfo, 12, 13
Diamond, 6
Diamond Ranch High School, 209, 327

Dinkeloo, Roche, 225
Disney Concert Hall, 217, 385
Disorganized fields, 46–47, 52–53
Dispersed fields
   defined, 5, 46–53
   disorganized, 46–47, 52–53
   organized, 46, 48–51
Distorted geometries, 301, 326–327
DMJ and Mendenhall, 479
Dominus Winery, 293, 457
Double-loaded
   architecture, 24–25
   illustrated, 18
   urban, 26–27
du Cerceau, Baptiste, 337
Duomo, 459

E
Eames, Ray and Charles, 41, 188
Eames House, 41, 188
East Wing of the National Gallery, 149, 198, 244, 293
Eberswalde Library, 441
Ecology, materials, 377
Eden Project, 409
Edinburgh International Conference Center, 315
Eiffel, Gustave, 365
Eiffel Tower, 365
Einstein Tower, 333
Eisenman, Peter, 32, 51, 191, 505, 529, 589
Elevational poche, 264–265
Elgin Marbles, 419
Ely Cathedral Dome, 159
Ennis House, 185
Entablature at the Parthenon, 435
Entablature of Basilica Vicenza, 467
Environment, perception, 515, 530–531
Esherick House, 190
Etruscan House, 88
EUR, 537
Exeter Cathedral, 78
Exeter Library, 9, 204, 235, 317, 381, 473
Experience Music Project, 408
Experiential sequence, 535, 540–541
Experimental formalism, 233, 248–251
Expo Pavilion, 305

F
Facades
   church, 72
   parallel evolutionary, 87
Fagus Works, 575
Fallingwater, 286
Farnsworth House, 39, 188, 387, 396
Fenestration, 384
Fenice Theater, 214
Ferrell, Sir Terry, 315
Figure / ground
   defined, 254
   Nolli map of Rome, 254, 258–259
   poche(s), 254–255, 260–281
   positive / negative, 256–257

Final Wooden House, 391
Flatiron Building, 148
Florence Baptistery, 154
Florence Duomo, 362
Florentine House, 89
The Forbidden City, 336, 355
Form
   for categorization, 232
   contextual formalism, 232, 238–239
   as essential, 233
   experimental formalism, 233, 248–251
   functional formalism, 232, 236–237
   geometric formalism, 233, 244–245
   grid, 31, 36–37
   material formalism, 233, 246–247
   organizational formalism, 233, 242–243
   performative / technological formalism, 232, 240–241
   platonic formalism, 232, 234–235
   sequential formalism, 233, 250–251
Form—geometry / system / pattern / ornament
   concrete, 382–383
   defined, 376
   glass, 386–387
   masonry, 380–381
   metal, 384–385
   plastic, 388–389
   wood, 378–379
Formal / geometric hierarchy
   architecture, 350–351
   defined, 348
   urban, 352–353
Formwork, 382
Fort McHenry, 165
Fort William, Calcutta, 166
Fort Worth Water Gardens, 527
Foster, Norman, 239
Fragnano Olona Elementary School, 208
Frank Gehry House, 379
Free plan, 42–45
   courtyard and, 59
   defined, 5
   example types of, 42–43
   types of, 42–43
   with voids, 59
Freeman House, 307, 455
Frequency, 445, 500–505
Ftown Building, 369
Fujimoto, Sou, 391
Fuller, Buckminster, 305, 475
Functional formalism, 232, 236–237
Functional poche(s), 255

G
Gallaratese Housing, 21
Gamble House, 399, 573
Garage de la Societe, 224
Garden House, 151
Garnier, Charles, 215, 540
Gaudi, Antoni, 304, 341, 463
Gehry, Frank, 33, 53, 62, 201, 217, 373, 379, 385, 408, 519
Geometric / form typology, 124–167

circle, 125, 134–141
defined, 123, 124
oval, 125, 142–147
polygon, 125, 154–161
spiral, 125, 162–163
square, 125, 126–133
star, 125, 164–167
triangle, 125, 148–153
Geometric formalism, 233, 244–245
Geometry / proportion
circle as figure / form, 299, 312–313
circle as space / void, 300, 316
circle-to-drum, 300, 314–315
distorted geometries and computational complexities, 301, 326–327
golden section, 300, 322–323
point / line / plane, 299, 302–303
root rectangles, 300, 320–321
three-dimensional volume / mass, 299, 304–305
triangle, 300, 318–319
two-dimensional and three-dimensional module form, 300, 324–325
two-dimensional module in plan, 299, 306–307
two-dimensional square module as space / void, 299, 310–311
two-dimensional square module in elevation as figure / form, 299, 308–309
use of, 298
German Architecture Museum, 62
The Getty Center, 417
The Getty Villa, 359, 547
Gigon and Guyer, 527
Gila Cliff Dwellings, 267
Ginzburg, Moisei, 21
Glasgow School of Art, 206, 338
Glass
detail, 406–407
form—geometry / system / pattern / ornament, 386–387
Glass House, 45, 187, 280
Glass Pavilion, 135, 387
Globalism, 553, 594–595
The Globe Theater, 214
Goff, Bruce, 57, 163
Goldberg, Bertrand, 395
Golden section geometry / proportion, 300, 322–323
Goody and Hamilton, 388
Goreme, Cappadocia, Turkey, 266
Goryokaku Fort, Hakodate, Japan, 167
Gothic, 551, 558–559
Graham, John, 365
Graves, Michael, 291, 585
Great Wall of China, 271
The Great Temple of Amun at Karnak, 38
Greek Cross, 7, 14–15
Greene and Greene, 399, 573
Grids, 30–41
defined, 5
form, 31, 36–37
module, 31, 40–41
position, 30, 32–35
relative reading of, 30
structure, 31, 38–39
uses for, 30
Grimshaw, Nicholas, 409

Gropius, Walter, 207, 575
Guggenheim Bilbao, 62, 201
The Guggenheim Museum, 194, 243, 350, 546
Guimard, Hector, 433

H
Habitat 67, 41, 489
Hadid, Zaha, 327
Hadrian, 271
Hadrian's Villa, 52
Hall of Records, 485
Harajuku Church, 265
Harrison, Abramovitz, and Abbe, 397
Harvard University, 227
Hauer, Erwin, 383
Haussmann, Georges-Eugène, 29, 357, 537
Heatherwick Studio, 303
Heidi Weber Pavilion, 397, 461
Herzog and de Meuron, 63, 143, 223, 293, 295, 303, 385, 425, 441, 451, 457, 469, 477, 503, 519
Hierarchical formalism, 233
Hierarchical poche, 276–277
Hierarchy
axial, 349, 354–357
color, 349, 370–371
formal / geometric, 348, 350–353
material, 349, 372–373
programmatic / functional, 349, 368–369
visual / perceptual, 349, 358–359
Hierarchy of monument, 349, 364–365
Hierarchy of scale
architecture, 360–361
defined, 349
urban, 362–363
Hierarchy of visual control, 349, 366–367
High Museum of Art, 199, 315, 547
The Hirschhorn Museum, 137
Historical context, 285, 290–291
Historical precedent, 67
Hi-Tech, 553, 582–583
HOK, 197
Holl, Steven, 25, 249, 517, 521
Holocaust, Berlin, 51
Homewood, 276
Hoover Dam, 279
Horizontal sequence, 535, 536–537
Horta, Victor, 571
Horta House, 571
Hotel de Beauvais, 150
Hotel Guimard, 112
House, 22, 49
House III, 32, 589
House of Education, 98–99
House typology, 168, 182–191
House X, 191
Houses of Parliament, 567
Howe and Lescaze, 222
Huang, Qin Shi, 271
Hybrid systems, 58–63
defined, 5
example types of, 59

I

Il Redentore, 73, 117, 173
Il Gesu, 72, 172
Il Teatro del Mondo, 216
Illinois Institute of Technology, 229
Imai Daycare Center, 319
Indeterminate poche, 274–275
Indianapolis, Indiana, 133
Industrialism, 552, 574–575
The Inn at Middleton Place, 343
Intellectual precedent, 67
Ito, Toyo, 127, 245

J

Jackson, Cywinski, 287
Jacobsen, Arne, 208
Jahn, Helmut, 145
Jefferson, Thomas, 84, 113, 115, 145, 228, 499, 521
John Soane House, 193
Johnson, Philip, 15, 45, 98–99, 102–103, 143, 163, 187, 195, 223, 280, 353, 406, 527
Johnson Wax Building, 221
Jones, Fay, 82, 287, 390
Judd, Donald, 49

K

K-25 Plant, 237
Kaaba, 373
Kahn, Louis, 9, 151, 190, 196, 204, 235, 237, 288, 317, 345, 359, 381, 403, 473, 526, 530
Kalkriese Archaeological Group, 527
Karlskirche, 174
Katsura Imperial Villa, 398
The Kennedy Center, 215
The Kimbell Art Museum, 196, 526
Kings Road House, 185
Kirche am Steinhof, 175
Knights of Columbus Building, 343
Koenig, Pierre, 189
Koolhaas, Rem, 57, 61, 63, 191, 200, 205, 217, 261, 263, 291, 317, 368, 511, 543, 595
Kunsthal, 200
Kunsthaus Bregenz, 9, 200, 247, 407
Kunsthaus Graz, 201
Kurokawa, Kisho, 48, 325, 493

L

L House, 389
La Bourse de Commerce, 137
La Madeleine, 175, 569
La Sainte-Chapelle, 516
Laban Dance Centre, 519
Labrouste, Henri, 111, 202, 439
Lakeuden Risti Church, 177
Las Vegas Strip, 27
Latin Cross, 7, 12–13
Lautner, John, 155
Le Corbusier, 23, 25, 44, 82, 100–101, 109, 119, 131, 176, 186, 246, 251, 263, 265, 310, 311, 322, 326, 351, 380, 382, 397, 402, 427, 461, 517, 543, 577
le Duc, Viollet, 421

le Pautre, Antoine, 150
Ledoux, Claude Nicolas, 17, 112
Lemercier, Jacques, 337
L'Enfant, Pierre, 28, 353, 357, 364
Lenin's Mausoleum, 270
Lequeu, Jean-Jacques, 313
Library typology, 169, 202–205
Light, 516–517
Light, perception, 515
Light Frames, 247
Lineages, 69–91
Linear systems, 18–29
    defined, 4
    double-loaded, 18, 24–27
    point-to-point, 18–19, 28–29
    single-loaded, 18, 20–23
    types of, 18–19
L'Institute du Monde Arab, 405
The Lion Gate of Mycenae, 274
Lipstick Building, 143
Lloyd's of London, 241, 281
Local symmetry, 331, 340–341
Looping experience, 55
Loos, Adolf, 56, 187, 251, 292, 309, 335, 341, 367, 545
Los Manantiales, 394
Louvre, 235, 319, 407
Lovell Beach House, 186
Loyola Law School, 519
Lutyens, Edwin, 276
Lynn, Greg, 409

M

Machado and Silvetti, 359
MacKay-Lyons, Bryan, 49
Mackintosh, Charles Rennie, 206, 338
Maderna, Carlo, 75
Main Temple at Chichen Itza, 278
Maison Carrée, 71, 171
Maison de Verre, 60, 277, 386
Maisons Jaoul, 380
Majolika House, 465
Malin Residence, 155
Mannerism, 551, 562–563
Mannheim, Germany, 156
Mansart, Jules Hardouin, 140
Manzana de Chinati, 49
Marie Tjibaou Cultural Center, 295, 391
Marika-Alderton House, 61, 593
Markets of Trajan's Forum, 218
Marne School of Architecture, 209
Marshall Fields Building, 219
Marshall House, 131
Masonry
    detail, 400–401
    form—geometry / system / pattern / ornament, 380–381
    structure, 392–393
Massachusetts State House Dome, 372
Mat building design, 63
Material
    construction—ornamental, 412, 414–417
    context, 285, 292–293

detail, 376, 398–409
ecology, 377
form—geometry / system / pattern / ornament, 376, 378–389
formalism, 233, 246–247
hierarchy, 349, 372–373
meaning of, 377
pattern, 444, 452–457
perception, 377, 515, 524–525
poche, 255, 278–280
precedent, 67
program, 377
structure, 376, 380–397
symmetry, 331, 342–343
typology, 123
McKim, Mead, and White, 203, 339
Meaning
color / association / history, 377
theory / themes / manifestos, 550–595
Medieval Wall, 275
Meduna Brothers, 214
Meier, Richard, 199, 315, 417, 547
Melnikov, Constantine, 135
Melnikov House, 135
Memory / history, 525, 528–529
Mendelsohn, Erich, 333
Mercedes Benz Museum, 149
Metal
detail, 404–405
form—geometry / system / pattern / ornament, 384–385
structure, 396–397
Metopes, 419
Milam House, 507
Military / defensive poche, 255, 268–269
Miralles, Enrique, 491
MIT Chapel, 393
Modernism, 552, 576–577
Module, 445, 494–499
Module, grid, 31, 40–41
Module / bay, 445, 482–487
Moller, Villa, 56
Monadnock Building, 219, 273, 392
Moneo, Rafael, 83, 177, 289, 381, 529
Monocoque, 388
The Monsanto House of Future, 388
Monument to Vittorio Emmanuele II, 363
Monumental poche, 255, 270–271
Moo Moo, 389
Morphosis, 209, 327, 405
Morris, William, 471
Mosque, 100–101
Mozart Place, 384
Mullion, 378
Munkegards School, 208
Murcia Town Hall, 289
Murcutt, Glenn, 61, 593
Museum, 104–105
The Museum of Modern Art, 238
Museum of Modern Literature, 487
Museum of Roman Art, 529
Museum typology, 169, 192–201

N
Nakagin Capsule Tower, 325, 493
Narkomfin, 21
National Air and Space Museum, 197
The National Mall, 353
The National Museum of Roman Art, 381
Natural context, 285, 286–287
Nelson-Atkins Museum of Art, 249
Neoclassicism, 552, 568–569
Neo-Gothic, 551, 566–567
Netsch, Walter, 318
Neue Nationalgalerie Museum, 126, 195, 307
Neue Staatsgalerie, 85, 139, 198
Neuf-Brisach, France, 157
Neutra, Richard, 485
New Haven Parking Garage, 225
New House, 415
New Museum of Contemporary Art, 361, 509
New Orleans House, 90
New Orleans, Louisiana, 34, 36
New York, New York, 37
Newton's Cenotaph, 234, 312, 316
Nolli map of Rome, 254, 258–259
Notre Dame, 421
Notre Dame du Haut, 82, 176, 263, 517
Notre Dame du Raincy, 81, 176
Nouvel, Jean, 405
Nuclear reactor, 100–101

O
Odeon of Herodes Atticus, 213
Office / high-rise / commercial typology, 169, 218–223
Office dA, 401
Oklahoma City Memorial, 528
Olivetti Showroom, 403
Olmstead, Frederick Law, 229
OMA, 511, 543, 595
Organizational formalism, 233, 242–243
Organizational systems
centralized, 4, 6–17
defined, 4
dispersed field, 5, 46–53
free plans, 5, 42–45
grid, 5, 30–41
hybrid, 5, 58–63
linear, 4, 18–29
methods of composition and design, 4
raumplan, 5, 54–57
types of, 4
Organizational typology, 123
Organized fields, 46, 48–51
Ornament
applied—historical, 413, 438–441
applied—performative—mechanical, 413, 426–429
applied—performative—structural, 413, 422–425
applied—referential—organicism, 413, 430–433
applied—referential—structural, 413, 434–437
applied—religious, 413, 418–421
defined, 412
material construction as, 412, 414–417
Ospedale degli Innocenti, 40, 483

Oval
   defined, 125
   as figure / form—architecture, 142–143
   as void / space—architecture, 144–145
   as void / space—urban, 146–148
Oxford, 226

P
PA Technology Center, 241
Palace, 104–105
Palace of Assembly, 100–101, 311
Palace of Parliament, 363
Palais-Royal, 337
Palazzo Cancelleria, 180
Palazzo Davanzati, 178
Palazzo dei Diamanti, 400
Palazzo del Te, 180, 501, 563
Palazzo delle Prigioni, 210
Palazzo Farnese, 106, 130, 306
Palazzo Massimo, 181
Palazzo Medici Riccardi, 179, 561
Palazzo Porto, 345
Palazzo Rucellai, 179
Palazzo typology, 168, 178–181
Palimpsest systems, 32
Palladio, Andrea, 8, 73, 117, 118, 173, 183, 213, 277, 332, 340, 345, 360, 467, 523
Palmanova, Italy, 160, 167
Palmonova, 17
Panopticon, 11, 211, 367
Pantheon, 84, 92–93, 96–97, 114, 138, 248, 272
Parc de la Villette, 51, 371, 591
Paris, 29, 357, 537
Paris Metro Entry, 433
Paris Opera House, 215, 540
Park Guell, 304, 463
Parking garage typology, 169, 224–225
Parthenon, 70, 170, 555
Partis, 4. *See also* Organizational systems
Pattern
   color, 444, 458–463
   defined, 444
   material, 444, 452–457
   referential, 445, 464–469
   repetition, 445, 470–493
   rhythm, 445, 494–511
   shape, 444, 446–451
Pavilion 9 at UVA, 113
Paxton, John, 404
Pazzi Chapel, 320, 324
Pei, I.M., 149, 198, 235, 244, 293, 319, 407
Pensacola Houses, 294
The Pentagon, 155
Perception
   color, 515, 518–519
   constructed focal point, 515, 520–521
   defined, 514
   environment, 515, 530–531
   light, 515, 516–517
   material, 377, 515, 524–525
   memory / history, 515, 528–529
   perspectives, 515, 520–523
   sound, 515, 526–527
Performative / technological formalism, 232, 240–241
Perrault, Dominique, 531
Perret, Auguste, 81, 176, 224
Persian Courtyard House, 89
Perspectives
   constructed focal point, 515, 520–521
   false, 515, 522–523
Peruzzi, Baldassare, 181
Pharaoh Ramses II, 354
Phidias, 435, 555
Phillips Pavilion, 326
Piacentini, Piccanato, Rossi, Vietti, 537
Piano, Renzo, 197, 199, 240, 289, 295, 371, 391, 429, 583
Piano nobile, 44
Piazza at Sant Ignazio di Loyola, 147
Piazza del Campidoglio, 497, 545
Piazza dell'Anfiteatro, 146
Piazza Pio II, 520
Piazza Rucellai, 152
Piazza San Marco, 22
Piece / unit, 445, 470–475
Pilotis, 44
Place Dauphine, 153
Place des Victoires, 140
Place des Vosges, 337
Plan for Algeria, 23
Plan poche, 255, 260–261
Plastic
   detail, 408–409
   form—geometry / system / pattern / ornament, 388–389
Platonic formalism, 232, 234–235
Poche(s)
   defensive, 255
   defined, 254
   elevational, 264–265
   functional, 255
   hierarchical, 255, 276–277
   indeterminate, 274–275
   material, 255, 278–279
   military / defensive, 255, 268–269
   monumental, 255, 270–271
   plan, 255, 260–261
   positional, 255
   programmatic, 280–281
   sectional, 255, 262–263
   service, 255
   structural, 255, 272–273
   urban, 255, 266–267
Point / line / plane, 299, 302–303
Point-to-point
   illustrated, 18–19
   urban, 28
Polshek, James, 313
Polygon
   defined, 125
   as figure / form—architecture, 154–155
   as figure / form—urban, 156–157
   as void / space—architecture, 158–159
   as void / space—urban, 160–161

Pompeii Amphitheater, 16
Pompeii House of the Faun, 182
Pope-Leighey House, 378
Porte-cochere, 205
Portion / panel, 445, 476–481
Portland Building, 585
Position, grid
   architecture, 32–33
   illustrated, 30
   urban, 34–35
Positional poche(s), 255
Positive / negative, 25–27
Positive negative, 59
Postal Office Savings Bank, 220
Postal Savings Bank, 453
Postmodernism, 553, 584–585
Post-Structuralism, 553, 588–589
Prada Headquarters, 223
Precedent
   assemblages, 92–105
   categories, 67
   combinations and hybridization, 68
   comparatives, 106–119
   defined, 66
   lineages, 69–91
   using, 68
Prince Toshihito, 398
Princeton University, 227
Prison typology, 169, 210–211
Program, materials, 377
Programmatic / functional hierarchy, 349, 368–369
Programmatic / functional typology
   campus, 169, 226–229
   defined, 123
   house, 168, 182–191
   library, 169, 202–205
   museum, 169, 192–201
   office / high-rise / commercial, 169, 218–223
   palazzo, 168, 178–181
   parking garage, 169, 224–225
   prison, 169, 210–211
   school, 169, 206–209
   temple/church, 168, 170–177
   theater, 169, 212–217
Programmatic poche, 280–281
Programmatic sequence, 535, 542–543
Programmatic symmetry, 331, 344–345
Prouve, Jean, 384
PSFS Building, 222
PTW Architects, 389
Pyramid at Cheops, 262

R
Radial, 6, 16–17
Radial geometry, 6
Raphael, 74
Rationalism, 552, 578–579
Raumplan, 54–57
   defined, 5
   looping experience, 55
   sectional visual control, 54
   volumetric acension, 55
The Red House, 471
Referential pattern, 445, 464–469
Regional Modernism, 553, 592–593
Renaissance, 551, 560–561
Repetition pattern
   chunk, 445, 488–493
   module / play, 445, 482–487
   piece / unit, 445, 470–475
   portion / panel, 445, 476–481
Rhythm pattern
   frequency, 445, 500–505
   module, 445, 494–499
   scale, 445, 506–511
Rice University, 228
Richards Medical Center, 237
Richardson, H.H., 203, 219
Ricola Storage Building, 469, 477
Rietveld, Gerrit, 370
Robie House, 184
Roche and Dinkeloo, 343
Rodeo Drive, 27
Rogers, Richard, 197, 240, 241, 281, 371, 429, 583
Roman Coliseum, 142
Roman Encampment, 50
Roman Triumphal Arch, 94–95
Roman Wall, 268
Romanesque, 551, 556–557
Romano, Giulio, 180, 501, 563
Rome, 29, 356
Root rectangles geometry / proportion, 300, 320–321
Rose, Thompson, 53
The Rose Center, 313
Rossellino, Bernardo, 520
Rossetti, Biagio, 400
Rossi, Aldo, 21, 208, 216, 308
Rothko Chapel, 15
Rotunda, 84
Royal Saltworks, 17
The Royal Circus, 141
The Royal Crescent, 23
Rudin House, 295
Rudolph, Paul, 189, 225, 281, 507

S
Saarinen, Eero, 305, 393, 395
Sacconi, Giuseppe, 363
Safdie, Moshe, 41, 489
Sainsbury Wing, National Gallery, 239
Saint Ivo Della Sapienza, 165
Salk Institute, 359, 403, 530
San Andrea, 495
San Carlo alle Quatro Fontane, 80, 173, 565
San Cataldo Cemetery, 308
San Giorgio Maggiore, 73, 360
San Giovanni de Fiorentini, 11
San Miniato al Monte, 557
San Sabastiano, 14, 334
San Spirito, 13, 172
SANAA, 245, 309, 361, 387, 509
Sant Abbondio, 77

Sant Andrea, 79, 94–95
Santa Croce, 12
Santa Maria del Fiore, 13
Santa Maria della Pace, 358
Santa Maria Novella, 72, 325
Santa Maria presso San Satiro, 522
Sant'Andrea al Qirinale, 144, 174, 261
Saynatsalo Town Hall, 249
Scala Regia, Vatican, 366, 523
Scale, 445, 506–511
Scamozzi, Vincenzo, 17
Scarpa, Carlo, 403
Scharoun, Hans, 33
Schindler, Rudolf, 185, 186
Schinkel, Karl Friedrich, 85, 108, 193, 335
School typology, 169, 206–209
Schröder House, 370
Scogin, Elam, and Bray, 205
Scottish Parliament Building, 491
Scusev, A.V., 270
Seagram Building, 102–103, 222, 243, 546
Seattle Public Library, 63, 205, 368, 543
Sectional poche, 255, 262–263
Sectional visual control, 54
Sendai Mediatheque, 127, 245
Sequence
   ceremonial, 535, 538–539
   circulation, 534–547
   experiential, 535, 540–541
   horizontal, 535, 536–537
   programmatic, 535, 542–543
   vertical, 535, 544–547
Sequential formalism, 233, 250–251
Service poche, 255, 276–277
Shadao, Shoji, 305, 475
Shape pattern, 444, 446–451
Sheldon Museum, 195
Sidney Opera House, 216
Signal Station, 503
Silvetti, Machado, 547
Simmons Hall, 25
Single-loaded
   architecture, 20–21
   illustrated, 18
   urban, 22–23
Smirke, Sir Robert, 194
Soane, John, 193
SOM, 237, 318, 351, 481
Sonsbeek Pavilion, 587
Sony Center, 145
Sound, 515, 526–527
Space Needle, 365
Spacelab, 201
Spandrels, 222
Spanish Courtyard House, 90
Speer, Albert, 355
Spiral
   defined, 125
   figure / form—architecture, 162–163
Square, 6, 8–9
   defined, 125
   as figure / form—architecture, 126–127

   as figure / form—urban, 128–129
   as void—architecture, 130–131
   as void—urban, 132–133
St. Benedict Chapel, 379
St. Ignatius, 517
St. Paul's Cathedral, 80
St. Peter's Basilica, 15, 74, 75, 79, 147, 333, 352, 536
Stanford University, 229
Star
   defined, 125
   figure / form—architecture, 164–165
   figure / form—urban, 166–167
Stata Center, MIT, 33
Steinhof, 81
Stirling, James, 85, 139, 198
Stockholm Library, 139, 314
Stone, Edward Durrell, 215, 238
Structural poche, 255, 272–273
Structuralism, 553, 586–587
Structure
   concrete, 394–395
   masonry, 392–393
   material, 376
   metal, 396–397
   wood, 390–391
Structure, grid, 31, 38–39
Sullivan, Louis, 220
Surface / skin, 377
Swiss Pavilion at Expo 2000 in Hanover, 399
Symmetrical formalism, 233
Symmetry
   asymmetry, 331, 338–339
   bilateral, 331, 332–337
   defined, 330
   local, 331, 340–341
   material, 331, 342–343
   programmatic, 331, 344–345

T
Takeshi Hosaka Architects, 151
Taliesin East, 541
Taut, Bruno, 135
Teatro Olimpico, 213, 523
Tectonic / technology, 377
Tempietto, 134
The Tempietto, 10
Temple / church typology, 168, 170–177
Temple Complex at Karnak, 354
Temple of Hera II at Paestum, 70
Temple of Jupiter, 71
Temple of Queen Hatshepsut, 538
Temple of the Earth, 313
Temple plan, 71
Ten Peachtree Place, 291
Terragni, Giuseppe, 39, 107, 129, 207, 221, 321, 323, 539, 579
Theater at Ostia Antica, 212
Theater typology, 169, 212–217
Thorncrown Chapel, 82, 287, 390
Thornton, Latrobe, and Bullfinch, 361
Three-dimensional volume / mass, 299, 304–305
Tomb of Augustus, 136
Tongxian Gatehouse, 401

Transco Tower, 223
Travertine, 417
Treasury at Petra, 264
Trenton Bath House, 345
Très Grande Bibliotheque, 57, 263, 317
Triangle
　defined, 125
　as figure / form—architecture, 148–149
　geometry / proportion, 310, 318–319
　as void / space—architecture, 150–151
　as void / space—urban, 152–153
Triglyphs, 435
Tristan Tzara House, 292, 309
Tschumi, Bernard, 51, 209, 371, 591
Turner, Major Reuben, 211
TWA Terminal, 305, 395
Twin Towers at Marina City, 395
Two-dimensional and three-dimensional module form geometry / proportion, 300, 324–325
Two-dimensional module in plan, 299, 306–307
Two-dimensional square module as space / void, 299, 310–311
Two-dimensional square module in elevation as figure / form, 299, 308–309
Typological formalism, 232
Typology
　categories, 122–123
　defined, 122
　geometric / form, 123, 124–167
　history as value system, 123
　material, 123
　organizational, 123
　programmatic / functional, 123, 168–229

U
The Uffizi, 20, 192
UK Pavilion Shanghai Expo, 303
UN Studio, 149
Ungers, O.M., 62
Unite d'Habitation, 25, 265, 382, 402, 427
United d'Habitation, 246
Unity Temple, 541
University of Houston, 98–99
University of St. Thomas, 353
University of Virginia, 228
University of Virginia Rotunda, 84
Urban axial hierarchy, 349
Urban context, 285, 288–289
Urban formal hierarchy, 348
Urban poche, 255, 266–267
U.S. Capital Building, 361
U.S. Pavilion at '67, 475
U.S. Steel Building, 397
Utzon, Jorn, 216
UVA Lawn, 499, 521
UVA Rotunda, 115, 145

V
van der Rohe, Mies, 39, 45, 126, 188, 195, 222, 229, 243, 302, 307, 342, 387, 393, 396, 437, 525, 546
Van Eyck, Aldo, 587
Vanna Venturi House, 190, 290, 311, 339
Vasari, Georgio, 20, 192
Vatican Museum Courtyard, 159
V.C. Morris Shop, 162
Venturi, Robert, 190, 290, 311, 339
Vernacular precedent, 67
Versailles Entry Court, 153
Vertical sequence, 535, 544–547
Vignon, Pierre-Alexandre, 175, 569
Villa Barbaro, 332
Villa Dall'Ava, 191, 291
Villa Farnese, 164
Villa Giulia, 86, 87, 183, 250
Villa Lante, 544
Villa Mairea, 524
Villa Malcontenta, 118
Villa Moller, 251, 335
Villa Müller, 56, 187, 341, 367, 545
Villa Rotunda, 8, 96–97, 183, 277
Villa Savoye, 44, 131, 186, 251, 543, 577
Villa Snellman, 184
Villa Stein at Garches, 44, 119, 310, 322
Visual / perceptual hierarchy, 349, 358–359
Vitra Fire Station, 327
Volumetric acension, 55
von Erlach, Johann Fischer, 174
Voussoirs, 272

W
W. G. Low House, 339
Wagner, Otto, 81, 175, 220, 453, 465
Wainwright Building, 236
Walker Guest House, 189
Warehouse, 104–105
Washington, D.C., 28, 357, 364
Water Cube, 389
Webb, Philip, 471
Whitney Museum, 581
Winn Memorial Library, 203
Winton Guest House, 53
Wittgenstein, Ludwig, 321
Wittgenstein House, 321
Wood
　detail, 398–399
　form—geometry / system / pattern / ornament, 378–379
　structure, 390–391
Wood, John the Younger, 23, 141
Wren, Sir Christopher, 80
Wright, Frank Lloyd, 162, 184, 185, 194, 221, 243, 286, 307, 350, 378, 455, 541, 546

Y
Yale Center for British Art, 196, 288
Yale Fine Arts Gallery, 151
Yale Parking Garage, 225
Yale School of Architecture, 281

Z
Zollverein School of Management and Design, 309
Zumthor, Peter, 9, 83, 200, 247, 279, 379, 399, 407